Reflective Cracking in Pavements

State of the Art and Design Recommendations

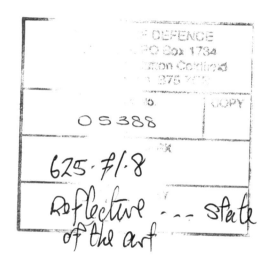

Related RILEM publications

Mechanical Tests for Bituminous Mixes
Characterization, Design and Quality Control
Edited by H.W. Fritz and E. Eustacchio

Bituminous mixes are used throughout the world for highway construction and surfacing. Traditional test methods for bituminous mixes are largely based on empirical procedures and have only a limited significance. Newer, more fundamental test methods can provide more reliable and significant results but they often require complex testing facilities and are not suitable for routine use. The traditional and modern methods need to be brought together to combine the advantages of both.

This book forms the proceedings of the International Symposium organized by RILEM in Budapest, Hungary in October 1990. The purpose of the Symposium was to bring together current developments and new research information in order to promote tests for the characterization, design and quality control of bituminous mixes which will improve the relevance and reliability of test methods.

The Symposium was the fourth on the testing of bituminous and asphalt materials and it focuses on three aspects of mechanical testing: specimen preparation; tests with unique loading (which are used for mix design and control of mechanical properties); and tests with repeated loading (which give information on fatigue, deformation and moduli, and are widely used for mix design).

RILEM Proceedings 8, 672 pages, 1990, ISBN 0 412 39260 7

Geomembranes
Identification and Performance Testing
Edited by A. Rollin and J.M. Rigo

Geomembranes are synthetic impermeable sheets placed in contact with soil to contain or restrain the movement of gases or liquids. The tremendous worldwide growth in the use of geomembranes in transportation, environmental and geotechnical applications has led to an urgent need for agreed standards covering their properties and use. Few standards exist and little harmonization at international level has yet been achieved.

This book puts forward for the first time recommended procedures for identification, characterization and testing of the materials themselves, and of the procedures for joining them. It has been prepared by RILEM Technical Committee 103-MGH. It will be of value to geotechnical and environmental engineers, manufacturers of geomembranes, those involved in research into and standardization of geomembranes, and testing organizations.

RILEM Report 4, 376 pages, 1990, 0 412 38530 9

A full list of RILEM publications available from E & FN Spon is given at the back of the book.

Reflective Cracking in Pavements

State of the Art and Design Recommendations

Proceedings of the
Second International RILEM Conference,
organized by the Belgian Research Centre
for Plastics and Rubber Materials (CEP) and
the Construction Materials Laboratory (LMC),
Liege University, Belgium

Liege, Belgium
March 10–12, 1993

EDITED BY

J.M. Rigo, R. Degeimbre

Construction Materials Laboratory, Liege University, Belgium

and

L. Francken

Belgian Road Research Centre, Brussels, Belgium

E & FN SPON
An Imprint of Chapman & Hall

London · Glasgow · New York · Tokyo · Melbourne · Madras

**Published by E & FN Spon, an imprint of Chapman & Hall,
2–6 Boundary Row, London SE1 8HN**

Chapman & Hall, 2–6 Boundary Row, London SE1 8HN, UK

Blackie Academic & Professional, Wester Cleddens Road, Bishopbriggs, Glasgow G64 2NZ, UK

Chapman & Hall Inc., 29 West 35th Street, New York NY10001, USA

Chapman & Hall Japan, Thomson Publishing Japan, Hirakawacho Nemoto Building, 6F, 1-7-11 Hirakawa-cho, Chiyoda-ku, Tokyo 102, Japan

Chapman & Hall Australia, Thomas Nelson Australia, 102 Dodds Street, South Melbourne, Victoria 3205, Australia

Chapman & Hall India, R. Seshadri, 32 Second Main Road, CIT East, Madras 600 035, India

First edition 1993

© 1993 RILEM

Printed in Great Britain at the University Press, Cambridge

ISBN 0 419 18220 9

Apart from any fair dealing for the purposes of research or private study, or criticism or review, as permitted under the UK Copyright Designs and Patents Act, 1988, this publication may not be reproduced, stored, or transmitted, in any form or by any means, without the prior permission in writing of the publishers, or in the case of reprographic reproduction only in accordance with the terms of the licences issued by the Copyright Licensing Agency in the UK, or in accordance with the terms of licences issued by the appropriate Reproduction Rights Organization outside the UK. Enquiries concerning reproduction outside the terms stated here should be sent to the publishers at the London address printed on this page.
  The publisher makes no representation, express or implied, with regard to the accuracy of the information contained in this book and cannot accept any legal responsibility or liability for any errors or omissions that may be made.

A catalogue record for this book is available from the British Library

Library of Congress Cataloging-in-Publication data

Publisher's note
This book has been produced from camera ready copy provided by the individual contributors in order to make the book available for the Conference.

Contents

Scientific Committee · xiii
Conference Organization · xiv
Preface · xv

PART ONE INTRODUCTION AND KEYNOTE PAPERS · 1

1 **General introduction, main conclusions of the 1989 Conference on Reflective Cracking in Pavements, and future prospects** · 3
J.M. RIGO
Chairman, RC93; Chairman, RILEM TC 97-GCR; Construction Materials Laboratory, University of Liege, Belgium

2 **Evaluation of pavement structure with emphasis on reflective cracking** · 21
A.A.A. MOLENAAR
Road and Railway Research Laboratory, Delft University of Technology, Netherlands

3 **Les procédés utilises pour maitriser la remontée des fissures: bilan actuel** (Retarding measures for crack propagation: state of the art) · 49
M.G. COLOMBIER
Laboratoire Régional des Ponts et Chaussées d'Autun, France

4 **Laboratory simulation and modelling of overlay systems** · 75
L. FRANCKEN
Belgian Road Research Centre, Brussels, Belgium

5 **Geotextile use in ashphalt overlays – design and installation techniques for successful applications** · 100
F.P. JAECKLIN
Geotechnik Consulting Engineers, Ennetbaden, Switzerland

PART TWO DESIGN MODELS FOR REFLECTIVE CRACKING IN PAVEMENTS 119

6 CAPA: a modern tool for the analysis and design of pavements 121
A. SCARPAS and J. BLAAUWENDRAAD
Structural Mechanics Division, Delft University of Technology, Netherlands
A.H. de BONDT and A.A.A. MOLENAAR
Road and Railroad Research Laboratory,
Delft University of Techology, Netherlands

7 Semirigid pavement design with respect to reflective cracking 129
I. GSCHWENDT
Slovak Technical University, Bratislava, Czechoslovakia
I. POLIAČEK
Viaconsult, Bratislava, Czechoslovakia
V. RIKOVSKÝ
Research Institute of Civil Engineering, Bratislava, Czechoslovakia

8 Numerical modelling of crack initiation under thermal stresses and traffic loads 136
A. VANELSTRAETE and L. FRANCKEN
Belgian Road Research Centre, Brussels, Belgium

9 Evaluation of crack propagation in an overlay subjected to traffic and thermal effects 146
J.M. RIGO, S. HENDRICK, L. COURARD and C. COSTA
Geosynthetics Research Centre, Materials Research Centre,
University of Liege, Belgium
S. CESCOTTO
M.S.M., University of Liege, Belgium
P.J. KUCK
Hoechst, Bobingen, Germany

10 Design of ashphalt overlay/fabric system against reflective cracking 159
B. GRAF
Consulting Engineer, Zurich, Switzerland
G. WERNER
Geosynthetics Consulting, Polyfelt GmbH, Linz, Austria

11 Design limits of common reflective crack repair techniques 169
T.R. JACOB
Owens Corning Fiberglas, Granville, Ohio, USA

PART THREE ASSESSMENT METHODS FOR REFLECTIVE CRACKING IN PAVEMENTS   177

12 Apparatus for laboratory study of cracking resistance   179
H. DI BENEDETTO and J. NEJI
Laboratoire Géomatériaux, ENTPE, France
J.P. ANTOINE and M. PASQUIER
Gerland Routes, Corbas, France

13 Investigations on the effectiveness of synthetic asphalt reinforcements   187
P.A.J.C. KUNST
Netherlands Pavement Consultants, Hoevelaken, Netherlands
R. KIRSCHNER
Huesker Synthetic GmbH, Gescher, Germany

14 A new test for determination of fatigue crack characteristics for bituminous concretes   193
S. CAPERAA, E. AHMIEDI and C. PETIT
Civil Engineering Laboratory, University of Limoges, Egletons, France
J-P. MICHAUT
Colas Society, Paris, France

15 Dynamic testing of glass fibre grid reinforced asphalt   200
M.H.M. COPPENS
Netherlands Pavement Consultants, Hoevelaken, Netherlands
P.A. WIERINGA
Chomarat, Mariac, France

16 On the thermorheological properties of interface systems   206
L. FRANCKEN and A. VANELSTRAETE
Belgian Road Research Centre, Brussels, Belgium

17 Influence of modulus ratio on crack propagation in multilayered pavements   220
C. PETIT and S. CAPERAA
Civil Engineering Laboratory, University of Limoges, Egletons, France
J-P. MICHAUT
Colas Society, Paris, France

18 Reflective cracking in asphalt pavements on cement bound road bases under Swedish conditions   228
J. SILFWERBRAND
Swedish Cement and Concrete Research Institute, Stockholm, Sweden

19 **Study on reflection cracks in pavements** 237
   K. SASAKI
   Civil Engineering Research Institute, Hokkaido Development Bureau, Japan
   H. KUBO
   Faculty of Engineering, Hokkaigakuen University, Japan
   K. KAWAMURA
   Civil Engineering Research Institute, Hokkaido Development Bureau, Japan

20 **Processes reducing reflective cracking: synthesis of laboratory tests** 246
   Ph. DUMAS and J. VECOVEN
   Public Road Laboratory of Autun, France

21 **Paving fabric specification for asphalt overlays and sprayed reseal applications** 254
   W. ALEXANDER
   Geofabrics Australasia Pty Ltd, Melbourne, Australia
   R. McKENNA
   Geosynthetic Testing Services, Albury, Australia

22 **New testing method to characterize mode I fracturing of asphalt aggregate mixtures** 263
   E.K. TSCHEGG
   Technical University of Vienna, Austria
   S.E. STANZL-TSCHEGG
   University of Agriculture, Vienna, Austria
   J. LITZKA
   Technical University of Vienna, Austria

PART FOUR   RETARDING MEASURES FOR REFLECTIVE CRACKING IN PAVEMENTS 271

23 **The precracking of pavement underlays incorporating hydraulic binders** 273
   G. COLOMBIER
   Laboratoire Régional des Ponts et Chaussées, Autun, France
   J.P. MARCHAND
   Cochery Bourdin Chausse, Nanterre, France

24 Treatment of cracks in semi-rigid pavement: cold
    microsurfacing with modified bitumen emulsion and
    fibers: Spanish experience   282
    R. ALBEROLA
    Ministry of Public Works and Transport, Madrid, Spain
    J. GORDILLO
    E.S.M. Research Center, Madrid, Spain

25 Application of geosynthetics to overlays in Cracow
    region of Poland   290
    W. GRZYBOWSKA, J. WOJTOWICZ and L. FONFERKO
    Cracow University of Technology, Cracow, Poland

26 Minimization of reflection cracking through the use of geogrid
    and better compaction   299
    A.O. ABD EL HALIM and A.G. RAZAQPUR
    *Center for Geosynthetic Research, Information and Development,
    Carleton University, Ottawa, Canada*

27 New system for preventing reflective cracking: Membrane
    Using Reinforcement Manufactured On Site (MURMOS)   307
    J. SAMANOS
    Screg Routes, France
    H. TESSONNEAU
    Screg Sud-Est, France

28 The use of a polypropylene bituminous composite overlay to
    retard reflective cracking on the surface of a highway   316
    A.R. WOODSIDE, B. CURRIE and W.D.H. WOODWARD
    *Department of Civil Engineering and Transport, University of Ulster,
    Jordanstown, Northern Ireland*

29 Fabric reinforced chip seal surfacing and resurfacing   323
    C.J. SPRAGUE
    Sprague and Sprague Consulting Engineers, Greenville, USA
    D.M. CAMPBELL
    Hoechst Celanese Corporation, Spartanburg, USA
    P.J. KUCK
    Hoechst AG, F+E Spunbond, Bobingen, Germany

30 Jute fibre for production of non-woven geotextiles to
    prevent reflective cracking in pavements   334
    S.N. PANDEY and A.K. MAJUMDAR
    J.T.R.L., (I.C.A.R.), Calcutta, India

31 Bituminous pre-coated geotextile felts for retarding
    reflection cracks   343
    M. LIVNEH, I. ISHAI and O. KIEF
    *Transportation Research Institute, Technion-Israel Institute
    of Technology, Haifa, Israel*

PART FIVE   CASE HISTORIES   351

32 Comparative sections of reflective crack-preventing systems:
    four years evaluation   353
    G. LAURENT
    C.E.T.E. de l'Ouest, Nantes, France
    J.P. SERFASS
    Screg Routes, Guyancourt, France

33 Assessment of methods to prevent reflection cracking   360
    M.E. NUNN and J.F. POTTER
    Transport Research Laboratory, Crowthorne, UK

34 Experience of Du Pont de Nemours in reflective cracking:
    site follow up   370
    G. KARAM
    Du Pont de Nemours, Luxembourg

35 Soft pavement wide interlayer in Japan   378
    K. INOUE
    Japan Seal Industries Co. Ltd, Osaka, Japan

36 The application of a geotextile manufactured on site on the
    Belgian motorway Mons–Tournai   384
    R. DUMONT
    Walloon Ministry of Infrastructure and Transport, Mons, Belgium
    Y. DECOENE
    Screg Belgium, Brussels, Belgium

37 Belgian applications of geotextiles to avoid reflective cracking
    in pavements   391
    Y. DECOENE
    Screg Belgium, Brussels, Belgium

38 Prevention of cracking progress of asphalt overlayer with
    glass fabric   398
    ZHONGYIN GUO
    Tongji University, Shanghai, P.R. China
    QUANCAI ZHANG
    Taiyuan Highway Administration, P.R. China

39 Long term performance of geotextile reinforced seals to control
shrinkage on stabilized and unstabilized clay bases   406
P. PHILLIPS
Geofabrics Australasia, Sydney, Australia

40 In situ behaviour of cracking control devices   413
M. LEFORT and D. SICARD
Laboratoire Régional de l'Ouest Parisien, Trappes, France

41 Asphalt overlay on crack-sealed concrete pavements using
stress distributing media   425
G. HERBST
*Central Road Testing Laboratory, Road Administration of Lower
Austria, St Pölten-Spratzen, Austria*
H. KIRCHKNOPF
*Road Construction – Department 2, Road Administration of Lower
Austria, Tulln, Austria*
J. LITZKA
*Institute for Road Construction and Maintenance, Technical
University of Vienna, Austria*

42 Experimental project on reflection cracking in Madrid   433
R. ALBEROLA
National Highway Department, Madrid, Spain
A. RUIZ
*Centro de Estudios y Experimentación de Obras Públicas,
Madrid, Spain*

43 Two kinds of mechanism of reflective cracking   441
SHA QING-LIN
Research Institute of Highways, Beijing, P.R. China

44 Movements of a cracked semi-rigid pavement structure   449
A.H. de BONDT
*Road and Railroad Research Laboratory, Delft University of
Technology, Netherlands*
L.E.B. SAATHOF
*Road and Hydraulic Engineering Division, Ministry of Public
Works, Transport and Water Management, Delft, Netherlands*

45 A crack-resistant surface dressing: the results of 8 years
of application, and future prospects   458
J.P. MARCHAND
Cochery Bourdin Chausse, Nanterre, France

**46 Thin overlay to concrete carriageway to minimise
reflective cracking** 464
I.D. WALSH
Engineering Services Branch, Kent County Council, Maidstone, UK

**47 Design and first application of geotextiles against reflective
cracking in Greece** 482
A. COLLIOS
Edafomichaniki Ltd, Athens, Greece

Author index 488

Subject index 489

Scientific Committee
(RILEM Technical Committee 97-GCR, Applications of Geotextiles to Crack Prevention in Roads)

J.M. Rigo, Liege University, Belgium (Chairman)
L. Francken, Belgian Road Research Centre (Secretary)
G. Colombier, Laboratoire Régional des Ponts et Chaussées d'Autun, France
A.H. de Bondt, Delft University of Technology, The Netherlands
R. Degeimbre, Liege University, Belgium
Ph. Delmas, BIDIM Geosynthetics, Bezons, France
G. Karam, Du Pont de Nemours, Luxembourg
R.L. Lytton, Texas A and M University, College Station, Texas, USA
A.A.A. Molenaar, Delft University of Technology, The Netherlands
P. Rimoldi, Tenax Plastotecnica, Como, Italy
J.P. Serfass, Screg Routes et Travaux Publics, St Quentin en Yvelines, France
N. Sprecher, Owens Corning Fiberglass, Brussels, Belgium
N. Thom, University of Nottingham, UK
G. Werner, Polyfelt GmbH, Linz, Austria
P. Wieringa, Chomarat, Mariae, France

Conference Organization

Belgian Research Centre for Plastics and Rubber Materials (CEP)
and
Construction Materials Laboratory (LMC)

Liege University
Civil Engineering Institute
Quai Banning, 6
B-4000 Liege
Belgium

Collaborating Organizations

Belgian Association for Materials Studies (ABEM)
Belgian Committee for Geosynthetics (BCG)
Belgian Road Research Centre (BRRC)
Bituminous Information Centre (CIB)
European Asphalt Pavement Association (EAPA)
Intercontinental Club for Plastics Use in Building and Engineering (ICP)
International Geotextile Society (IGS)
International Road Federation (IRF)
Permanent International Association of Road Congress (PIARC)
Internation Union of Testing and Research Laboratories for Materials and Structures (RILEM)

Organizing Committee

Professor Mst. Dr Ir J.M. Rigo, Liege University, Belgium (Chairman)
Dr L. Francken, Belgian Road Research Centre (Scientific Secretariat)
Professor Dr Ir. R. Degeimbre, Liege University, Belgium (Management)

Preface

The rehabilitation of cracked roads by overlaying is rarely a durable solution. In fact, the cracks rapidly propagate through the new asphalt layer. This phenomenon is called 'reflective cracking' and is widespread over many countries. With current financial restrictions, road maintenance authorities have to find solutions with a good cost:benefit ratio.

Many solutions have been proposed to meet this challenge:

- placing a stress-absorbing membrane interlayer between the cracked support and the overlay;
- modifying the overlay composition, principally by the use of modified bitumen (bitumen plus polymer or elastomer);
- a combination of the two above proposals.

These solutions are supported by numerous studies dealing with:

- analytical evaluation of the reflective cracking phenomena and of the anti-reflective cracking systems (elastic and viscoelastic stress analysis, fracture mechanics, etc);
- experimental evaluation, including laboratory studies and/or field installation completed by follow-up operations.

In spite of these efforts, it seems that universal crack repair treatment with good durability is still lacking.

The First Conference on Reflective Cracking (March 1989)

RILEM Technical Committee TC 97-GCR was set up in 1986 to study the problems of the use of geotextiles and related products in preventing crack propagation in pavements. The objectives of the TC are:

- to establish a state of the art on the subject;
- to recommend on site and laboratory test methods to determine relevant parameters.

This Technical Committee decided on the organization of the first conference which was held in Liege in March 1989, with the title:

'Reflective Cracking in Pavements: Assessment and Control'

Very valuable contributions were presented on: laboratory and full scale experiments; modelization; case histories.

The time has come to ensure these objectives are followed up and to draw attention to:

- the state of the art in this field;
- the design recommendations for field applications.

These are the two main topics on which the 1993 Conference will be based.

Professor J.M. Rigo,
Conference Chairman

PART ONE
INTRODUCTION AND KEYNOTE PAPERS

1 GENERAL INTRODUCTION, MAIN CONCLUSIONS OF THE 1989 CONFERENCE ON REFLECTIVE CRACKING IN PAVEMENTS, AND FUTURE PROSPECTS

J.M. RIGO
Chairman, RC93; Chairman, RILEM TC 97-GCR; Construction Materials Laboratory, University of Liege, Belgium

Abstract
A R.I.L.E.M. Technical Committee was set up in 1986 to study the problems related to the use of geotextiles and geotextile-related products in preventing cracks propagation in pavements : TC 97-GCR.
The objectives of this Technical Committee are :
- to establish a state of the art on the subject;
- to recommend on site and laboratory test methods to determine relevant parameters.
This Technical Committee decided the organization of a first conference which was held in Liège, March 1989, with the title : "Reflective Cracking in Pavements : Assessment and Control".
During this conference, very valuable contributions were presented on :
- laboratory and full scale experiments;
- design models for reflective cracking in pavements;
- case histories.
Introductory papers gave overviews on :
- nature, origin of the cracks and remedial measures;
- design aspects.
This second conference has as objectives :
- to insure the follow up on the three above mentionned topics;
- to draw attention on the design recommendations for field applications.
This conference is coming at a time where International bodies like C.E.N. (and specially the C.E.N. Technical Committee TC 189 on geotextiles and geotextile-related products) are strongly interested in the development of design and testing standards on the use of geotextiles and geotextile-related products in asphalt pavements.
<u>Keywords</u> : reflective cracking, geotextiles, geosynthetics, design, testing, assessment, case histories, standardization.

Reflective Cracking in Pavements. Edited by J.M. Rigo, R. Degeimbre and L. Francken.
© 1993 RILEM. Published by E & FN Spon, 2–6 Boundary Row, London SE1 8HN. ISBN 0 419 18220 9.

1 Introduction

The rehabilitation of cracked roads by simple overlaying is rarely a durable solution. In fact, the cracks rapidly propagate through the new overlay. Indeed, in general, the newly overlayed bituminous layer does not have the deformation ability to bridge living cracks without damage. This phenomenon is called "reflective cracking" and is widespread over many countries. With the current financial restrictions, the road maintenance authorities have to use solutions with a good cost/benefit ratio.

Cracks in the top layer of a pavement cause numerous problems :
- discomfort for the users;
- reduction of the safety;

but also
- intrusion of water and subsequent reduction of the soil bearing capacity;
- pumping of soil particles through the crack;
- progressive degradation of the road structure in the neighbourhood of the cracks due to local overstresses;
- ...

Cracks origins are numerous and they differ by their shape, configuration, movement mode (I, II, III), amplitude, deformation velocity... It means that at lot of remediations technics were developed in the past and are still in development now.

These remedial measures may be roughly classified in three classes :
- modification of the overlay characteristics in order to absorb stresses or strains without damage;
- placement of a stress or strain-absorbing membrane interlayer between the old structure and the new overlay;
- combinations of the two above counter-measures.

These solutions are supported by numerous studies dealing with :
- analytical evaluation of the reflective cracking phenomena and of the anti-reflective cracking systems;
- experimental evaluations, including laboratory studies and/or field trials completed by follow-up operations.

On the other hand, field experience is growing and one of the main objectives of this conference is to review the accumulated knowledge on this matter in order to prepare a synthesis useful for the designer.

No miracle solution does exist and an engineering approach of each specific problem is necessary :
- identification of the problem;
- analysis of the possible solutions and their limitations;
- feasability in terms of cost effectiveness;
- analysis of the laydown and construction considerations.

The problem of the free circulation of goods and services throughout the European Community and the European Free Trade Associated Countries is typical. This requires a unification of standards on design and testing of materials and/or structures.
A C.E.N. Technical Committee on the uses of geotextiles and geotextile-related products in building and civil engineering was created in 1989 : CEN TC 189. It is a Belgian initiative and the author acts as chairman of this committee.
Among other topics, harmonization of design and testing of geosynthetics used in pavements will be developed in the near future. R.I.L.E.M. and C.E.N. have signed a memorandum of understanding by which R.I.L.E.M. work should be taken as prenormative documents for the C.E.N. standardization work. This conference takes place at the right moment in this respect.
This paper will review the main conclusions of the 1989 conference taking into consideration what will be presented by the four other introductory papers :
- evaluation and assessment methods [Molenaar, 1993];
- retarding measures [Colombier, 1993];
- laboratory simulations and modelling [Francken, 1993];
- existing recommendations and specifications [Jaecklin, 1993].

2 Cracks origin and propagation

It is first necessary to make a distinction between two different types of problems [Bonnot, 1989] :
- cracking occuring in flexible roads structure;
- cracking occuring in rigid or semi-rigid structures resulting from cement treated layers shrinkage or thermal movement or resulting from the movements of the joints between concrete slabs.

Most of the communications presented at the first conference in 1989 were devoted to the second type of problem.
Three different mechanisms were pointed out as origin of the cracks in rigid and semi-rigid pavements :
- <u>thermal stresses</u> (or thermal fatigue) : daily temperature variations induce openings and closures of the cracks induced by the shrinkage in the cement treated base layers of the structure. If the bituminous concrete overlay perfectly adheres on the cement treated layer, the thermal movements induce stress concentrations in the overlay. If this one resists, a debonding between layers may occur. If not, a crack will propagate in the overlay from bottom to the top;
- <u>thermal stresses</u> may also be initiated by a rapid cooling down of the top layer which induces important tensile stresses and cracks. This type of observation was reported by Nunn (1989) and some others (figure 1).

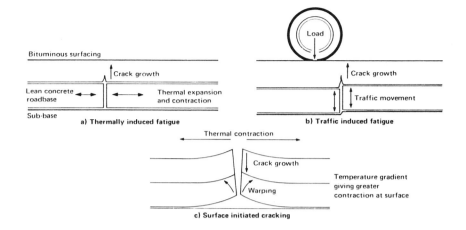

Figure 1 : mechanisms of reflective cracking [Nunn, 1989]

- **traffic loads** : the cyclic application of traffic loadings induces an additional distress of the overlay and the propagation of cracks originated by thermal effects.

Authors did not agree on the predominant mechanism by which traffic loads are acting on the crack propagation :
- some authors considered that the most important effect is obtained by the opening of the cracks (mode I) resulting from the structure bending;
- others considered that the vertical shear stresses resulting from a poor load transfer between the two edges of the crack (mode II) is the more detrimental;
- top **surface micro-cracking** due to bad compaction procedure was also mentionned [Halim, 1989];
- **settlement of the soil** (hydric variation...) underneath the pavement may also create cracking;
- **abrupt transitions between different types of foundations** (road widening, local repair, trenches...) were also reported as cracks origins.

The two last items received less attention during the 1989 conference.

Agreement was found on the fact that :
- thermal stresses are supposed to induce the cracks and take part to the initial propagation of the cracks;
- traffic stresses take part to the second step of the crack propagation.

It is interesting to note that practically nobody spoke about the healing phenomenon that occurs during hot periods under traffic effects where the integrity of the overlay is partially or totally restored.

Cracks differ in shape, aspects orientation, opening, configuration... and are function of a numerous number of parameters. These will be reviewed later by G. Colombier (1993).

3 Anti-reflective cracking measures

3.1 Remediation technics
Remediation technics for the above described phenomenons are numerous and may be classified in three main classes :
- remediation by <u>modifications of the overlay characteristics</u> in order to improve its ability to resist to the strains or stresses imposed by the crack movements.
  This may be obtained by the use of modified bitumen with or without short fibers as a binder for the overlay [Walsh, 1989]. This may also be obtained by the reinforcement of the overlay by means of steel wires [Molenaar, 1989] or glass fibers [Schuster, 1989]. Increase of the overlay thickness is also a solution;
- remediation by the placement of <u>a stress or strain-absorbing membrane interlayer</u> adhering to the old structure and the new overlay : S.A.M., S.A.M.I., bituminous mortars, bitumen impregnated geotextiles, bitumen impregnated in situ projected fibers, textiles or plastic geogrids... [Colombier, 1989];
- remediation by <u>acting on the origin of the cracks</u> when it is possible : stabilization of slabs joints, modification of the road foundation...

It clearly appears that the resistance to crack propagation involves all the pavements components :
- overlay;
- interlayer (if any);
- old pavement structure;
- soil.

Each of these must be treated in view of the global system efficiency. It is a case by case decision.

3.2 Interlayer between the old pavement and a new overlay
The interlayer functions are as follow :
- <u>stress release</u> : basement and/or overlay differential movements of thermal origin may be absorbed by the interlayer, having an important stress release effect on the structure.
  On the other hand, traffic stresses must be transmitted correctly between the pavement layers. All the interlayer actually developed are based on bitumen or modified bitumen. The visco-elasto-plastic behavior of this material fits very well to the above mentionned requirements;
- <u>waterproofing</u> : in case of cracks, one must avoid any intrusion of water in the soil underneath the pavement. Else, soil bearing capacity will decrease and some pumping effect is to be expected under traffic loads, inducing extra stresses into the pavement and additional distress.

Bitumen is the basic interlayer componant. In order to decrease the interlayer thermal susceptibility important modifications based on elastomers or plastomers are done. Perfetti (1989) and some others reported on the use of ethyl-acrylate (E.V.A.) copolymer modified bitumen in order to decrease the thermal susceptibility and to optimize the ability for the textile saturation without bleeding and the adhesion to the old structure and the new overlay.
The use of styrene-butadiene-styrene (S.B.S.) copolymer modified bitumen permits to reach higher performances but is more expensive. This type of binder may be used for more difficult cases : very cold climate, important cracks movements...

The interlayer bitumen is mechanically stabilized by means of fibers acting as a container. These may be presented as manufactured fabrics or as short or long fibers projected on site on the structure to be repaired. Rigo et al (1989) reported that the majority (about 2/3) of the mechanical effect of the stress release function is insured by the interlayer bitumen. The remaining part is due to the fibers or the fabric structure.

It is evident that the interlayer mechanical effectiveness is highly dependent on its adhesion to the old pavement and the new overlay [Fock, 1989]. This adhesion must be permanent and uniform. It is essential to the bearing capacity, stability and durability of the new road structure. The use of a paving felt as a bitumen carrier offers the possibility of applying a defined and regular bitumen film as the bonding agent or bonding bridge between the individual composite strata.

The waterproofing function may be obtained by the complete saturation of the interlayer fibers mass. It is why this requirement is fulfilled in case of nonwoven (needle punched, thermobonded) fabrics and in situ projected fibers. Open structures like grids are not able to fulfil this requirement.
Recently developed composites made of textile grids and nonwovens arrived on the market.

Important remark : the use of an interlayer as a way to react to reflective cracking is possible as long as the vertical movements of the crack edges (modes II and III) are strictly limited. Most of the RC 89 conference authors supposed that the shear modes were under control. It is clear that the shear movements may not be supported by such a thin layer without severe damage to the overlay.

3.3 Modifications of the overlay characteristics
Other ways to resist to crack propagations are to inject sufficient :
- deformability or
- tensile resistance
in the overlay.

Excessive deformability must be avoided because it induces rutting effect under traffic loading. Walsh (1989) reported on the successful use of E.V.A. (Ethyl Vinyl Acetate) and S.B.S. (Styrene Butadiene Styrene) polymers modified bitumen as an overlay binder used on concrete carriage ways in the U.K.

Improvement of the overlay bituminous concrete tensile resistance may be obtained by means of :
- the use of short fibers spread in the whole mass;
- the use of reinforcing elements located in the lower part of the overlay.

The use of short fibers mixed and spread into the hot mass before laying down is of recent development. Courard L. et al (1991) and some others during this conference are reporting on the use of short steel, glass, polyester, polyacrylonitril (PAN), polyvinylalcohol (PVA)... fibers in bituminous concretes.

If the tensile improvement is expected, the decrease of the rutting effect is also obtained.

Table 1 presents a compilation of these fibers characteristics. Some bitumens characteristics are also reported.

The bitumen, even if modified, is highly temperature and frequency sensitive. Polymeric, glass and steel are less sensible to these parameters.

Reinforcement, understood on the classical way, requires that the modulus of elasticity of the reinforcing material be higher than the one of the material to be reinforced. Looking at the E modulus raw in table 1, it may be observed that an effective reinforcement by means of thermoplastic fibers (short, long), fabrics or grids is to be expected mostly for thermal sollicitations. Glass fibers and steel wires have higher moduli.

It may also be remarked that short fibers in the overlay mass inject quasi isotropic characteristics.

Reinforcement of the overlay may also be obtained by means of "reinforcing" fabrics, grids or meshes in the lower part of the overlay. Taking into account the above remark on the relative modulus between the reinforcing material and the bituminous concrete, it seems that steel and glass are able to act as reinforcing elements.

It must be noticed that these reinforcing elements generally have two principal directions due to fibers or wires orientations. Orthotropic behavior is to be expected from the reinforced overlay. It also means that diagonal directions are weaker and more subjected to failure.

Finally, the placement conditions are very important. No wrinkles may occur. This is why these reinforcing elements are placed under tension (by nailing for instance).

Finally, geosynthetics may also be used as a reinforcement of <u>microsurfacing</u> layers (fabric reinforced chip seal). This was reported by Perrier (1989) and will developed later during this conference.

Table 1 : some characteristics of fibers, wires and bitumens

Characteristics	PET	PP	PVA	PAN	GLASS	STEEL	BITUMEN
Softening point (°C)	250	160	140	250	860	1600	30→100
Glass transition T (°C)	80	−15	85	90	−	−	−
Degradation T (°C)	300	280	240	250	2000	−	−
Tensile (N/mm^2)	1000	600	1000	500	3500	1900	0.5→10
Elongation (%)	15	15	10	35	4	1,5	−
E modulus (N/mm^2)	12500	12000	25000	6500	73000	210000	10000→30000 (traff.) 100→1000 (therm.)
Density	1.38	0.91	1.29	1.18	2.60	7.9	−
Chemical resistance . solvents	+	+	+	+	+	+	−
. fuels	+	+	+	+	+	+	−
. de-icing	+	+	+	+	+	−	+
Biological resistance	+	+	+	+	+	+	±
Water resistance	+	+	±	±	±	−	+

3.4 Temperature limitations to the use of fibers in asphalt overlaying

Bituminous materials (concrete, slurry,...) are laid down at temperatures ranging between 130°C and 200°C.
When fibers are injected in a hot bitumen during the mixing it is clear that the fibers raw materials must have a certain heat resistance (see table 1).
On the other hand, when used as a componant of an interlayer, the maximum temperatures occur when the overlay is laid down. Courard et al (1989) reported that temperatures really occuring at the level of the textile are, in such cases, ranging between 90 and 110°C.

Data obtained from field trials showed the same values even in cases of overlay overheated at 220°C.
Realistic requirements are to be formulated with that respect.

4 Design models and laboratory simulation

Design models are excellent tools to predict the structure behavior under various types of loads provided that :
- the loads simulation is realistic. Basically this problem is to be solved in 3D. For simplification reasons most of the developments were done in 2D. Correspondance between 2D and 3D solutions are generally limited to an area very near to the crack;
- traffic and thermal loads are correctly combined. This is one of the actual key issues;
- the constitutive equations representing the materials behaviors are available. Still a lot of information is missing : fatigue law with temperature changes, rupture criteria, effect of rest periods and cracks healing, simulation of the crack pattern considering the asphalt concrete heterogeneity, bituminous visco-elasto-plastic behavior, the fiber-bitumen composite interlayer characteristics;
- ...

In spite of this, these simplifications permit to progress in the knowledge in the field but one must be aware that oversimplified problems lead to unrealistic solutions. In any case, tendancies may be identified and solutions may be compared. L. Francken (1993) will review this problem later.
    One way to reduce the design models approximations is to calibrate these by means of laboratory and full scale experiments.
Many test methods were developed in the past and most of these were reported during the RC 89 conference.
L. Francken (1993) will review these later.

Test methods in this field may be classified as follow :
- <u>index tests</u>
  - on individual componants (geotextiles, bitumen...);
  - on combined materials (interlayer, bituminous concrete...);
- <u>performances tests</u>
  - on combined materials;
  - on structures.

Tables 2 and 3 give examples of test methods to be used in that respect.

Table 2 : index test methods on materials used for reflective cracking applications

---

Index tests on individual components :
- Geotextile and geotextile-related products
  - definitions
  - mass per unit area
  - thickness
  - classification
  - tensile
  - modulus of elasticity (F)
  - C.B.R.
  - perforation
  - melting temperature (F)
  - bitumen retention   (F)
  - chemical resistance (F)
  - ...   (F) = also for short and long fibers
- Bitumen
  - ring/ball
  - penetration
  - viscosity
  - thermal susceptibility
  - elastic modulus
  - shear modulus
  - modifier content
  - ...

Index tests on combined materials :
- Interlayer
  - modulus of elasticity (vertical)
  - shear modulus
  - thermal susceptibility
  - thickness at various pressures
  - ...
- Asphalt overlay
  - Marshall
  - stiffness modulus
  - thermal expansion
  - void content
  - ...

Table 3 : performance tests on materials and structures used for reflective cracking in pavements

Performance tests on combined materials :
- interlayer
  . modulus of elasticity in function of temperature and frequency
  . shear modulus in function of temperature and frequency
  . tensile resistance
  . elongation
  . adhesion
  . thermal resistance of the componants
  . damage during installation
  ...

- asphalt overlay
  . fatigue law
  . crack propagation law
  . healing of cracks
  ...

Performance tests on structures :
- crack propagation simulation test
  . under traffic loading
  . under thermal loading
  . under combined effect (with or without waterhead) followed by an interlayer adhesion test

- in situ non destructive testing

- ...

Important developments are to be expected in the near future for the preparation of such a list of test methods (if not already existing). The European Committee for Standardization of the design and test methods on geotextiles and geotextile-related products (CEN TC 189) is willing to take this in hands. CEN TC 189 will also have to fix minimum values of well selected characteristics in relation with safety and stability to permit the CE Mark labelling and as a consequence to permit the free circulation of products in Europe. Finally, design procedures will be recommended.

5 Case histories

More than 25 papers of the RC 89 conference described field trials or applications covering all kind of structures mostly rigid and semi-rigid, only a few cases of flexible pavements were described.

5.1 Fibers and wires based materials

Fibers or wires based materials are used for :
- interlayer underneath an asphalt concrete overlay;
- interlayer underneath a cement concrete overlay [Hellenbroich, 1989];
- the reinforcement of surface dressing [Perrier, 1989] or as reinforcement of microsurfacing with chip seal;
- local bridging of cracks [Kubo, 1989].

The fibers or wires based materials are presented as :
- long or short fiber needle-punched fabrics (thermoplastics);
- nonwoven thermally bonded (thermoplastics);
- woven fabrics (thermoplastics or glass);
- woven/nonwoven composites;
- extruded grids (thermoplastics);
- honey comb shaped mesh (galvanized steel);
- short fibers projected on site (glass or thermoplastic);
- continuous filaments projected on site (glass or thermoplastic);
- short fibers reinforcing the overlay (glass, steel or thermoplastics).

It is important to remind that for the temperature ranges met in road structures :
- thermoplastic exhibit a visco-elasto-plastic behavior;
- glass and steel are elastic.

5.2 Described case histories

The main conclusions to be mentionned from these experiences are dealing with :
- the assessment of the case to be treated;
- the treatment before the interlayer and overlay applications;
- the obtained results.

The assessment of the cases to be treated generally start by a visual inspection. In some cases, the cracks pattern is recorded. Unfortunately in the great majority of the presented applications the assessment is limited to this visual analysis.

Cases were reported where concrete slab joints were stabilized prior to the treatment.

Cracks movements are generally not measured and in very few cases, for instance Van de Griend (1989), non-destructive evaluations of the road structure were done.

Taking into consideration the very strong limitation of use of thin interlayers in case of excessive cracks shear movements, one must draw the attention of the users and site trial researchers that a correct structure assessment is indispensable to draw correct conclusions.

Prior to the interlayers applications some minor but important treatments are realized. Concrete slab joints exhibiting "too big" movements are stabilized, cracks larger than 2 to 3 mm are filled and potholes are also treated.

In some cases an intermediate regulating layer is to be applied to obtain a flat surface.

The repair layers may then be applied. The placement procedures are function of the applied solution.

In case of use of geotextiles (woven, nonwoven composites) the application technics vary :
- lay down of the interlayer bitumen (hot bitumen or emulsion);
- lay down of the fabric (when the emulsion is broken);
- some authors reported a second bitumen application on top of the geotextile;
- lay down of the overlay.

In case of the use of chopped fibers or continuous filaments projected on site, the 3 first above mentionned steps are generally assembled in one : simultaneous projection of bitumen and fibers.

The fabric placement is a critical step :
- wrinkels in this fabric have been reported as cracks initiations;
- roads turns induce wrinkels or necking in very stiff products;
- the fabric must generally be protected by some sand or fine gravel to avoid undesirable adhesion between the fibers and the trucks or finishing machines tires.

Geogrids are generally applied under tension. Gilchrist (1989) reported a tension corresponding to 0.5 % strain in the grid for the application of the overlay. The grid is place on a flat surface :
- lay down;
- tensioning;
- placement of a protective surface dressing (chip seal or hot mix pad coat);
- laying down of the overlay.

Steel wire meshes are nailed on the old structure without particular tension and are directly covered by the overlay.

Abd El Halim (1989) focussed the attention on micro-cracking created in the top layer of the overlay by incorrect compaction. He suggested the use of compactor rolls with soft surface.

Obtained results were generally reported by means of the observation of the reflected cracks. Many ways were used to characterize these cracks :
- there are cracks or not (without any additional informations);
- number of cracks per kilometer;
- length of cracks;
- percentage of reflected cracks by comparison to the cracks existing in the old structure;
- percentage of reflected cracks [Lefort, 1989] :
  . if the total length, weight factor = 1;
  . if the reflected crack is discontinuous, weight factor = 0.5;
  . if the crack just appears, weight factor = 0.25;
- degree of cracks = total length or cracked area/m^2; coefficient of cracks = area containing cracks * 100 (%)/total area [Fukuoka, 1989];
- number, length, difference in level between the cracks edges [Bernard, 1989];
- cracking index 0 when
  . no cracks appear;
  . no rutting effects;
  . no deformations;
  cracking index 100 when
  . completely cracked;
  . deformations > 1.5 cm [Bernard, 1989];
- inspection and cracks survey.
Levelling with instrument.
Measurements with falling weight.
Rutting measurement with laser [Johansson, 1989].

Finally, Lefort (1989) presented a classification of shrinkage cracks in function of geometrical criteria (table 4).
Common language is not easy to find. It is clear that a unification of these classification systems is required. This classification must be easy to handle, give numbers or classes to the engineer.
It must be based on accurate and realistic observations.
In that respect it is clear that the development of non destructive testing methods in order to quantify the road structure deterioration on site and in lab is highly necessary. Indeed, cracks are observed on site when they appear at the top surface. This observation gives no idea about the initiation and the propagation phases.
Some case histories were reported without any reference section which does not permit any analysis.
Some others were devoted to a single technic. Comparisons views between technics are necessary and it is why site trials of sufficient length and including various technics in strictly similar conditions are necessary.

Table 4 : shrinkage cracks classification in function of geometrical criteria [Lefort, 1989]

Parameters	Classes of cracks			
Spacing between cracks	Classification in function of the mean and standard deviation values			
Relative width of the cracked zone (versus the pavement width)	A < 1/3	B 1/3 < l/L < 2/3	C > 2/3	
Opening and aspect of the cracks edges	A fine cracks	B cracks with spaling around aggregates	C cracks with spaling (in blocs)	
Ramification	A No ramif.	B Part. doubl.	C Part. triple	D Crocodile skin
Tortuosity l = width of the fictive zone including the crack	A l<20 cm	B 20<l<40	C 40<l<60	D l>60
Discontinuity Number of blocks constituting the crack	A 1	B 2	C 3 and more	

It is not my intention to compare technics but some observations can be pointed out from the case histories presented during the RC 89 conference :
- the use of thin interlayer including fiber type materials have a retarding effect on the crack initiation and propagation. The most significant effects are obtained for flexible structure. Success are met for rigid and semi-rigid structures when shear modes are avoided;
- the use of overlay reinforcing elements seems to be the more efficient when steel wires and fiber glass grids are used. Polymeric grids uses are more discussed;
- overlay bitumen modifications by means of thermoplastic or elastomeric modification and/or inclusion of short fibers (thermoplastic, glass, steel...) are promising field;
- in all presented cases, placement conditions are reported to be essential for the success of the technic. Wrinkels, overlaps, lack of saturation, remaining emulsion water, adhesion to the machines tires... induce damages which may cause failures;

- solutions were used outside of their applications
  limits. It is a challenge for the future to identify
  these limits of use.

6 Conclusions

In terms of the designer, the reflection cracking problem should be approached from the combined perspective of :
- if the problem exists, what are the factors and
  mechanisms and what are the alternatives available;
- how can the alternatives be screened and evaluated to
  find the best solution and how should it be implemented.

This is our basic challenge.
The RC 89 conference gave a first occasion to review the first item of this challenge. Time is now to screen the available solutions and to define, even roughly, their applications limitations. This is the objective of the RC 93 conference.
In a near future, technical bodies like CEN will review this work and definitively start a standardization work on testing and design methods. This conference comes at the right time.

7 Bibliography

Abd El Halim A. A new approach towards understanding the problem of reflection cracking. Reflective Cracking in Pavements. Assessment and Control. Liège, March 1989.

Bernard et al. Geotextile imprégné; une solution aux problèmes de remontée des fissures pour le climat nordique canadien. Reflective Cracking in Pavements. Assessment and Control. Liège, March 1989.

Bonnot. Oral presentation as reporter of the session on laboratory and full scale experiment. Reflective Cracking in Pavements. Assessment and Control. Liège, March 1989.

Colombier G. Fissuration des chaussées : nature et origine des fissures; moyens pour maîtriser leur remontée. Reflective Cracking in Pavements. Assessment and Control. Liège, March 1989.

Colombier G. Retarding measures for crack propagation. State of the art. Reflective Cracking in Pavements : State of the Art and Design recommendations. Liège, March 1993.

Courard L., Rigo J.M. The Use of fibers in bituminous concrete as a solution to decrease rutting. 5th Eurobitume Congress, Stockholm, 1993.

Courard et al. Compatibility between fibers and modified bitumen. Reflective Cracking in Pavements. Assessment and Control. Liège, March 1989.

Fock G. The use of paving felts to influence the life expectancy and permanent adhesion of asphalt road surfaces. Reflective Cracking in Pavements. Assessment and Control. Liège, March 1989.

Francken L. Laboratory simulation and modelling of overlay systems. Reflective Cracking in Pavements : State of the Art Report and Design Recommendations. Liège, March 1993.

Fukuoka. Some old case histories on reflection cracks in Japan. Reflective Cracking in Pavements. Assessment and Control. Liège, March 1989.

Gilchrist A. Control of reflection cracking in pavements by the installation of polymer geogrids. Reflective Cracking in Pavements. Assessment and Control. Liège, March 1989.

Hellenbroich Th. Cement concrete on geotextiles. Reflective Cracking in Pavements. Assessment and Control. Liège, March 1989.

Jaecklin F. Geotextile use in asphalt overlays. Design and installation techniques for successful applications. Reflective Cracking in Pavements : State of the Art Report and Design Recommendations. Liège, March 1993.

Johansson S. Reinforcement of bituminous wearing course with geotextiles. A research project on road 588 in Soderhamn, Sweden. Reflective Cracking in Pavements. Assessment and Control. Liège, March 1989.

Kubo et al. Site experiments on reflective crack retarding measures in asphalt pavements. Reflective Cracking in Pavements. Assessment and Control. Liège, March 1989.

Lefort M. et al. Maîtrise de la remontée des fissures d'assises traitées aux liants hydrauliques dans les chaussées. Résultats obtenus sur chantiers expérimentaux en France. Reflective Cracking in Pavements. Assessment and Control. Liège, March 1989.

Molenaar A. et al. Steel reinforcement for the prevention of reflective cracking in asphalt overlays. Reflective Cracking in Pavements. Assessment and Control. Liège, March 1989.

Molenaar A.A.A. Evaluation of pavement structures with Emphasis on reflective cracking. Reflective Cracking in Pavements : State of the Art and Design Recommendations. Liège, March 1993.

Nunn M. An investigation of reflection cracking in composite pavements in the United Kingdom. Reflective Cracking in Pavements. Assessment and Control. Liège, March 1989.

Perfetti et al. Système anti-fissure textile-liant modifié. Concept, évaluation, application. Reflective Cracking in Pavements. Assessment and Control. Liège, March 1989.

Perrier H. Enduits renforcés par géotextile sur routes en terre. Expérience française en Guyane. Reflective Cracking in Pavements. Assessment and Control. Liège, March 1989.

Rigo et al. Laboratory testing and design methods for reflective cracking interlayers. Reflective Cracking in Pavements. Assessment and Control. Liège, March 1989.

Schuster A. et al. Polyester and glass-fibre geogrids for the prevention of reflective cracking. Reflective Cracking in Pavements. Assessment and Control. Liège, March 1989.

Van de Griend et al. Practical investigations concerning geotextiles and reinforcing grids in asphaltic overlays. Reflective Cracking in Pavements. Assessment and Control. Liège, March 1989.

Walsh I.D. Overbanding and polymer modified asphalt in overlay to concrete carriage-ways. Reflective Cracking in Pavements. Assessment and Control. Liège, March 1989.

2 EVALUATION OF PAVEMENT STRUCTURE WITH EMPHASIS ON REFLECTIVE CRACKING

A.A.A. MOLENAAR
Road and Railway Research Laboratory, Delft University of Technology, Netherlands

Abstract
This paper describes the evaluation of pavement structures with emphasis on characterization of the reflective crack potential, pavements have in relation to future maintenance strategies.
It is shown that precise characterization of the movements across the cracks and joints is essential as well as a careful characterization of the various interfaces that can be recognized in overlayed cracked pavements. Visual condition surveys, coring, deflection measurements, joint movement measurements and material characterization are the most important steps in the evaluation process. Attention is also paid to the mathematical modelling of cracked pavements as well as engineering approaches that are available to do so. Furthermore it is indicated how the crack resistance of overlay materials can be analyzed as well as the effect of reinforcement and interlayers.
Keywords: Pavement Evaluation, Visual Condition Surveys, Deflection Measurements, Crack Movements, Fracture Mechanics, Pavement Modelling, Overlay Design.

1 Introduction

When pavements have been in service for a certain number of years, defects will appear at the pavement surface. These defects can be related to ageing of the surface layer (e.g. raveling, surface cracking) or can be due to lack of deformation resistance of the structure (e.g. rutting), environmental effects (e.g. transverse cracking) or can be a result of lack of structural bearing capacity

Reflective Cracking in Pavements. Edited by J.M. Rigo, R. Degeimbre and L. Francken.
© 1993 RILEM. Published by E & FN Spon, 2–6 Boundary Row, London SE1 8HN. ISBN 0 419 18220 9.

of the structure (fatigue cracking).

When these defects have developed to a certain severity and extent, maintenance is needed in order to provide sufficient service to the road user for a substantial period of time. In order to be able to select the most appropriate maintenance strategy, a detailed evaluation of the condition of the pavement structure should be carried out. This evaluation should reveal the reasons for the development of the observed defects.

Furthermore the evaluation should provide input for the predictions that should be made with respect to future deterioration as well as input for predictions that should be made with respect to the effectiveness of the various maintenance strategies that are possible.

In order to be able to select the appropriate evaluation procedures one should recognize the possible causes and consequences of the defects observed. If e.g. transverse cracking is the main problem of a flexible pavement one should question whether or not this cracking is low temperature cracking of the asphalt or whether this cracking is initiated by movements of the underlying cement treated base.

In order to be able to select the most appropriate maintenance strategy for those situations, it is not sufficient to determine the horizontal movements across the crack due to temperature variation.

Also the vertical movements across the crack due to the passage of a wheelload is of importance. Lack of aggregate interlock across the crack requires a rather high resistance of the maintenance treatment to shear forces which would restrict the number of possible maintenance strategies. If such load transfer information is not available, a less appropriate or even wrong maintenance strategy could be selected.

In this paper attention will be paid to the evaluation of pavement structures in relation to the design of maintenance strategies, with emphasis on reflective cracking. The paper will start with a short description of possible reasons for cracking in asphalt pavements followed by a discussion on how these cracks might propagate through overlays and other maintenance strategies. This discussion will result in a number of aspects that should be taken into account in pavement evaluation.

Furthermore the paper describes a number of measurement techniques that can be used for pavement evaluation with emphasis on reflective cracking.

Finally the paper will describe some mechanistic models that can be used in conjunction with the evaluation procedures in order to be able to make predictions on future deterioration and the effectiveness of various maintenance strategies.

2 Cracking of pavements

Cracking of pavements can be due to a very large number of reasons. An excellent discussion on this is given by Colombier (1989). It is not intended to repeat this discussion here, only some general remarks with respect to the causes and development of cracking will be given.

Basically four main crack categories can be discriminated which are:

- transverse cracking
- block cracking
- longitudinal cracking
- alligator cracking

Transverse and block cracking (figure 1) are in general related to environmental effects. A well known example is the development of shrinkage cracks in cement treated bases due to hydratation of the cement and subsequent seasonal thermal variations. These cracks tend to propagate through the upper pavement layers which can be either asphalt or concrete. It should however be noticed that this type of cracking can also occur without the presence of a cement treated base. In cold regions low temperature cracking of asphalt pavements as well as low temperature fatigue cracking is a well known phenomena.

Although the horizontal movements perpendicular to the crack or joint are of main importance, one should be aware of the fact that also vertical movements and especially the difference in movements across the cracks is of importance.

In figure 2a it is indicated that the primary fracture mode of a layer above a cement treated base is the so called mode I cracking. Crack growth is generated through forces which are acting perpen-dicular to the crack face; these are forces which are generated due to shrinkage of the cement treated base because of a drop in temperature.

One should however be aware of the fact that also other modes of cracking are responsible for the growth of the crack through the upper layer. Due to a wheel passage shear forces will be generated as well as bending moments. The shear forces are responsible for mode II cracking while the bending moments will develop mode I cracking. It is however shown (Molenaar, 1989) that mode II cracking is dominating in almost all cases (figure 2b).

The extent to which traffic loads will influence crack propagation is dependent to a fairly large extent on the load transfer across the crack or joint.

Due to the horizontal movements because of temperature variations and due to the shear forces generated by the traffic loads, the cracks in the cemented base will in general propagate from bottom to top.

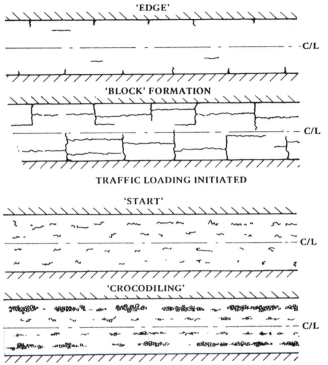

Fig.1. Thermal and traffic initiated cracking (source: Dickinson, 1984).

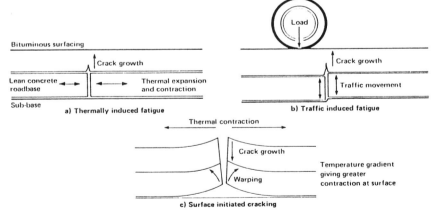

Fig.2. Cracking modes in pavements with cement treated bases (source: Nunn, 1989).

One should however be aware of the fact that due to warping of the cement treated base, because of the temperature gradient that exists over the thickness of the base, also reflective cracks might develop in the upper layer which in this case however will grow from top to bottom (figure 2c). Evidence for this type of crack propagation, which is of the mode I type, is given by Nunn (1989).

All in all it is quite clear that the reflective cracking process in pavements with cement treated bases is a complicated one. This immediately implies that a careful analysis should be made of the various influence factors in order to determine their relative importance. This is essential in order to be able to determine the most appropriate maintenance strategy.

When the environmental effects and the characteristics of the pavement are such that block cracking develops, the mode I cracking due to temperature variations becomes less predominant. In those cases the influence of the traffic loads will become increasingly important because of the reduction in bending stiffness of the entire structure due to the block cracking.

Longitudinal cracking (figure 1) in the wheelpaths is most likely influenced by traffic. It is clearly indicated by Molenaar (1983), Gerritsen e.a. (1987) and Jacobs e.a. (1992) that the horizontal shear forces that occur under both, free rolling and driven wheels are responsible for very high tensile stresses and strains at the pavement surface acting perpendicular to the direction of travel. These strains occur near the tire edges and produce longitudinal cracks. An example of the stress distribution in the pavement as a result of the shear forces and the vertical contact pressure is given in figure 3.

As one will observe from figure 3, the influence of the horizontal shear forces is neglectable at a depth of approximately 50 mm. This is in agreement with many observations in practice where it was determined that many longitudinal cracks only propagated through the wearing course.

These cracks will propagate in the longitudinal direction of the pavement because of tearing which is commonly called mode III type cracking (figure 4).

Of course this type of cracking is not a serious one from a structural point of view. Nevertheless they tend to propagate through an overlay if no proper treatment has taken place in advance.

Alligator cracking (figure 1) is commonly related to fatigue of the pavement and according to the commonly used design models, it is related to the tensile strain at the bottom of the asphalt layer. According to the commonly

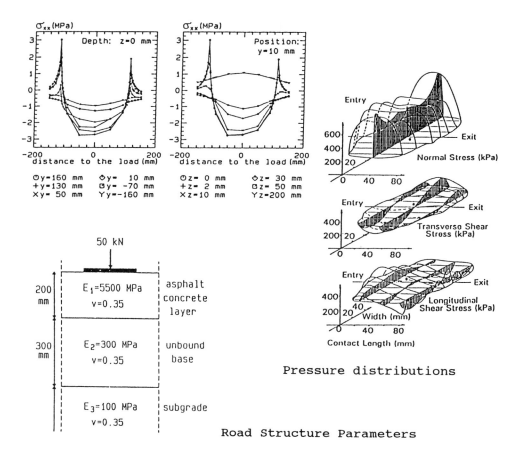

Fig.3. Effects of non-uniform stress distributions (x-axis: perpendicular to moving direction; y-axis: moving direction; z-axis: depth) (source: Jacobs e.a., 1992).

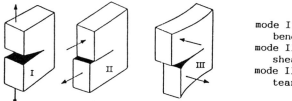

mode I
    bending mode
mode II
    shear mode
mode III
    tear mode

Fig.4. Modes of cracking.

used linear elastic multi layer analysis, fatigue cracks are developed in a direction transverse to the direction of travel. These transverse cracks are only developing over a limited length given the fact that the width of the loaded area is limited to the wheelpaths. Propagation of these cracks is generated by the bending (mode I) and shearing (mode II) action of the traffic loads. Also fatigue cracks will propagate through layers which will be placed on top of the cracked layer.

3 Evaluation of cracked pavements

From the discussion given in the previous sections it is obvious that a careful analysis should be made on the causes of the observed cracking and of the consequences the cracks might have on the performance of the maintenance strategy to be applied.

In analyzing the condition of cracked pavements special attention should be given to characterization of the movements that take place in the vicinity and across the cracks and joints. In the sections here-after some analysis techniques will be described.

However, before going into detail with respect to the evaluation procedures to be used, some remarks will be made with respect to pavement evaluation procedures related to maintenance management of an entire road network.

3.1 Evaluation of pavement networks, the network and project level approach

As has been indicated in the introduction, pavement maintenance is normally due to the presence of a number of defects. It seldomly happens that only one type of defect is apparent on the pavement surface.

In order to be able to make effective use of the available resources, pavement management systems are used which consist of the following aspects:

- network inventory: gives information on the size of the network, types of structures involved, length and width of the individual roads etc.;
- condition survey: gives information on the quality of the roads within the network;
- prioritization: gives information on which roads should be maintained first given their current condition and expected future deterioration;
- strategy selection: gives information on which types of maintenance strategies are suitable to upgrade the pavement condition given the types, severity and extent of the various defect types that are observed on the roads;

- optimization: determines which roads are going to be maintained as well as the most appropriate type of maintenance strategy given the present and expected future degree of deterioration, the possible maintenance strategies, the costs involved and the benefits which are obtained. Normally the benefit of a maintenance strategy is defined as the ratio increase in pavement life over costs associated with the particular maintenance strategy.

Since performing all these steps in great detail and depth on network level is an enormous task, usually a discrimination is made between the analysis on network and project level. The main difference between these two types of analyses is that the network level approach should result in information on where the pavement condition is insufficient, how fast the condition of the network as a whole deteriorates and what the budget requirements are to keep the network on a sufficient level of service.

The project level approach then concentrates on the sections with insufficient condition. By means of a detailed analysis the reasons and causes for the observed deterioration are analyzed finally resulting in the design of a maintenance strategy for the particular road sections under consideration.

This approach implies that the condition information collected on network level is rather general in nature.

In the Netherlands e.g. the visual condition of a pavement is rated by means of a 1 - 5 ranking scale (1 = good, 5 = bad) (Koning e.a., 1987). An example of this ranking system is given in figure 5; this figure indicates that at best also some general information is obtained on the extent and the severity of the observed damage.

Furthermore one should realize that in general deflection measurements are not a part of the pavement evaluation on network level.

With respect to cracking and reflective cracking this all means that the information on network level is not sufficient to make any useful analysis what so-ever with respect to reflective crack performance of maintenance treatments placed on cracked pavements. At best estimates on future condition can be made using normalized performance models which are shown in figure 6. These performance curves can be written as:

$$P = 1 - (t/T)^a \tag{1}$$

where P = 1- present amount of damage/maximum possible amount of damage
      t = period between time of construction or last major maintenance and time of inspection

road name : EXAMPLE		road nr. : 101				WEATHER		SURFACE	
from : HERE		section : A/B				☐ cloudless		☑ dry	
to : ETERNITY						☐ partially clouded		☐ drying	
length = 200 m						☑ overcast		☐ wet	

part			TL	BL	PP	PL	TL	BL	PP	PL
lane			L	R			L	R		
pavement type			ASPHALT				ASPHALT			
length / area			L = 200 m		A = 720 m²		L = 200 m		A = 720 m²	

		ext\deg	very slight	slight	mod.	severe	very slight	slight	mod.	severe
TEXTURE	raveling (%)	≤ 15		2	3	5		2	3	5
		16 - 30	(1)	3	4	5	(1)	3	4	5
		> 30		3	5	5		3	5	5
	bleeding (%)	≤ 15				5				5
		16 - 30			4	5			4	5
		> 30			5	5			5	5
DEFORMATION	rutting (%)	≤ 15		2	3	5		2	3	5
		16 - 30	(1)	3	4	5	(1)	3	4	5
		> 30		3	5	5		3	5	5
	unevenness (nr. / 100m)	≤ 7		2	3	5		2	3	5
		8 - 15	(1)	3	4	5	(1)	3	4	5
		> 15		3	5	5		3	5	5
STRUCTURAL CONDITION	long. cracking (m¹ / 100m)	≤ 25		2	3	5		(2)	3	5
		26 - 100	1	(3)	4	5	1	3	4	5
		> 100		4	5	5		4	5	5
	alligator cracking (%)	≤ 10		(2)	3	5		(2)	3	5
		11 - 20	1	3	4	5	1	3	4	5
		> 20		4	5	5		4	5	5
	joints (%)	≤ 15				5				5
		16 - 30			4	5			4	5
		> 30			5	5			5	5
EDGE DEFECTS	edge defects (m¹ / 100m)	≤ 15		2	3	5		2	3	5
		16 - 30	1	(3)	4	5	(1)	3	4	5
		> 30		4	5	5		4	5	5
VARIOUS	drainage	> 5	p k b		4	5	p k b		4	5
	verge		- + to		4	5	- + to		4	5
	parking lane				4	5			4	5
	busstop				4	5			4	5
			1	2	3	4 5	1	2	3	4 5
MEASUREMENTS	skid resistance									
	unevenness									
	deflection									
IMMEDIATE REPAIR	transverse cracking					m¹	5			m¹
	transverse joints					m¹				m¹
	long. joints					m¹				m¹
	potholes					nr.				nr.

Fig.5. Broad visual condition survey sheet.

T = period between time of construction or last major maintenance and moment when P reaches a value of 0
a = curvature parameter

The value of a depends on the type of structure but also to a very large extent on the type of defect considered. Some indicative values for a are given in table 1.

Table 1. Typical values for 'a'

Defect	a
rutting in asphalt pavements	0.6
alligator cracking in asphalt pavements	3 - 4
longitudinal cracking in asphalt pavements	3 - 4
transverse cracking in asphalt pavements	0.3 - 0.5
raveling of asphaltic surface layers	3 - 4

When from the network analysis, a pavement is selected for an analysis on project level, then the following measurements are recommended with respect to the analysis of a cracked pavement.

3.2 Visual condition surveys and coring
Visual condition surveys are a useful tool to quantify and qualify the extent and severity of the cracking visible at the pavement surface. In order to let the visual condition survey to be effective on project level, the survey should preferably be done by making maps of the observed cracking. Certainly at project level such a detailed survey is worthwhile. The result of such a survey could be as shown in figure 7.

Knowing the extent and severity of the cracks is not enough. As will be shown later on, knowledge on the depth of the cracks is also important in order to be able to perform an adequate analysis on the reflective crack potential of possible maintenance strategies. It is for that reason that it is strongly recommended to take cores on cracks. From these cores one can examine the depth to which the crack is propagating and whether or not it is likely that the crack has propagated from bottom to top.

3.3 Deflection measurements
Of course deflection measurements are needed in order to be able to evaluate the condition of the existing pavement in terms of cracking. Basically deflection measurements are used to determine the bending stiffness of the structure and to calculate the stiffness moduli of the various pavement layers. The commonly used procedure

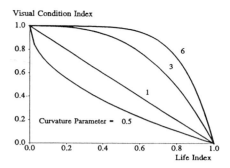

Fig.6. Visual condition performance curves (source: Koning e.a., 1987).

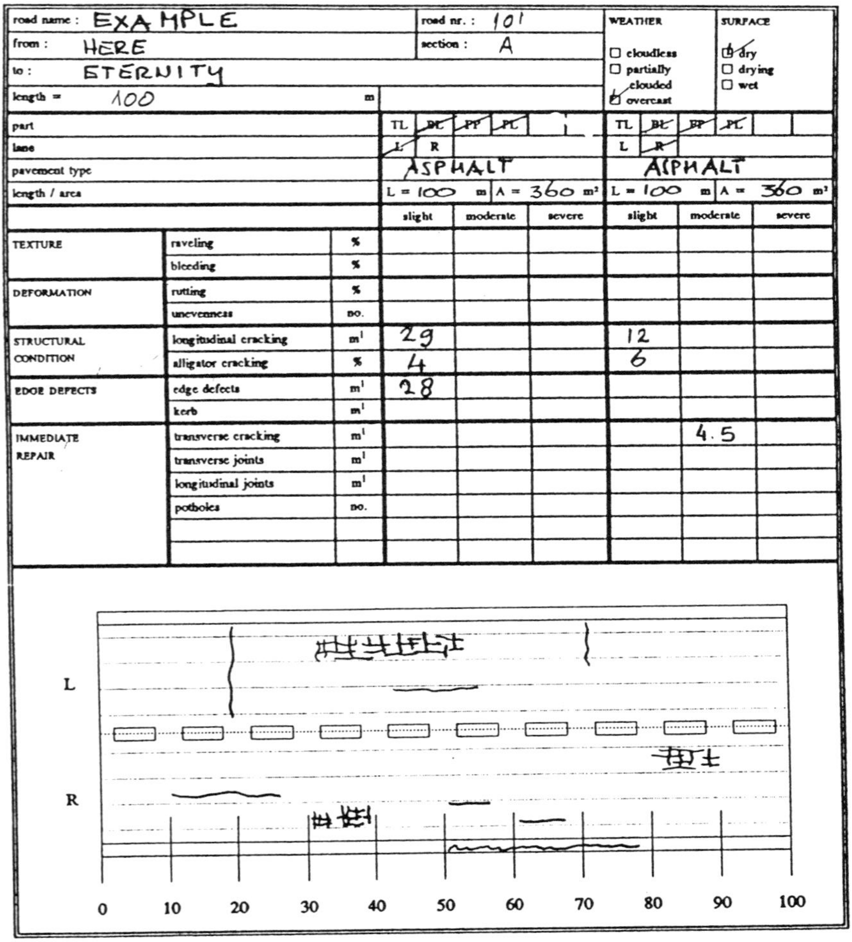

road name : EXAMPLE				road nr. : 101			WEATHER	SURFACE
from : HERE				section : A			☐ cloudless	☑ dry
to : ETERNITY							☐ partially clouded	☐ drying
length = 100			m				☑ overcast	☐ wet
part				TL ☐ BL ☐ PP ☐ PL ☐			TL ☐ BL ☐ BP ☐ PL ☐	
lane				L ☐ R ☑			L ☐ R ☑	
pavement type				ASPHALT			ASPHALT	
length / area				L = 100 m A = 360 m²			L = 100 m A = 360 m²	
				slight	moderate	severe	slight	moderate	severe
TEXTURE	raveling	%						
	bleeding	%						
DEFORMATION	rutting	%						
	unevenness	no.						
STRUCTURAL CONDITION	longitudinal cracking	m¹	29			12		
	alligator cracking	%	4			6		
EDGE DEFECTS	edge defects	m¹	28					
	kerb	m¹						
IMMEDIATE REPAIR	transverse cracking	m¹					4.5	
	transverse joints	m¹						
	longitudinal joints	m¹						
	potholes	no.						

Fig.7. Result of a detailed visual condition survey.

consists of the use of the falling weight deflectometer and a computer program for the analysis of multi layer linear elastic systems.

Although a cracked pavement certainly not fulfils the requirements of being a homogeneous, isotropic and linear elastic system, practice has shown that in most cases stiffness moduli can be back calculated, even for the cracked layer, using these programs.

Even for concrete block pavements, which can be considered as the ultimate stage of a cracked pavement, moduli for the block layer can be achieved through a good fit between the measured and calculated deflection profiles (Shackel, 1992; De Groot e.a., 1983).

As indicated before, not only the stiffness of the various layers is of importance in analyzing the structural condition of pavements but also the measurement of movements near cracks and joints.

When concrete pavements need to be rehabilitated e.g., it is of importance to know whether or not the overlay can be placed directly on top of the existing pavement or additional measures need to be taken like e.g. crack and seat procedures.

An interesting example of the use of deflection measurements for such conditions is given by Breyer e.a. (1990). A description has been given there of the evaluation of concrete pavements in Austria resulting in guidelines for deciding on crack and seat procedures.

Crack and seat procedures where considered to be necessary if the back calculated modulus for the concrete slabs was less than 30000 MPa and if poor support conditions or poor load transfer conditions where observed near the joints. If these later conditions prevailed, high shear forces in the overlay would be generated through traffic (mode II cracking) and the overlay would probably show excessive reflective cracking within a relatively short period of time.

Cracking and seating the existing concrete pavement, would result in a high quality unbound base, which is fully supported and where the chance in reflective cracking is reduced drastically.

Figure 8 shows the moduli values that were obtained on the original concrete pavement as well as on the cracked and seated concrete.

Figure 9 shows the ratio of deflection measured at the joint (DR) to the deflection measured at the slab centre (DM). Also the load transfer efficiency as determined by means of load transfer measurements using the falling weight deflectometer is shown in this figure.

As one will notice the slow lane, indicated by index 1 shows lower load transfer capabilities than the fast lane. The measured deflection ratios can also be used to calcu-

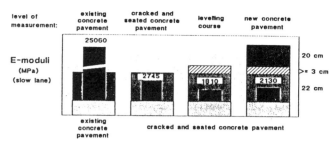

Fig.8. Moduli of a cracked and seated concrete pavement (source: Breyer e.a., 1990).

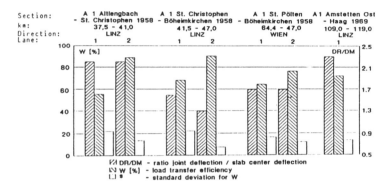

Fig.9. Load transfer at joints of several concrete pavements (source: Breyer e.a., 1990).

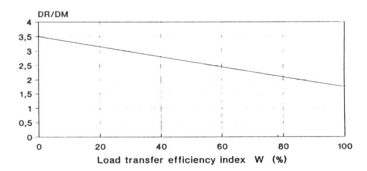

Fig.10. Load transfer efficiency chart (source: Breyer e.a., 1990).

late a load transfer efficiency. For this, figure 10 was developed; when the measured load transfer efficiency is less than the load transfer efficiency determined from the deflection ratio and figure 10, than the poor load transfer was most probably due to lack of support of the concrete slab near the joint. In other cases poor load transfer was most probably due to poor aggregate interlock and low dowel efficiency.

It is believed that this example nicely shows the capabilities of the falling weight deflectometer for the evaluation of cracked or jointed pavements and the analysis of the risks of future reflective cracking.

3.4 Crack and joint movement measurements

In the previous sections the importance of measurement of the movements across a crack or joint has already been indicated. As was shown earlier and will be shown later on, the movements across a crack in pavements containing cement treated bases is not only because of temperature variations resulting in mode I cracking; also mode II cracking, because of traffic, can be of very significant influence. It is therefore of importance to measure the movements across the crack or joint due to both environmental effects and traffic loads.

An effective way for making these measurements is by means of techniques like the Crack Activity Meter (CAM) as developed by the National Institute for Transport and Road Research in South Africa (Rust, 1986; also De Bondt e.a., 1992).

The principal of the CAM and an example of the results one obtains with it are shown in figure 11. This figure nicely shows the combination of both horizontal and vertical movements that takes place when a wheel load is passing the crack. It will be quite obvious that the magnitude of these displacements is a clear indication of the reflective crack potential of the structure under consideration.

Also it gives a clear indication of the possible maintenance strategies and needed pretreatment of the structure before the actual maintenance treatment can be applied.

3.5 Summary on evaluation of cracked pavements

In this chapter it has been shown that various measurements should be made in order to evaluate properly the behaviour of cracked pavements and to asses the reflective crack potential of the pavement.
These measurements are:

- visual condition surveys; in addition to that cores on cracks should be taken
- deflection measurements

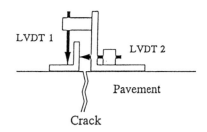

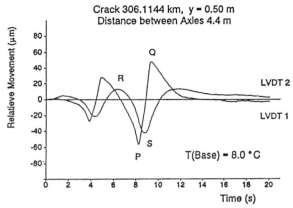

Fig.11. Principle of CAM and example of measurement result (sources: Rust and de Bondt e.a., 1992).

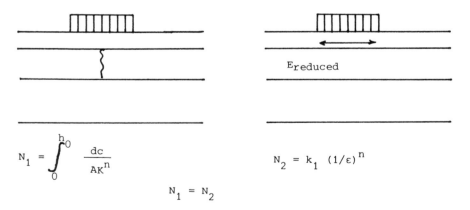

$$N_1 = \int_0^{h_0} \frac{dc}{AK^n} \qquad N_2 = k_1 (1/\varepsilon)^n$$

$$N_1 = N_2$$

Fig.12. Principle of simulation of reflective crack process by means of layered theory.

- measurements of movement of cracks subjected to loads.

4 Data analysis

The main goal in the analysis of data obtained during the evaluation measurement process on a cracked pavement should be the assessment of the reflective crack potential of the structure. By this it is meant the assessment of the forces to which any maintenance strategy will be subjected when applied over the cracked pavement under consideration.

From the description given in the previous sections it will be clear that this assessment becomes more complex if one has to deal with asphalt pavements having cement treated bases or concrete pavements because in that fact various factors are influencing the movements at the cracks and joints.

One should however be aware of the fact that also cracked asphalt pavements on unbound bases will have a certain reflective crack potential. Both types of pavements will be discussed here.

4.1 Engineering approaches

Since cracked pavements cannot be analyzed using the well known linear elastic multi layer theory (BISAR), finite element codes should be used to evaluate the measuring data. One major disadvantage of FEM applications still is that they are considered to be too cumbersome (time consuming and complicated) and therefore not suited for every days practice.

In time a number of engineering approaches have been developed which are based on FEM analyses of cracked pavements, application of fracture mechanics principles and simulations of the behaviour of cracked pavements by means of linear elastic multi layer systems.
An example of such an approach has been given by Van Gurp e.a. (1989).

They simulated the crack propagation through an overlay by using a traditional fatigue analysis of the overlay. The principle of the procedure is shown in figure 12.

In order to be able to allow fatigue damage to occur, the stiffness of the existing cracked asphalt layer needed to be reduced in order to be able to take into account the reflective crack potential of that cracked layer.

The effective modulus of the cracked pavement layer that needed to be used in the multi layer analysis was shown to be mainly dependent on the overlay thickness, the thickness of the existing cracked asphalt layer (it was assumed that this layer was completely cracked from bottom to top), the amount of aggregate interlock across the existing crack and the stiffness of the overlay mix.

Figure 13 shows the chart developed for the assessment of the effective modulus of the cracked existing asphalt layer. From this figure one can observe that in case of rather thin overlays (thinner than 50 mm), a relatively low effective modulus needs to be adopted (500 - 1000 MPa).

This indicates that in those cases the cracked existing asphalt layer should be rated in the reflective crack analyses as a good quality stone base!

A simular approach has been reported by Molenaar e.a. (1990). In this study the equivalent stress intensity factor at the top of the crack reflecting from the existing pavement through the overlay was shown to be dependent on the surface curvature of the deflection bowl, the stiffness of the overlay mix and the ratio of the thickness of the overlay over the thickness of the cracked existing asphalt layer (figure 14).

The deflections were assumed to be measured by means of a falling weight deflectometer (F= 50 kN, t = 25 ms) and the surface curvature index was defined as the difference between the maximum deflection and the deflection measured at 600 mm from the loading centre.

The equivalent stress intensity factor is the stress intensity factor that takes into account the combined effect of K_1 (bending mode) and K_2 (shearing mode) due to traffic loads. The value for K_{1eq} was used in the calculation of the number of load cycles needed to have the crack propagated completely through the overlay. This calculation was done by means of Paris' law being

$$dc/dN = AK_{1eq}^n \qquad (2)$$

where dc/dN = increase of crack length c per load cycle N
A, n = material constants

$$N = h_0 / dc/dN \qquad (3)$$

where N = number of load repetitions to complete reflection of the crack
h_0 = overlay thickness

This analyses results only in a crude estimation of the number of load repetitions to complete reflection of the crack through the overlay since it does not take into account the increase of K_{1eq} when the crack propagates through the overlay.

Through additional work reported by Jacobs e.a. (1992), predictive equations were developed which take care for this phenomenon.

One of the main questions of these engineering approaches is of course the validity of the predictions

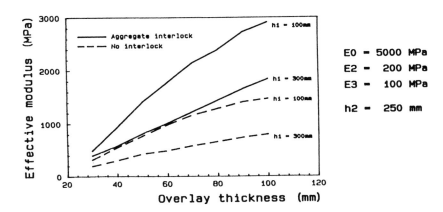

Fig.13. Effective modulus of cracked existing asphalt layer in relation to its thickness and the overlay thickness; reference modulus value of the cracked layer in 3000 MPa (source: Van Gurp, 1989).

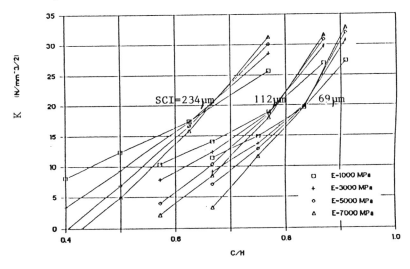

Fig.14. Equivalent stress intensity factor at tip of crack entering the overlay in relation to the surface curvature index of the existing pavement and the ratio c/h: thickness cracked asphalt thickness over total asphalt thickness (including overlay) (source: Molenaar, 1990).

made for conditions which are different from those for which they have been developed.

Another important question is of course how to arrive to the crack growth characteristics A and n of Paris crack growth law without the necessity of complicated testing. This last aspect will be discussed later on this paper.

4.2 Fundamental approaches

The fundamental approach to the analysis of cracked pavements and especially the analysis of the reflective crack potential, involves the use of finite element programs. These programs should allow a proper modelling of the load transfer that takes place across cracks due to aggregate interlock.

This means that special interface elements are needed since in ordinary finite element programs only smooth cracks can be modelled.

Special interface elements are also needed in order to be able to characterize the effects of reinforcing materials.

Up till recently no finite element programs were available that were fitted for the purpose of analyzing cracked pavements and the effects of various maintenance strategies. The available programs all suffered from lack of proper modelling of the various interfaces in the pavement structure (especially existing cracks, interlayers etc.).

Quite recently however a powerful tool has become available through research efforts made at the Delft University of Technology in the Netherlands (Scarpas, 1991; Scarpas e.a., 1993; De Bondt e.a., 1992). In this computer code CAPA, allowance is made for special interface elements that model the shear transfer across a crack or joint (figure 15).

Also special attention is given to the modelling of reinforced overlays (figure 16). This is believed to be an important aspect since the pressure of an adequately anchored reinforcement prevents unrestrained crack opening and therefore supports the development of a more efficient friction/aggregate interlock. For these reasons a special reinforced interface element has been developed which consists of the superposition of two ordinary interface elements and a truss element (figure 17).

Two interfaces elements, one on each side of the reinforcement, are considered because of the fact that due to construction difficulties the bond between the reinforcement and the existing pavement might be quite different from the bond between the reinforcement and the overlay! It is quite obvious that proper material testing should be done to obtain characteristics for the different elements involved. A testing program is currently undertaken at the Delft University to obtain these characteristics.

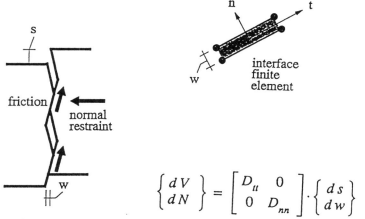

Fig.15. Interface element to model load transfer at crack (source: Scarpas, 1991).

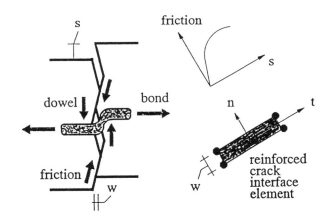

Fig.16. Reinforced crack interface element (source: Scarpas, 1991).

It is believed that such analysis tools will allow in depth evaluations to be made on the reflective crack potential of pavements and the effects of various possible maintenance strategies including the use of reinforcements and interlayers.

5 Material characterization

As has been stated before, proper analysis of cracked pavements and potential maintenance strategies involves the use of finite element codes and fracture mechanics principles.

In order to be able to do so a relatively large number of material characteristics should be known as for instance crack growth characteristics of the materials used.

As indicated in the previous section also information on the interface conditions should be available.

5.1 Crack resistance of overlay materials

For pavement evaluation purposes it is not feasible to rely on sophisticated testing to obtain these characteristics.
Time and budget constraints will normally not allow accurate testing of materials. The question then is how to obtain in a simple and effective way these important parameters.

In this section some attention will therefore be paid to the assessment of the constants A and n in Paris' crack growth law in case of asphalt concrete overlays.

According to Schapery (1973, 1981) crack growth in visco elastic materials can be described using Paris' law which is:

$$dc/dN = AK^n \tag{4}$$

The constants A en n are highly dependent on material characteristics like tensile strength, fracture energy, elastic stiffness. According to Schapery A and n can be determined using:

$$A = \frac{\pi}{6\sigma^2 \, I^2} \left\{ \frac{(1-\mu^2) \, D1}{2\Gamma} \right\}^{1/m} \int_0^{\Delta t} w(t)^{2(1+1(m)} dt \tag{5}$$

$$n = 2(1 + 1/m)$$

where σ = tensile strength at temperature and loading time conditions which are prevailing in the field

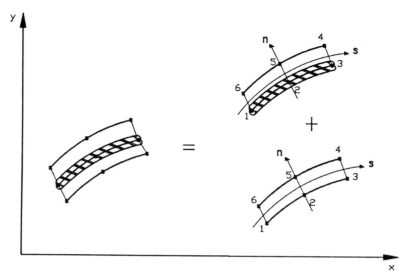

Fig.17. Reinforcement interface element (source: Scarpas, 1991).

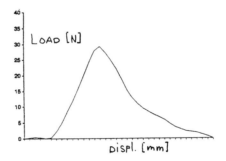

Fig.18. Fracture energy determined using indirect tensile test; the fracture energy is the area enclosed by the load vs vertical displacement graph.

I	= constant taking values between 1 and 2; 1.5 is a reasonable assumption
D_1	= 1/E at loading time of 1 sec
E	= modulus at the given temperature
Γ	= fracture energy (see figure 18)
m	= slope of the relation between D and loading time given the prevailing temperature conditions
w(t)	= shape of the stress pulse as a function of time
μ	= Poisson's ratio

From an extensive testing program (Molenaar, 1983), it was shown that n is also dependent on the void content of the asphalt mix following

n_{cor} = n / CF
CF = -0.93 + 0.65 Va
Va = void content in %

The elegance of this approach is that relative simple tests can be used to obtain the constants from Paris' crack growth law.
The tensile strength and fracture energy can be calculated from the results of e.g. indirect tensile tests.
The constants D_1 and n can be determined from modulus testing or, if these tests cannot be performed, estimated from available nomographs and formulas for the prediction of Smix.
In order to be able to use this approach for every days engineering purposes, rather simple predictive equations have been developed for the assessment of A. It has been shown by Molenaar (1983) that A can be estimated using

$$\log A = -0.977 - 1.628\, n_{cor} \qquad (6)$$

This equation was developed using the test results obtained on a number of reference mixes. It shows clearly the importance of knowledge on the slope of the compliance curve (log D vs log t) as well as knowledge on the void content of the mix used.
In case polymer modified mixes are used, this relation cannot be used. In those cases the ratio A of the polymer modified mix over A of the reference mix should be calculated first of all from the ratios σ of the polymer modified mix over σ of the reference mix, Γ of the polymer modified mix over Γ of the reference mix, m of the polymer modified mix over m of the reference mix and D_1 of the polymer modified mix over D_1 of the reference mix. Then the A of the polymer modified mix can be calculated from the A of the reference mix (calculated from the equation given above) and the ratio of A of the

polymer mix over A of the reference mix as determined following the procedure mentioned here.

It is believed that in this way valuable estimates can be obtained in a rather simple way on the crack growth characteristics of various possible overlay materials.

5.2 Interface conditions

As indicated in the previous chapter, proper characterization of the interface conditions is essential in order to be able to model the movements across a crack or joint and in order to be able to model the effect of various interlayers.

In order to be able to arrive to these characteristics a test set up as shown in figure 19 can be used; this test set-up is developed at the Delft University of Technology (de Bondt e.a., 1992).

Although this test reveals important fundamental characteristics, it is too complicated to be used in every days evaluation practice. Therefore it was investigated to what extent a simple shear test as indicated in figure 20 can be used for rating and quality control purposes, related to the use of reinforcing materials.

Some examples of test results obtained in this way are shown in figure 21. As one will observe there is a remarkable difference in shear performance of the various reinforcing materials which will undoubtedly have a significant influence on their performance with respect to retardation of reflective cracking.

More work is certainly needed in order to be able to draw firm conclusions on the applicability of this test for the purposes mentioned above.

6 Conclusions

In this paper attention has been paid to the evaluation of cracked pavements with emphasis on reflective cracking.

It has been shown that reflective cracking can be a major problem on any overlaid cracked pavement; it is certainly not restricted to pavements with cement treated bases and concrete pavements.

It has also been shown that much work should be done on the assessment of the movement that take place across a crack or joint since this will have a major effect on the maintenance strategies to be selected.

Furthermore it has been shown that much attention should be given to the characterization of the crack resistance of possible maintenance strategies. This not only means characterization of the crack resistance of the asphaltic overlay material but also of the effects of the possible reinforcing materials and interlayers.

Special attention should be given to proper modelling

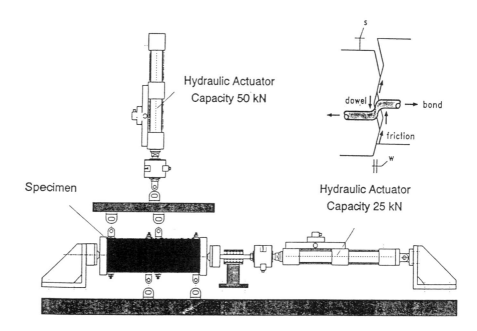

Fig.19. Test set-up the characterize reinforced cracked asphalt layers (source: De Bondt e.a., 1992).

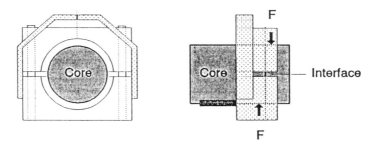

Fig.20. Simple shear test frame to be used under Marshall press for testing shear resistance of interfaces.

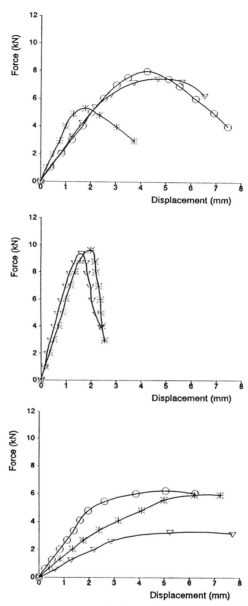

Fig.21. Typical examples of shear test results obtained on three different materials using test frame shown in figure 20 (source: De Bondt e.a., 1992).

of the various interfaces that can be recognized in the overlayed cracked pavement.

From the material presented here it will be obvious that a lot of work still has to be done before an adequate in depth analyses of cracked pavements and their reflective crack potential is possible.

7 References

Bondt, A.H. de, Saathof, L.E.B., Steenvoorden, M.P. (1992) **Test sections asphalt reinforcement A50 Friesland**, Report 7-92-209-21, Road and Railroad Research Laboratory, Delft University of Technology, Delft.

Bondt, A.H. de, Scarpas, A. (1992) CAPA: a modern tool for the design of pavements (in Dutch), in **Proceedings Wegbouwkundige Werkdagen,** Ede, CROW Vol II pp. 781-792.

Breyer, G., Fuchs, M., Litzka, J., Molenaar, A.A.A., Nievelt, G. (1990), Survey of reconstruction methods of worn-out (aged) rigid pavements, in **Proceedings Third International Conference Bearing Capacity of Roads and Airfields,** Trondheim, Vol II pp. 1147-1157.

Colombier, G (1989) Fissuration des chaussees, nature et origine des fissures, moyens pour maitiser leur remontee, in **Proceedings Conference on Reflective Cracking in Pavements,** Liege, pp. 3-22.

Dickinson, E.J. (1984) **Bituminous Roads in Australia**, Australian Road Research Board, Vermont South, Victoria.

Gerritsen, A.H, van Gurp, C.A.P.M., van der Heide, J.P.J., Molenaar, A.A.A., Pronk, A.C. (1987) Prediction and prevention of surface cracking in asphalt concrete pavements, in **Proceedings Sixth International Conference Structural Design of Asphalt Pavements,** Ann Arbor, Vol I pp. 378-391.

Groot, E. de, Houben, L.J.M., Molenaar, A.A.A. (1983) **Analysis of the behaviour of concrete block pavements using linear elastic multi layer theory,** Report 7-83-200-7, Road and Railroad Research Laboratory, Delft University of Technology, Delft.

Gurp, C.A.P.M. van, Molenaar, A.A.A. (1989) Simplified method to predict reflective cracking in asphalt overlays, in **Proceedings Conference on Reflective Cracking in Pavements,** Liege, pp. 190-198.

Jacobs, M.J.J., Bondt, A.H. de, Molenaar, A.A.A., Hopman, P.C. (1992) Cracking in Asphalt concrete pavements, in **Proceedings Seventh International Conference on Asphalt Pavements,** Nottingham, Vol I pp. 89-105.

Koning, P.C., Molenaar, A.A.A. (1987) Pavement management system for municipalities with emphasis on planning and cost models, in **Proceedings Sixth International Conference Structural Design of Asphalt Pavements,** Ann

Arbor, Vol I pp. 1041-1049.

Molenaar, A.A.A. (1983) **Structural performance and design of flexible road constructions and asphalt concrete overlays**, Dissertation, Delft University of Technology, Delft.

Molenaar, A.A.A. (1989) Effects of mix modifications, membrane interlayers and reinforcements on the prevention of reflective cracking of asphalt layers, in **Proceedings Conference on Reflective Cracking in Pavements,** Liege, pp. 225-232.

Molenaar, A.A.A., Stet, M.J.A. (1990) Surface curvature index of a deflection profile, a measure for the assessment of pavement life, overlay design and mix selection (in Dutch), in **Proceedings Wegbouwkundige Werkdagen,** Ede, CROW Vol I pp. 129-142.

Nunn, M.E. (1989) An investigation of reflective cracking in composite pavements in the United Kingdom, in **Proceedings Conference on Reflective Cracking in Pavements,** Liege, pp. 146-153.

Rust, F.C. (1986) **A detailed description of the working of the crack-activity meter (CAM),** Research Report RP/36, National Institute for Transport and Road Research, Pretoria.

Scarpas, A. (1991, 1992) **Personal communication on finite element analyses on cracked and reinforced pavements,** Structural Mechanics Department, Delft University of Technology, Delft.

Scarpas, A., Blaauwendraad, J., de Bondt, A.H., Molenaar, A.A.A. (1993) CAPA: a modern tool for the analysis and design of pavements, **Paper to be presented at the Int. Conference on Reflective Cracking in Pavements,** Liege.

Schapery, R.A. (1973) **A theory of crack growth in visco-elastic media**, Report MM 2764-73-1, Mechanics and Materials Research Centre, Texas A&M University, Texas.

Schapery, R.A. (1981) On visco-elastic deformation and failure behaviour of composite materials with distributed flaws. in **"1981 advances in aerospace structures and materials AD-01"** (editors S.S. Wang and W.J. Renton), American Society of Mechanical Engineers, New York.

Shackel, B. (1992) Computer based procedures for the design and specification of concrete block pavements, **Proceedings Fourth International Conference on Concrete Block Pavements**, Auckland, Vol I pp. 79-88.

3 LES PROCÉDÉS UTILISES POUR MAITRISER LA REMONTÉE DES FISSURES: BILAN ACTUEL
(Retarding measures for crack propagation: state of the art)

M.G. COLOMBIER
Laboratoire Régional des Ponts et Chaussées d'Autun, France

Résumé
De très nombreuses techniques ont été mises au point et sont utilisées pour ralentir ou empêcher la remontée des fissures dans les couches de chaussées. Le texte ci-après présente, par grandes familles, tous les procédés actuellement connus avec leurs différentes variantes. Pour chaque type de procédés ou de produits, on essaie de faire un bilan du comportement de manière à faire ressortir les avantages de chacun d'eux et de définir leurs domaines d'emploi privilégiés.
<u>Mots clés</u> : Chaussées, Fissures, Fatigue, Retrait, Géotextiles, Membranes, Enduits, Enrobés bitumineux, Bitumes élastomères, Géogrilles.

1 Introduction

Les interventions du précédent colloque " Reflective Cracking in Pavements - Assesment and Control ", qui avait eu lieu à LIEGE en Mars 1989, avaient permis de montrer l'extrême diversité des fissures que l'on peut rencontrer sur les chaussées. Cette diversité, qui s'explique par les causes multiples qui peuvent être à l'origine du phénomène de fissuration, se traduit par des fissures extrêmement variées dans :

* leur forme
* leur configuration
* leur ouverture

Ces différentes fissures se transmettront donc, selon des schémas et avec des vitesses eux-mêmes très différents, à une couche supérieure.
" <u>A chaque type de maladie son ou ses remèdes</u> ", c'est ce qui explique les très nombreuses techniques qui ont été imaginées et mises au point pour empêcher ou ralentir la remontée des fissures dans les chaussées. La panoplie des techniques proposées était déjà très large en 1989. Depuis 1989, les entreprises routières, les fabriquants de produits et de liants, les services techniques des

Reflective Cracking in Pavements. Edited by J.M. Rigo, R. Degeimbre and L. Francken.
© 1993 RILEM. Published by E & FN Spon, 2–6 Boundary Row, London SE1 8HN. ISBN 0 419 18220 9.

administrations et les laboratoires de recherche universitaires ont poursuivi leurs recherches, ce qui a élargi la gamme des solutions disponibles pour lutter contre la remontée des fissures et a permis d'améliorer l'efficacité des techniques les plus anciennes. L'objectif de ce texte est de dresser un catalogue le plus exhaustif possible des techniques actuellement connues, de décrire leurs avantages et leurs inconvénients et de proposer pour chacune de ces techniques des domaines d'emploi privilégiés.

2 Pourquoi vouloir empêcher la remontée des fissures dans les chaussées

L'apparition de fissures à la surface d'une chaussée est toujours considérée par les responsables de la gestion et de l'entretien des routes comme un phénomène contre lequel il faut se prémunir car il aura des conséquences néfastes sur la bonne tenue de cette chaussée.

Rappelons brièvement les conséquences défavorables des fissures de surface :

* perte d'étanchéité et risque de pénétration d'eau dans le corps de la chaussée et le sol support,
* accroissement des contraintes, au droit de la fissure, sur le sol support et dans le corps de la chaussée, ce qui réduit la durée de vie de la structure,
* dégradations de la couche de roulement au voisinage de la fissure sous l'effet du trafic, de l'eau, du gel etc..

Face à ces conséquences, les responsables de la gestion des routes peuvent avoir des exigences différentes selon l'importance économique et stratégique et le trafic des routes, le niveau de service qu'ils veulent offrir aux usagers, les crédits dont ils disposent.

Trois buts peuvent alors être visés :

* ne pas avoir de fissures visibles en surface,
* ralentir suffisamment la remontée des fissures pour qu'elles n'apparaissent pas avant un renouvellement normal de la couche de roulement,
* limiter la remontée des fissures et freiner leur évolution pour qu'elles ne nécessitent pas de travaux d'entretien et soient sans conséquences graves sur la tenue de la chaussée.

C'est en fonction des types de fissures en cause et des objectifs définis ci-dessus que les techniques, qui vont être présentées, pourront être choisies.

3 Les Méthodes pour maitriser la remontée des fissures dans les chaussées

Les solutions actuellement connues pour maitriser la remontée des fissures peuvent se classer en 3 grandes catégories.

a) On peut intervenir sur la cause de la fissuration, soit pour la supprimer, soit pour la maitriser de telle manière que les fissures soient moins actives. Rendre une fissure moins active vis à vis de sa remontée dans une couche placée au dessus d'elle consiste à rendre les mouvements des bords de cette fissure moins amples, moins rapides et moins fréquents.

b) On peut agir sur la nature et les caractéristiques de la couche qui sera mise en oeuvre sur la chaussée fissurée. Il s'agit dans ce cas d'utiliser un matériau plus résistant ou plus déformable dans lequel la vitesse de propagation de la fissure sera la plus faible possible.

c) On peut enfin introduire, entre la couche fissurée et la couche supérieure, un matériau ou un complexe qui assurera un découplage partiel entre les 2 couches, vis à vis des sollicitations lentes, tout en maintenant des liaisons suffisantes entre ces deux couches vis à vis des sollicitations rapides induites par le trafic. C'est à ce principe que font appel la plupart des solutions actuellement proposées pour lutter contre la remontée des fissures.

Rappelons que l'on peut envisager 3 types de fonctionnement pour justifier de telles solutions (Fig.1)

* La fissure existante se transforme, dans le matériau intermédiaire, en de multiples microfissures (Fig.1.1.) La phase d'initialisation de la fissure à la base de la couche non fissurée sera alors plus longue et le temps de remontée de la fissure, dont les mouvements à la base seront de faible amplitude, sera fortement allongé.

* La matériau intermédiaire se déforme sans se rompre sous l'effet des mouvements de la fissure (Fig.1.2.). La période d'amorçage de la fissure dans la couche supérieure sera alors très longue et le temps d'apparition de la fissure à la surface de cette couche sera rallongé d'autant.

* La fissure se propage dans le matériau intermédiaire ou à son interface selon un trajet long qui augmente le temps d'apparition en surface (Fig.1.3.)

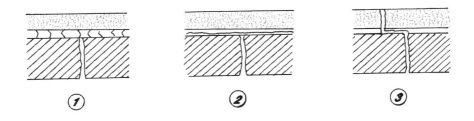

Fig. 1. Principes du découplage partiel

Il existe des moyens de réparer les fissures en les pontant sur une largeur de 10 à 20 cm par un film de bitume élastomère de 1 à 2 mm d'épaisseur. Cette méthode est efficace vis à vis de l'étanchéité qu'elle rétablit, mais elle est sans effet sur les conséquences mécaniques de la fissuration. Par ailleurs, le pontage des fissures rend beaucoup plus visible pour l'usager un phénomène dont il n'avait pas conscience et l'esthétique très discutable de ces réparations est très mal perçue par les maîtres d'ouvrages (Fig. 2).

Fig. 2. Fissures réparées par pontage

4 Intervenir sur la cause des fissures

Il existe deux possibilités pour agir sur les causes de la fissuration dans les chaussées. On peut agir sur le choix des matériaux et des structures pour éviter que des fissures se forment mais l'on peut aussi agir lorsque des fissures se sont formées pour supprimer ou réduire le phénomène qui les a créées.

4.1 Empêcher les fissures de se créer

Il s'agit essentiellement de respecter le code de bonne pratique en matière de construction routière dont on peut rappeler quelques points essentiels souvent à l'origine de phénomènes de fissuration.

4.1.1 Choisir des matériaux adaptés pour résister aux contraintes auxquelles ils devront résister pendant leur durée de vie. Il faut, par exemple, utiliser des liants bitumineux qui restent suffisamment déformables aux plus basses températures auxquelles ils seront soumis. Pour les matériaux traités par des liants hydrauliques, il faut éviter de cumuler tous les facteurs aggravant le retrait thermique (utilisation d'un granulat à fort coefficient de dilatation, module d'élasticité élevé, prise rapide du liant, mise en oeuvre par température élevée).

4.1.2 Dimensionner correctement les chaussées pour les niveaux de trafic auxquels elles seront soumises. C'est évidemment une des bases de la construction d'une chaussée sinon on verra apparaître rapidement une fissuration de fatigue.

4.1.3 Prévoir les dispositions constructives nécessaires. Dans ce domaine, il faut rappeler que les défauts de drainage et une protection insuffisante contre les phénomènes de retrait hydrique sont souvent à l'origine de fissures très importantes.

4.1.4 Appliquer les règles de l'art en matière de mise en oeuvre. Les défauts de collage entre couches, les joints longitudinaux et les reprises de répandage mal exécutés sont la cause de fissures qui peuvent être évitées.

4.2 Maitriser l'emplacement des fissures inévitables

Pour les chaussées à assises traitées par un liant hydraulique ou pouzzolanique, une fissuration de retrait thermique est inévitable pour le climat des régions tempérées dès lors que l'on vise des résistances et des modules d'élasticité supérieur à certains seuils. Lorsque les fissures se produisent naturellement, leur espacement est souvent voisin de 10 m et leur ouverture de quelques millimètres aux basses températures. Des techniques dites de PREFISSURATION ont été mises au point pour créer des fissures avec un espacement plus faible que celui des fissures naturelles afin de réduire l'ouverture et les mouvements des fissures ainsi créés.

Trois techniques de préfissuration sont actuellement opérationnelles. Il s'agit dans les 3 cas de créer, dans le matériau avant prise du liant, un joint ou une amorce de fissure à des espacements réguliers généralement compris entre 2 et 3 m.

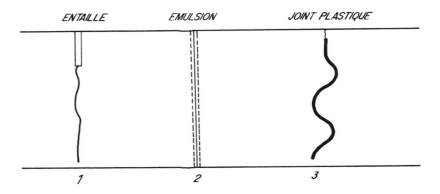

Fig. 3. Procédés de préfissuration

4.2.1 Découpage d'entailles à la partie supérieure de la couche (Fig. 3. 1.)
La méthode consiste à entailler la partie supérieure de la couche à préfissurer, après compactage. Les entailles, d'une profondeur de 5 à 10 cm, sont réalisées, soit à l'aide d'une plaque vibrante munie d'un couteau, soit avec un rouleau vibrant à double billes dont une des billes est équipée d'un disque de coupe.

4.2.2 Joint à l'émulsion de bitume (Fig. 3. 2.)
Le procédé consiste à créer un sillon, sur toute l'épaisseur de la couche à traiter avant son compactage, à l'aide d'un soc qui permet de projeter sur les parois du sillon une émulsion de bitume à rupture rapide. Le sillon est immédiatement refermé et le compactage de la couche traitée réalisé normalement. Par sa phase aqueuse à faible pH l'émulsion crée une zone de résistance plus faible favorable à la localisation de la fissure. Le film de bitume crée une discontinuité nette et rend insensibles à l'eau les lèvres de la fissure.

4.2.3 Joints ondulés (Fig. 3. 3.)

Le procédé consiste à pratiquer une saignée sur toute l'épaisseur de la couche répandue et légèrement compactée. On y introduit un élément de joint vertical à profil ondulé en matière plastique; on referme la saignée et on termine la mise en oeuvre normalement. L'élément de joint large de 2 mètres est placé transversalement dans l'axe de chaque voie. Sa hauteur est environ les 2/3 de l'épaisseur de la couche et il est placé en fond de couche. Le joint constitue une amorce de fissure et par sa forme ondulée permet un transfert de charges par engrènement des lèvres de la fissure quel que soit le matériau d'assise.

4.2.4 Domaine d'emploi des techniques de préfissuration

Destinées à traiter le cas des chaussées dont une couche au moins est traitée par un liant hydraulique, les techniques de préfissuration citées précédemment conduisent effectivement à des fissures :

* fines et rectilignes
* régulièrement espacées
* qui se dégradent moins que les fissures naturelles

La préfissuration n'évite pas la remontée des fissures dans les couches de roulement, mais elle retarde cette remontée. Associée à un traitement de la couche de roulement (enrobé bitumineux spécial, interposition d'un mortier bitumineux ou d'un géotextile imprégné), la préfissuration permet, soit d'éviter toute apparition de fissure en surface, soit de ne voir apparaitre que quelques fissures fines ne nécessitant aucun entretien.

4.3 Traiter la cause de fissures existantes avant mise en oeuvre d'une nouvelle couche de roulement.

Avant de mettre en oeuvre une nouvelle couche sur une chaussées fissurée, il est quelquefois possible d'intervenir sur la cause de la fissuration existante.
Pour illustrer cette possibilité, on peut rappeler quelques exemples qui avaient déjà été cités dans la conférence introductive du Colloque de Mars 1989.

* les fissures de fatigue de la couche de roulement sont dues au fait qu'elle est décollée de la couche inférieure. On peut, dans ce cas, raboter la couche jusqu'à l'interface et mettre en place une nouvelle couche de roulement bien accrochée à son support.

* les fissures de la couche de roulement bitumineuse sont dues au vieillissement excessif du bitume qui le rend fragile aux basses températures. On peut envisager de retraiter la couche en place en incorporant un régénérant qui rendra au bitume ses caractéristiques d'origine.

* les fissures sont dues à une perte de portance du sol support par excès d'eau. Il est possible, dans ce cas, d'assainir le sol par drainage et d'éviter que le phénomène se reproduise en étanchant la surface de la chaussée.
* les fissures sont dues à une fatigue généralisée de la structure. Un renforcement structurel bien dimensionné permettra alors de supprimer le problème.

A partir de cette série d'exemples, on voit qu'il existe des solutions permettant, en fait, de supprimer la cause d'un certain nombre de fissures d'une chaussée. Lorsqu'elles sont techniquement et économiquement possibles, ces solutions devront être préférées à toutes autres car elles sont évidemment les plus efficaces.

5 Intervenir sur la nature et les caractéristiques de la couche de roulement

5.1 Influence de l'épaisseur de la couche bitumineuse

Augmenter l'épaisseur de la couche bitumineuse mise en oeuvre sur une couche fissurée constitue une solution efficace pour retarder la réapparition des fissures en surface de la chaussée. En effet, l'augmentation de l'épaisseur de la couche supérieure réduit fortement les contraintes créées par le trafic au droit des fissures existantes (phase d'initialisation). En ce qui concerne le retrait thermique, l'augmentation de l'épaisseur constitue une protection qui réduit les écarts de température dans la couche fissurée diminuant ainsi l'amplitude des mouvements des bords de fissures (phase d'initialisation). Enfin, l'augmentation de l'épaisseur de la couche augmente directement la longueur du trajet que devra suivre la fissure initialisée pour atteindre la surface (phase de propagation).

De nombreuses études ont été faites pour modéliser le phénomène de remontée des fissures et calculer le temps de propagation de la fissure jusqu'à son apparition à la surface de la chaussée. Cette modèlisation est difficile car il faut apprécier l'importance respective du trafic et du retrait thermique qui sont les causes principales de propagation des fissures. Des constatations à caractère statistique réalisées en France sur un grand nombre de chaussées semi-rigides ont permis de définir le pourcentage de fissures remontant à travers une couche d'enrobé bitumineux en fonction de son épaisseur après 5 ans de trafic (Fig.4)

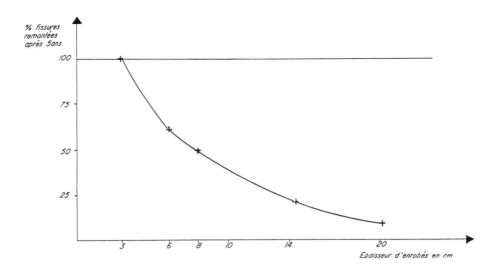

Fig. 4. Influence de l'épaisseur de la couche de roulement

Augmenter l'épaisseur de la couche bitumineuse est donc une technique efficace mais qui, pour éviter toute apparition de fissure en surface, dans le délai normal de renouvellement d'une couche de roulement, nécessite, pour les chaussées semi-rigides, des épaisseurs très importantes (environ 20 cm). C'est donc une solution très coûteuse.

5.2 Influence de la formulation et des caractéristiques de la couche bitumineuse

Les couches de roulement bitumineuses sont plus ou moins sensibles à la fissuration. D'une manière générale, les fissures se propageront d'autant moins facilement, dans ce type de matériaux, qu'ils seront capables d'accepter des déformations avant rupture importante même à basse température. La résistance à la fissuration d'un enrobé dépend essentiellement :

    * de la nature du granulat
    * de la richesse en liant bitumineux
    * des caractéristiques du liant

La nature du granulat intervient par l'intermédiaire de son coefficient de dilatation (sensibilité aux variations de température) et par la qualité des liaisons entre le granulat et le liant (adhésivité). On peut cependant considérer qu'il s'agit de facteurs secondaires en face de la richesse en liant et des caractéristiques du liant. Ces derniers facteurs jouent un rôle prépondérant sur la résistance à la fissuration des enrobés car ils interviennent très directement sur :

　　　　* la capacité de déformation élastique reversible, notamment aux basses températures, donc sur la possibilité qu'a l'enrobé d'encaisser les contraines transmises au droit des fissures,
　　　　* la capacité d'autoréparation de l'enrobé, c'est à dire la possibilité qu'ont les fissures de se fermer et se ressouder en été sous l'effet du trafic,
　　　　* la résistance au vieillissement, c'est à dire la capacité qu'a l'enrobé de conserver, dans le temps, les deux caractéristiques citées ci-dessus.

Pour obtenir ces diverses caractéristiques il faut choisir un liant peu visqueux et une teneur en liant élevée. Ce choix est malheureusement limité car il se fait au détriment de la résistance à l'orniérage et de la conservation des qualités d'antidérapance.

Pour résoudre cette contraction des objectifs, plusieurs solutions sont proposées.

5.2.1 Utilisation de bitumes polymères

Les principaux polymères utilisés sont les styrène-butadiène-styrène (SBS) et les éthylène-acétate de vinyle (EVA). Les premiers ont un caractère élastomérique, les seconds sont des plastomères. Ajoutés en quantité suffisante dans des bitumes sélectionnés, ils donnent des liants dont la viscosité est notablement supérieure à celle des bitumes purs normalement employés. Leur dosage dans l'enrobé peut alors être supérieur et atteindre 6,5 %.

5.2.2 Utilisation de bitumes modifiés par ajout de poudrette de caoutchouc récupéré

Cette technologie conduit à des liants extrêmement visqueux, qui peuvent être incorporés à de forts dosages, allant jusqu'à 7,5 % en couche très mince.

5.2.3 Addition de fibres

Dans cette technique la modification n'intervient pas au niveau du liant mais à celui du mastic. Les fibres utilisées, qui sont très fines et courtes, fixent une quantité importante de liant tout en " armant " le mastic. Les fibres minérales (roche, verre) ou organiques (cellulose) sont employées le plus souvent en conjonction avec un bitume pur. Les dosages en liant peuvent atteindre 7 % et plus.

5.2.4 Constatations

Les enrobés modifiés à forte teneur en liant ont une très bonne résistance en fatigue et au vieillissement. Vis à vis de la fissuration, les enrobés modifiés sont généralement considérés comme très efficaces lorsqu'ils sont appliqués sur des chaussées souples ou bitumineuses fissurées par fatigue. Appliqués sur des structures à assises traitées aux liants hydrauliques à forte fissuration de retrait, ces enrobés retardent mais n'empêchent pas la remontée des fissures. Il semble, par contre, que les fissures réapparues en surface restent, grâce à la richesse et au comportement du mastic d'enrobage, fines et qu'elles ne se dégradent pas ou peu. Certaines se referment en période chaude.

6 Interposition d'un produit ou d'un complexe entre la couche fissurée et la couche de roulement

Comme cela a été signalé au paragraphe 3, la plupart des solutions proposées comme SYSTEME ANTI FISSURES consistent à interposer, entre la couche fissurée et la couche bitumineuse que l'on envisage de mettre en oeuvre, un produit ou un complexe auquel on demande de jouer un triple rôle :
* dissiper, réduire ou répartir les contraintes extrêmement localisées qui apparaissent à la base de la nouvelle couche au droit des fissures existantes dans l'ancienne chaussée,
* assurer un collage suffisant entre la couche fissurée et la nouvelle couche de roulement de façon à ce que les contraintes liées au trafic se répartissent dans l'ensemble de la structure et ne conduisent pas à une fatigue trop rapide de la nouvelle couche de surface,
* maintenir l'étanchéité de la structure même dans le cas où des fissures sont réapparues en surface.

De tels procédés sont bien sûr utilisés dans le cadre de l'entretien de chaussées fissurées mais ils sont aussi quelquefois utilisés lors de la construction de chaussées neuves. Lorsqu'ils sont utilisés à titre préventif, les systèmes antifissures visent à ralentir ou empêcher la remontée de fissures que le constructeur de la route juge inévitables dans la structure de la chaussée. C'est notamment le cas des chaussées dont les couches d'assises sont traitées par un liant hydraulique et qui seront obligatoirement l'objet d'une fissuration de retrait thermique.

Pratiquement tous les systèmes antifissures utilisés actuellement peuvent se classer en 3 grandes catégories :

a) Les mortiers bitumineux riches en liants et en fines

b) Les membranes bitumineuses

c) Les fibres textiles imprégnées par un liant bitumineux

6.1 Mortiers bitumineux

C'est aux U.S.A que, dans les années 70, les premières applications d'un enrobé fin traité par un bitume fluidifié ou un bitume peu visqueux (200 - 300) sont signalées pour retarder la remontée de fissures d'origine thermique ou liées à la fatigue. Très rapidement la technique s'est développée en Europe (Grande Bretagne et France notamment).

6.1.1 Principe de fonctionnement

Il s'agit de réaliser une couche de mortier bitumineux de faible épaisseur (1 à 2 cm) souple et à forte teneur en liant qui sera interposée entre la couche fissurée et la nouvelle couche de roulement. L'objectif fixé à ce mortier est de dissiper les contraintes qui se concentrent au droit des fissures existantes et d'absorber, par glissement interne ou microfissuration, les déformations provoquées par les mouvements d'ouverture et de fermeture des fissures.

Par référence aux principes cités au paragraphe 3 (Fig.1) les mortiers bitumineux fonctionnent selon le schéma 1.

L'interposition d'un mortier bitumineux retarde l'amorçage de la fissure à la base de la nouvelle couche. De plus le temps de propagation de la fissure à l'intérieur du mortier est beaucoup plus long que dans un enrobé traditionnel. Les essais de laboratoire confirment cette hypothèse de fonctionnement (Fig. 5).

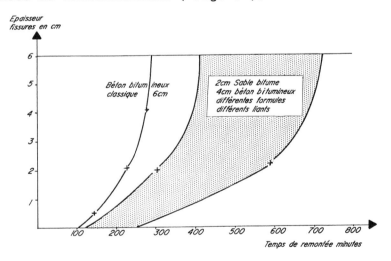

Fig. 5. Efficacité comparée d'un enrobé traditionnel et d'un enrobé bicouche d'après L.R.P.C. AUTUN.

6.1.2 Composition des mortiers antifissures

Ces mortiers sont constitués de sables d'origines diverses (matériaux éruptifs, calcaires durs, alluvionnaires). Les granulométries utilisées sont comprises entre 0/2 mm et 0/6 mm. Des fines calcaires sont toujours ajoutées. Si des bitumes purs sont quelquefois utilisés (pénétration 80/100 ou 180/200) l'emploi de liants modifiés est le plus souvent cité. Les additifs au bitume sont très divers (copolymères S.B.S. ou E.V.A., Styrène - Butadiène réticulé, poudrette de caoutchouc) . Des fibres (roche, verre, cellulose) sont quelquefois incorporées aux mortiers bitumineux pour fixer une quantité supplémentaire de liant, ce qui augmente l'efficacité de ce mortier vis à vis de la remontée des fissures.

Les formules les plus courantes répondent aux caractéristiques suivantes :

* Pourcentage de fines (< 0,08 mm)   10 à 15 %
* Dosage en liant   8 à 12 %
* Module de richesse   5,5 à 6 %

6.1.3 Couches de roulement

Les mortiers bitumineux de faible granularité et riches en liant ont une surface trop fermée pour servir de couche de roulement. Ils doivent donc être recouverts par un enrobé de surface qui assure les caractéristiques d'adhérence, répartit les charges dues au trafic et limite aussi les risques d'orniérage par fluage. Les enrobés de roulement utilisés sur des mortiers anti-fissures sont généralement de formulation classique et les épaisseurs les plus fréquentes comprises entre 3 cm et 8 cm. L'utilisation d'enrobés drainants en couche de surface est quelquefois citée.

Les ensembles mortier d'interposition + enrobé de surface sont souvent appelés " Enrobés - bicouches "

6.1.4 Comportement

L'ensemble des constatations citées dans la littérature montre que les mortiers bitumineux sont très efficaces pour lutter contre la remontée des fissures. De nombreux auteurs signalent qu'un tel système a évité toute remontée de fissures même dans des cas difficiles (trafic élevé, fissures nombreuses). Les mortiers bitumineux sont considérés comme efficaces vis à vis de la remontée des fissures d'origine thermique malgré la variation d'ouverture, dans le temps, de ces fissures.

Vis à vis de la résistance à l'orniérage, des défauts de comportement sont signalés pour les trafics les plus intenses lorsque l'on a utilisé des bitumes purs.

L'utilisation, pour le mortier bitumineux, d'un liant modifié semble éviter tout risque d'orniérage. Dans le cas de mise ne oeuvre du mortier sur une chaussée fortement déformée, ce qui conduit à des surépaisseurs locales importantes, les mortiers bitumineux peuvent se déformer par fluage sous la circulation.

6.1.5 Conclusions - domaines d'emploi

Les complexes **mortiers bitumineux - couche de roulement** sont le plus souvent considérés comme très efficaces pour éviter la remontée des fissures. Il s'agit par contre d'une solution coûteuse (1 cm de mortier au liant polymère coûte le prix de 2 cm d'enrobé bitumineux traditionnel). Pour les trafics importants, l'utilisation d'un bitume modifié par des élastomères est nécessaire pour éviter les risques de fluage. Les mortiers bitumineux ne doivent pas être employés sur des chaussées déformées. En conclusion, les mortiers bitumineux riches en liant sont à réserver, compte tenu de leur coût, au traitement de cas difficiles (trafics élevés, fissures actives).

6.2 Les enduits épais

Les enduits épais dénommés S.A.M. (" Stress absorbing membranes ") ont été développés aux U.S.A. à partir des années 1970; ils sont apparus un peu plus tardivement en Europe. La technique consiste à réaliser un film épais de liant bitumineux d'une épaisseur voisine de 3 mm (au lieu de 1 à 1,5 mm pour un enduit superficiel classique). De ce film épais de liant gravillonné, on attend qu'il supporte sans se rompre des déformations importantes même à basse température. Pour améliorer la susceptibilité thermique, augmenter les qualités élastiques et permettre une mise en oeuvre en couche épaisse sans coulage du liant avant gravillonnage, on utilise systématiquement des bitumes modifiés.

6.2.1 Les liants purs pour enduits épais

Deux grandes catégories de liant sont utilisées :

* Bitumes caoutchouc

Ces liants sont obtenus par ajout à un bitume pur de poudrette de caoutchouc obtenue par traitement de pneumatiques usagés. Plusieurs procédés permettent, par l'intermédiaire d'une huile compatibilisante, d'obtenir des mélanges homogènes de bitume et de caoutchouc. La quantité de caoutchouc introduite dans le bitume est importante (15 à 20 %), ce qui conduit à l'obtention d'un liant très visqueux (800 à 1000 cP à 200 °) et à cohésivité élevée.

* Bitumes polymères

L'utilisation de bitume modifié par ajout de polymères " neufs " (S.B.S., E.V.A., mélange S.B.S.-E.V.A.) est aussi fréquemment citée. Les quantités de polymères utilisés (de l'ordre de 10 %) conduisent à des liants moins visqueux que les bitumes caoutchouc. Les conditions d'application des bitumes polymères sont sensiblement les mêmes que celles des bitumes caoutchouc, leur viscosité élevée et leur faible mouillabilité nécessitent l'emploi d'épandeuses adaptées et le recours obligatoire à des gravillons laqués ou chauffés.

6.2.2 Comportement - Bilan des réalisations

Les jugements portés sur ce type de technique, quant à son efficacité à empêcher la remontée des fissures, sont très nuancés.
De manière générale, l'efficacité des enduits épais est jugée médiocre ou nulle vis à vis de fissures dites actives (fissures de retrait en particulier). Le jugement est plus favorable en ce qui concerne le traitement des fissures de fatigue des chaussées bitumineuses. Cependant la difficulté de réalisation de ces enduits, qui conduit à des échecs dès la mise en oeuvre, a fortement nui au développement de cette technique.

6.2.3 Conclusions - domaines d'emploi des enduits épais

Le coût élevé des enduits épais (équivalent à celui de 3 cm d'enrobé bitumineux), leurs conditions de réalisation délicates et leur inefficacité vis à vis de la remontée des fissures actives limitent fortement l'intérêt de cette technique dont l'emploi est stoppé dans de nombreux pays.

6.3 Les membranes bitumineuses

Les membranes bitumineuses ou S.A.M.I. (Stress Absorbing Membrane Interlayer) utilisent les mêmes liants que ceux des enduits épais (voir 6.2.1.) et avec des dosages voisins ou un peu plus élevés. Le gravillonnage est par contre réalisé avec des gravillons plus petits car la membrane est peu circulée avant la mise en oeuvre d'une couche de roulement en enrobés bitumineux de natures diverses et dont l'épaisseur varie généralement entre 4 et 6 cm.

6.3.1 Bilan de comportement des membranes bitumineuses

Le bilan de comportement des membranes bitumineuses est bon vis à vis du traitement des fissures de fatigue des chaussées souples. Avec cette technique on évite les risques d'échecs immédiats des enduits épais . Cette solution reste cependant toujours d'un coût élevé.

* Dans le cas du traitement des fissures " actives " c'est à dire ayant des mouvements importants de leurs bords sous l'effet du trafic ou des phénomènes thermiques, le bilan est beaucoup plus mitigé.
* Testées au laboratoire (Fig. 6) les membranes bitumineuses ont une efficacité proche de celle des mortiers bitumineux décrits en 6.1. Cette apparente contradiction avec le comportement sur la route peut s'expliquer par l'effet négatif du gravillonnage qui, sur chantier, par poinçonnement sous l'effet du trafic, créait un engrênement mécanique entre la couche fissurée et la couche de roulement.

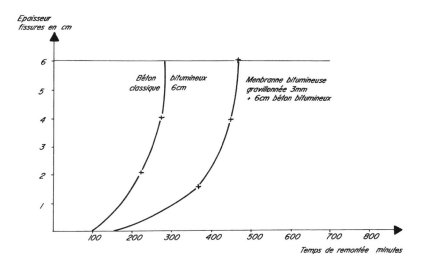

Fig. 6. Efficacité d'une membrane bitumineuse d'après L.R.P.C. AUTUN

Deux solutions ont été proposées pour pallier à cet inconvénient.
* Protéger la membrane par une mince couche d'enrobé à froid avant mise en oeuvre de la couche de roulement bitumineuse à chaud. Cette technique supprime le gravillonnage et empêche la migration du liant de la membrane dans l'enrobé de surface. Le recul est encore faible mais cette technique paraît très efficace.

* Ne réaliser la membrane qu'au voisinage des fissures en utilisant un liant très chargé en élastomère et légèrement sablé. En fait cette solution consiste à ponter les fissures préalablement à la mise en place d'une couche de roulement en enrobés bitumineux, selon une technique identique à celle citée au paragraphe 3 comme technique de réparation. Cette solution paraît efficace mais elle ne peut être utilisée que pour traiter des fissures en nombre limité et très localisées.

6.3.2 Conclusions - domaines d'emploi

Plus efficaces que les enduits épais, les membranes gravillonnées (S.A.M.I.) sont d'un coût élevé. Le recours à un gravillonnage préalablement à la mise en oeuvre de la couche de roulement fait perdre, par ailleurs, beaucoup de son efficacité à cette technique. Certains procédés proposés depuis peu pourront permettre de relancer l'utilisation des membranes bitumineuses peu employées actuellement.

6.4 Les géotextiles imprégnés de bitume

6.4.1 Historique

C'est vers les années 1970 que l'on utilise aux Etats-Unis, pour la première fois, des géotextiles non tissés en polypropylène imprégnés de bitume comme membranes antifissures sous une couche de roulement en enrobés bitumineux. La technique s'est très vite développée aux Etats-Unis, puis en Europe, au Japon et dans de nombreux autres pays. Le nombre important de communications consacrées à cette technique, présentées lors du précédent Colloque de LIEGE en 1989 montrait bien la très large diffusion des géotextiles imprégnés de bitume utilisés comme membranes antifissures. Depuis 1989, des enseignements ont pu être tirés des chantiers réalisés depuis les années 1980, ce qui a permis, en s'appuyant sur des études de modèlisation et des essais de simulation en laboratoire, de perfectionner et d'optimiser la technique dans toutes ses composantes. C'est ainsi que des progrès ont été réalisés dans le choix des liants et des dosages, la mise au point de géotextiles adaptés, le perfectionnement des matériels et des méthodes de mise en oeuvre et dans la définition des domaines d'emploi.

6.4.2 Principe de la technique

La technique consiste à réaliser, en place, sur la couche fissurée ou susceptible de se fissurer, un complexe constitué :
* d'un liant bitumineux assurant le collage entre la couche fissurée et la couche de roulement et dans l'épaisseur duquel on espère dissiper les mouvements horizontaux des lèvres des fissures. Du liant bitumineux on attend aussi une réduction des infiltrations d'eau.

* d'un géotextile dont le rôle essentiel est de permettre la mise en oeuvre d'une quantité importante de bitume sans risque de migration de ce bitume et de fluage de la couche de roulement. Le géotextile constitue donc un réservoir.

Le complexe ainsi formé fonctionne, théoriquement, selon le schéma 3 de la figure 1. Avec un tel système le temps d'initialisation de la fissure à la base de la couche supérieure doit être allongé.

6.4.3 Liants bitumineux - Nature et dosages

Les liants bitumineux sont mis en oeuvre, soit sous forme d'émulsion, soit de liants chauds. Les émulsions offrent l'avantage d'une mise en oeuvre facile quelles que soient les conditions climatiques. Par contre, les émulsions n'assurent un collage efficace du géotextile que si on attend leur rupture complète avant mise en oeuvre de la couche d'enrobé. Par temps très chaud, les émulsions (qui utilisent des bitumes relativement mous) ne garantissent pas contre un collage aux pneumatiques des camions. Lorsque les conditions climatiques le permettent, les liants répandus à chaud sont donc souvent préférés aux émulsions.

Si les premiers chantiers utilisaient des bitumes purs 80/100 ou 180/220, les essais de laboratoire réalisés dès 1980 ont montré tout l'intérêt des liants modifiés par des élastomères. Les liants élastomères offrent, par rapport aux bitumes purs, des capacités de déformation avant rupture plus importante. Par ailleurs la faible susceptibilité thermique de certains liants élastomères leur confère de meilleures performances aux basses températures.

Les dosages en liant doivent permettre de remplir totalement les vides du géotextile et d'assurer le collage de la couche fissurée à la couche supérieure. Selon la nature du géotextile, et à titre d'exemple, les quantités de liant utilisées avec un géotextile de masse spécifique de 120 à 150 g/m² se situent entre 800 g et 1 Kg/m².

6.4.4 Les géotextiles

De très nombreux types de géotextiles ont été utilisés. Si les non tissés aiguilletés en fibres de polypropylène semblent les plus employés, on utilise aussi des non tissés thermosoudés, des tissés et des fibres de polyester ou de polyéthylène.

L'utilisation de produits plus complexes tels que non tissés sur grille de verre, non tissé enduit de bitume est aussi citée.

Compte tenu du rôle de réservoir que l'on veut faire jouer au géotextile le portrait idéal d'un tel produit serait en fait le suivant :

* pourcentage de vide élevé mais faible compressibilité,
* bonne résistance à la température et faible coût,
* une certaine rigidité pour éviter les plis à la mise en oeuvre mais possibilité de déformation transversale pour s'adapter aux virages.

Toutes ces exigences sont souvent contradictoires. Les produits utilisés visent donc, en général, à proposer un compromis acceptable.

6.4.5 Essais en laboratoire

De très nombreux complexes liant bitumineux + géotextiles ont été testés par le Laboratoire Régional des Ponts et Chaussées d'AUTUN en France sur une machine qui avait été présentée dans le cadre du Colloque de 1989.
Dans des conditions expérimentales données, l'essai consiste à mesurer le temps de remontée d'une fissure utilisant le système à tester sous 6 cm d'un enrobé témoin et à le comparer au temps de remontée à travers 6 cm d'enrobé ou 2 cm de sable bitumineux recouvert de 6 cm d'enrobé. Les quelques exemples représentés sur la figure 7 montrent la grande diversité des efficacités constatées selon la nature des géotextiles et surtout des liants testés.

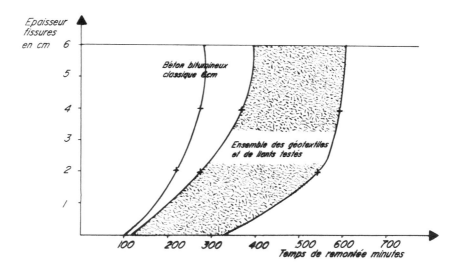

Fig. 7. Efficacité des géotextiles selon L.R.P.C. AUTUN

De manière générale, on peut cependant constater que la mise en place d'un géotextile imprégné retarde systématiquement l'amorçage de la fissure dans la couche de roulement. Les liants élastomères sont toujours plus efficaces que les bitumes purs.
Enfin les expérimentateurs n'ont jamais constaté de rupture du géotextile après apparition de la fissure en surface même pour des ouvertures de fissures plus larges que celles que l'on rencontre habituellement sur les chaussées.

6.4.6 Réalisation des chantiers

La méthode de réalisation la plus classique consiste à épandre en une seule couche la quantité de liant prévue (liant chaud ou émulsion) à l'aide d'une épandeuse pour enduits superficiels. Immédiatement , si l'on utilise un liant chaud, ou après rupture de l'émulsion, le géotextile est mis en place à l'aide d'une machine (porte rouleau maroufleur) (Fig. 8). Cet engin permet la pose du géotextile sans plis à des cadences élevées (jusqu'à 15 ou 20 000 m²/jour) . Les joints sont réalisés par recouvrement de 10 à 20 cm. La couche de roulement prévue (enduit superficiel ou enrobé bitumineux) est ensuite réalisée de manière classique.

Fig. 8. Mise en place d'un géotextile

6.4.7 Bilan du comportement

A la suite du développement rapide des années 1970, de nombreux constats d'échecs ont été faits aussi bien aux U.S.A. qu'en Europe. On peut expliquer ces échecs par l'utilisation de produits souvent mal adaptés (géotextiles et liants), sans études préalables et dans des contextes où l'échec était inévitable (fissures larges, fortement actives).

A partir de 1985, les produits et les moyens de mise en oeuvre étant mieux maitrisés, un constat plus positif et plus nuancé peut être fait.

* Cas des chaussées souples

Les géotextiles imprégnés de bitume sont considérés comme assez efficaces pour éviter la remontée des fissures de fatigue et des fissures liées à un vieillissement excessif du bitume de la couche de roulement.

Pour ces dernières, on constate souvent que les fissures ne réapparaissent pas en surface. Les fissures de fatigue, selon l'importance de cette fatigue, peuvent réapparaitre après quelques années de circulation. Dans ce cas, l'avis est généralement unanime, en ce qui concerne le maintien de l'étanchéité. On préconise généralement des géotextiles relativement absorbants et des teneurs en bitume d'imprégnation supérieures ou égales à $1kg/m^2$.

Lorsque la couche de roulement est constituée par un enduit superficiel, de bons résultats sont obtenus lorsque l'on utilise un liant élastomère pour imprégner le géotextile et réaliser l'enduit superficiel.

* Cas des chaussées semi-rigides

L'utilisation de géotextiles imprégnés de bitume pour traiter les fissures de retrait thermique des chaussées semi-rigides (possèdant des couches d'assises traitées par un liant hydraulique) s'est beaucoup développée dans tous les pays utilisant ce genre de structure (France, Espagne ...)

Pour ces applications, on constate généralement que l'interposition d'un géotextile imprégné ralentit la vitesse de remontée des fissures en surface mais n'évite pas, à terme, cette remontée. Lorsque le procédé est mis en oeuvre selon les règles, on constate, par ailleurs, que les fissures réapparues en surface sont plus fines que les fissures non traitées et qu'elles ne se dégradent pas ou peu dans le temps. Enfin, comme pour les chaussées souples, on constate que le géotextile ne se déchire pas si la fissure remonte, ce qui réduit très fortement le risque de pénétration d'eau dans le corps de la chaussée.

* Cas des chaussées béton
De nombreuses utilisations de géotextiles ont été faites pour éviter la remontée de joints ou de fissures de dalles bétons dans des couches d'entretien bitumineuses. La technique a beaucoup été utilisée sur des pistes d'aérodrome mais aussi sur des chaussées autoroutières.
Pour ces applications, on dispose rarement d'une section témoin, sans géotextile, permettant de montrer l'efficacité de ce dernier.
Les résultats semblent très variables en particulier selon les situations climatiques des sites concernés. De manière générale, les géotextiles imprégnés de bitume sont sans effets sur la remontée des fissures ou des joints dès lors qu'il y a battement, c'est à dire mouvement de cisaillement des bords des fissures sous l'effet du trafic.
Le procédé semble plus efficace vis à vis des fissures de fatigue que vis à vis des joints de construction.

6.4.8 Conclusions - Domaines d'emploi
Les géotextiles imprégnés de bitume constituent une solution relativement économique pour éviter ou retarder la remontée des fissures. L'emploi de liants élastomères semble augmenter fortement l'efficacité du procédé qui, même si les fissures réapparaissent en surface, maintient une certaine étanchéité. Les géotextiles imprégnés de bitume sont bien adaptés au traitement des chaussées souples fatiguées ou à celui des fissures de retrait des chaussées semi-rigides.

6.5 Fibres textiles imprégnés de bitume
Le principe consiste à mettre en place, entre la couche fissurée et la couche de roulement, une membrane constituée par un liant bitumineux armé de fibres courtes ou continues.
Selon la nature et le dosage en fibres et la quantité de bitume mise en oeuvre, les procédés se rapprochent des membranes bitumineuses ou des géotextiles imprégnés de bitume.

On peut citer deux procédés:

6.5.1 Utilisation de fibres continues
Le principe consiste à réaliser en place un complexe comprenant une membrane bitumineuse suivie par la projection d'une armature de fils continus. Un gravillonnage permet la circulation des camions et des engins utilisés pour la réalisation de la couche de roulement (enduit ou béton bitumineux).
Le liant est généralement une émulsion de bitume élastomère (1,5 à 1,7 Kg/m²).

Les fils synthétiques sont projetés à grande vitesse par une machine spéciale (Fig. 9) à un dosage voisin de 100 g/m². Un tel procédé s'apparente aux géotextiles imprégnés dont il a le même domaine d'emploi.

6.5.8 Utilisation de fibres courtes

Une entreprise propose la réalisation d'un enduit épais (S.A.M.) ou d'une membrane antifissure armée de fibres de verre (S.A.M.I.).

Le principe consiste à noyer une nappe de fibres de verre de 5 à 6 cm de long, entre 2 couches de liant (de préférence un bitume modifié) sous forme d'émulsion. Le complexe peut ensuite être gravillonné selon le procédé S.A.M. ou recouvert par une couche d'enrobés bitumineux procédé S.A.M.I.

La mise en place des fibres se fait à partir de bobines de grande longueur d'où les fils sont tirés puis coupés à grande vitesse. La répartition des fibres sur le sol est assuré par un dispositif à pression d'air (Fig. 10).

Le procédé testé en laboratoire parait efficace. Un recul insuffisant ne permet pas de connaitre son efficacité réelle sur chaussée.

Fig. 9. Machine à projeter les fibres

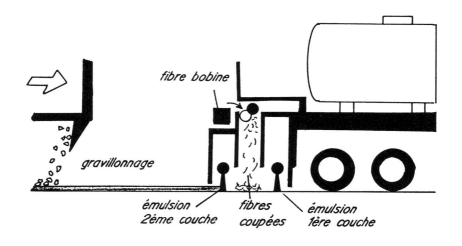

Fig. 10. Réalisation d'un enduit fibres courtes

6.6 Armatures et géogrilles

Les grilles et armatures sont utilisées depuis plus de 20 ans pour renforcer les enrobés bitumineux dont on souhaite améliorer les résistances en traction et en flexion, augmenter le module de rigidité et la résistance à la rupture par fatigue. Ces produits sont quelquefois proposés pour ralentir ou empêcher la remontée des fissures.

6.6.1 Les produits
De nombreux types de grilles et d'armatures sont proposés:

* grilles métalliques
* grilles en fibres de verre
* grilles en fibres de polyester à haute ténacité enduites ou non de bitume

6.6.2 Mise en oeuvre
Les grilles sont mises en oeuvre, soit directement sur la couche support sur laquelle on a épandu une couche d'accrochage bitumineuse, soit entre deux couches d'enrobés bitumineux.
Les grilles sont maintenues en place, avant mise en oeuvre de la couche de roulement, par clouage ou par des pelletées d'enrobé.

6.6.3 Comportement

L'efficacité des grilles est en général reconnue quant à leur amélioration de certaines performances des enrobés qu'elles arment. Dans ces conditions, leur utilisation pour l'entretien de chaussées souples fissurées par fatigue est courante dans plusieurs pays.
Les grilles sont considérées, par contre, comme inefficaces pour lutter contre la remontée des fissures de retrait thermique des chaussées semi-rigides. Leur efficacité est contestée pour le traitement des chaussées béton.

7 Conclusions

Comme l'avait déjà mis en lumière, le Colloque RILEM de LIEGE de 1989, la nature et les causes des fissures des chaussées sont extrêmement nombreuses et variées. Cela explique, en partie, le grand nombre de procédés ou de produits proposés pour éviter ou limiter les conséquences défavorables, pour la tenue de la route, de l'apparition de ces fissures en surface.

Mais cette grande diversité des fissures à traiter explique aussi qu'il n'existe pas de solutions aptes à résoudre tous les problèmes. A chaque procédé un domaine d'emploi privilégié. La gamme des produits disponibles en 1989 était déjà très grande; elle s'est encore élargie en même temps que les procédés les plus anciens se perfectionnaient.
Face à tous ces procédés, le maître d'oeuvre est donc confronté à un choix difficile.
Pour choisir il lui faut déjà savoir répondre à ces deux questions :

* Quel type de fissures je souhaite traiter ?

* Avec quel objectif ?

Ce n'est qu'ensuite que des comparaisons techniques et économiques lui permettront de retenir le procédé le mieux adapté à son problème.

BIBLIOGRAPHIE

COLOMBIER G. et all. " Fissuration de retrait des chaussées à assises traitées aux liants hydrauliques " Bulletin de liaison des L.P.C. FRANCE - N° 156 et 157 - JUILLET - SEPTEMBRE 1988.

Reflective cracking in Pavements - Colloque RILEM - LIEGE 8 - 9 - 10 MARS 1989

" Techniques pour limiter la remontée des fissures "
Note d'Information S.E.T.R.A. - L.C.P.C. (FRANCE)
N° 57 - MARS 1990

" Géotextiles et Technique routière " - Revue Générale des Routes et des Aérodromes - N° 679 - NOVEMBRE 1990

Fabrics in asphalt overlays and pavement maintenance
BARKSDALE R.D. - Transportation Research Board -
WASHINGTON - JULY 1991

" La lutte contre les Fissures " - Journée Ecole Nationale des Ponts et Chaussées - PARIS - 26 NOVEMBRE 1991

" Des chaussées semi-rigides sans fissures " - LEFORT M.
Revue " Laitiers sidérurgiques " - FRANCE - N° 74 -
JANVIER 1992

" Fisuracion de las mezclas con cemento causas y control "
VAQUERO J. - " Carreteras " - JUILLET - AOUT 1992

4 LABORATORY SIMULATION AND MODELLING OF OVERLAY SYSTEMS

L. FRANCKEN
Belgian Road Research Centre, Brussels, Belgium

Abstract

The paper recalls the basic information needed for describing and understanding the behaviour of an overlay system. It reviews models which are now proposed for analysis and design and summarizes the basic experimental procedures which have been used to evaluate the behaviour of such systems. It is concluded that the understanding of the problem has considerably improved during the last years. Further research remains necessary in the field of material characterization and design procedures including functional interlayers, anisotropy and non elastic components.

Key words: Reflection cracking, overlays, laboratory testing, models, design procedures, reinforcement, temperature, traffic loading.

1 Introduction

Reflective cracking is a major concern for engineers facing the problem of road maintenance and rehabilitation.
  There is evidence after several years of research and practice that materials and procedures having potential to improve the situation do really exist. But it must be emphasized that there is no standard solution suited in any case. Assessment has clearly revealed the complexity of the problem and also the fact that the terms "reflective cracking" cover a wide variety of phenomena.
  An efficient retarding measure does not simply consist in the use of a miraculous thin functional interlayer between the cracked pavement and a bituminous overlay. The success of innovative solutions depends on the correct choice of all the components of an overlay, on their combination and on their implementation in function of the loading conditions to which they will be exposed for a future design life.

Although no clear terminology exists yet we propose to call such a combination an **anti-reflective system**.

The justification of any innovation in this field will in most cases depend on its cost effectiveness in the long term, which includes initial investments and maintenance costs over the expected service life.

Designing a structure and evaluating its cost effectiveness is a current engineering task for which the essential starting elements are :

1. The characteristics of the basic materials
   They are determined by experimental test methods.

2. Analytical modelling of the behaviour of the system.
   Modelling starts from the input data describing the loading conditions and the materials. The essential part of a model is a computation tool to determine the stress-strain distribution in the structure. Another important part is the physical law used to describe the evolution of the damage.

3. Simulation of the system behaviour.
   Experimental simulation will allow the estimation of how a system behaves in reality and the verification of the agreement between prediction and actual behaviour in the field.

We will devote the present contribution to an overview of the elements which are required to carry out such evaluations.

2 Characterization of the basic components

The characteristics of the basic components of an anti-reflective cracking system are needed to allow :

1. the initial choice of a product according to standards;
2. the correct modelling of the pavement structure and the evaluation of its performance.

Table 1 : Characteristics for modelling and assessment

Bituminous layers	Stiffness modulus Poisson's ratio Thermal expansion coefficient Fatigue law Crack propagation law
Binders and Tack coat	Penetration, softening point, etc... Viscosity Shear modulus Temperature susceptibility
Interface layers	Modulus of elasticity Stiffness modulus Tensile resistance Thickness

The different components concerned are :

. the overlay, considered here to be a bituminous layer;
. the tack coat material;
. the interlayer product.

2.1 Bituminous overlay materials.

In most cases, the main part of the anti-reflective cracking system will be the bituminous layer itself.
The behaviour of this material in traditional types of overlays is the first information we need in order to have a reference to assess the future performance of more complicated structures, because any evaluation of the contribution of additional functional layers must be made in comparison with the traditional solution (so far the cheapest one for what concerns the construction costs).

What is of first importance in any case is the fact that we have to deal with a material displaying wide variations with temperatures and loading time in all its mechanical properties.

2.1.1 The modulus of bituminous materials

The dynamic modulus of bituminous mixes is a function of temperature and loading time. It can be represented under the form of a master curve built up from measurements carried out at different combinations of temperatures and frequencies by using shifting factors of the form :

$$\log \alpha_T = H \cdot (\frac{1}{T} - \frac{1}{T_S}) \tag{1}$$

in which H = 10920 K
T and Ts are respectively the test temperature and the reference temperature expressed in °K.

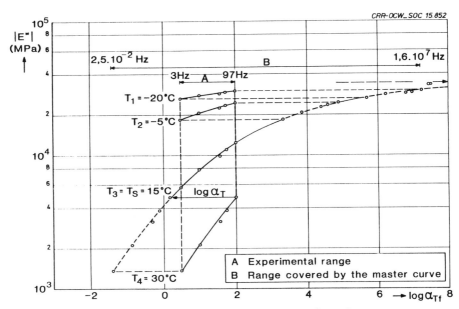

Figure 1 : Modulus master curve of a bituminous concrete

This modulus master curve can be estimated from the composition of the material and the binder characteristics for the case of normal bituminous binders (ref 1,2).

For the case of materials containing modified bitumens these evaluations are not correct any more and use must then be made of experimental evaluations.

It is worth mentioning here that the experimental procedures used in the determination of these basic characteristics is under study in another RILEM Committee (ref 3). For practical applications, modulus values of a given structure can be evaluated in the assessment phase by carrying deflection tests. Measured deflection profiles are then used to calculate in situ stiffness of the pavement layers using back-calculation techniques (ref. 6, 7).

2.1.2 Thermal properties

It is well known that thermal stresses can induce spontaneous surface cracking under severe winter conditions. Although this effect is not directly relevant to what is considered here as reflective cracking, the presence of such stresses superposed to the other loading conditions can have a strong influence on the resistance

of a pavement structure to cracks generated in the bottom layers (ref 4).
The magnitude of these stresses can be estimated from the thermal expansion coefficient and the stiffness modulus master curve. The resistance of the material to tensile stresses at low temperature can be determined either by tensile tests at low loading rates or by cooling tests on samples maintained at constant length. The critical temperature at which brittle fracture occurs may be determined in this way.

2.1.3 Performance laws
Complementary to the above mentioned data, bituminous materials must be characterized by performance laws which are used to define acceptance criteria for design purposes.
Each type of damage affecting the material can be described by a performance law. For the onset and propagation of cracks in bituminous overlay materials the most important ones are the fatigue law and the crack propagation law.

2.1.3.1 The fatigue law
This law allows the estimation of the number of loads N needed to initiate a crack resulting from the repetition of different strain (or stress) levels.

$$N = (\frac{C}{\varepsilon})^m \quad (2)$$

The parameters C and m can be experimentally determined on the basis of repeated bending tests.
For straight-run bitumen they can be readily estimated from parameters defining the volumetric mix composition and the binder rheological characteristics (ref 1,2).

2.1.3.2 The crack propagation law
The rate of crack growth in asphalt concrete can be predicted using the empirical power law developed by Paris and Erdogan (ref 5) :

$$\frac{dc}{dn} = A(\Delta K)^n \quad (3)$$

where :
ΔK = stress intensity factor amplitude
A, n = fracture parameters of the material
c = crack length
N = number of loading cycles

The stress intensity factor K is depending on the geometry of the specimen, on the mode of opening of the crack (mode 1 or 2) and on the crack length c.
For thermal movements of pavement materials only the where

opening mode 1 is most likely to occur. Traffic loads will introduce more complicated situations involving combinations of modes 1 and 2 (shear mode).
The Paris equation allows an estimation of the number of load repetitions Nf needed to propagate a crack through the overlay thickness h by integration (ref. 5, 6):

$$N_f = \int_0^h \frac{dc}{A \cdot (\Delta K(c))^n} \qquad (4)$$

The normal way to experimentally determine the material parameters A and n is to examine the stable crack growth through asphalt beam specimens under repeated loading conditions (ref 6,7).
Different set ups are possible, but what is essential is to measure the crack length c during the test over a sample geometry which is simple enough to allow an accurate estimation of the stress intensity factor K.
Although this approach is an excellent way to describe the reflection crack problem, it remains rather unpopular for the reasons that the input data needed for its implementation are very scarce and difficult to obtain by experimental means.
Some equations derived by Shapery (ref 8) can be used to determine the fracture mechanics parameters without performing expensive fracture tests when certain material properties are known.
However, there is no unanimity as to the validity of the basic theory in the case of heterogeneous materials such as bituminous mixes.
Paris crack propagation law assumes that once a small crack is created, for instance through fatigue, it will propagate as a simple plane discontinuity through the material considered as a homogeneous medium.
There is now evidence that this situation, which is more easy to handle as a concept in accordance with fracture mechanics theories, does not entirely correspond to reality. On the other hand a complete explanation can not be given unless the influence of temperature is included.
Observations by Jacobs (ref 9) have shown that a microcrack zone precedes the onset of macrocracks and that more fundamental principles such as the rate theory (ref 10) could be applied.

2.2 Binders and tack coat material

Before dealing with the case of the interlayer products it must be emphasized that the structural contribution of an overlay system is primary dependent on the way different interlayers interact. When the connection of the interface is purely mechanical (nailed or by granular interlock) the effect is governed by granular friction and tensile resistance of the links.

For most cases however the bond will be insured by a tack coat or a viscoelastic medium. In the case where the interlayer has a low contribution to the structural stiffness of the structure (SAMI or non woven interlayers) the effect of this component will be more and more important, and the properties like binder viscosity and resistance to low temperature fracture will be more importance than the properties of the fabric.

Experimental research has clearly shown to what extent the nature of the binder can influence the rate of crack propagation (ref 11).

2.3 Interlayer materials

The variety of interface products available on the market is very wide and a large diversity can be found in their mechanical properties, in their laying procedure, as well as in their cost price. Inquiries and surveys of actual implementations have shown that many field engineers consider an interface product, whatever it is, as a strengthening material to be applied for whatever purpose. Instead of this oversimplified -and wrong- statement, a thorough knowledge of the working mechanisms of these products can help in the choice of the proper solution to a specific type of damage. One important step in this choice relies on the identification of the product based on its physical and mechanical properties.

Most of the suppliers provide technical information allowing such a first choice but this type of information may also be different according to the type of product.

There is a real need for a rational classification of the products, and some base elements to do this can be found in specifications and state of the art reports (ref 12) but there is still a lack of homogeneity and consensus in the proposals so that much needs to be done in order to normalize the experimental characterization procedures . This is the type of issue this conference might help to address.

2.3.1 Thin interlayers

The interlayer thicknesses of foils, fabrics or grids range from some tenth of a mm to less than 2mm. In comparison with overlay thicknesses of several cm this is to be considered as an almost 2 dimensional foil. For this reason the stiffness of such a product is generally expressed in a force per unit of length which is equal to the component's modulus of elasticity times the thickness. The two dimensional stiffness allows a first classification to be made on the basis of the potential contribution of the interlayer to the strength of the overlay system . Table 2 gives an example of a tentative classification of fabrics and geosynthetics based on experimental data.

Table 2 : Example of tentative classification of fabrics and thin interlayers (ref 12)

Stiffness	Stiffness		Tensile strength		Failure elongation	
Units	kN/m		kN/m		% length	
Description	from	to	from	to	from	to
Very low	---	<140	9	26	10	100
Low	140	263	11	35	10	60
Stiff	263	701	15	175	10	35
Very stiff	701	1139	>61	--	5	15

NB : These values are purely indicative.

This characteristic, generally derived from tensile tests allows to determine the type of function the product is able to fulfil.

Very low or low stiffness products will of course not be suited as a strengthening material.
Such is the case of most of the non woven geosynthetics which thanks to their high porosity can be used as binder containers and play an important role in reducing stress concentrations.
In this case the stiffness of the product will be of less importance than its ability to absorb a given quantity of binder.
Very stiff interlayers on the contrary will be suited for strengthening purposes.
A given material will act as a reinforcement if its overall stiffness modulus is higher than that of the upper layer of the system.
Owing to the fact that the bituminous overlay material is temperature susceptible, the ability of a given interlayer to reinforce will depend on the temperature. This means that most of them will act as such only at the highest temperatures. One may conclude from this that the reinforcement function is generally effective for traffic loads in the medium and high temperature range but that it is inefficient under winter conditions.
For some of these products such as grids or woven

fabrics these properties are strongly anisotropic so that for two main directions different stiffness values have to be attributed to a same product. Although many information is now available on the characteristics of the basic products of interlayer systems, it is noticeable that very little information is to be found on the characteristics of the interface once it is in place. This cannot be obtained by simply combining the properties of the individual components (for example tack coat binder + membrane or grid), but it is the kind of information which is required in order to get correct evaluations.

2.3.2 Thick interlayers

The functional layer of the system can also consist of an improved mix composition, a sand asphalt or a bituminous mix containing fibre reinforcing materials (ref 13).
The characteristics needed in these cases are the same as for the asphaltic overlay (moduli, fatigue and crack propagation law). But in some cases these characteristics can display a stronger anisotropy. A classification of these products may be based on their stiffness modulus and performance criteria such as fatigue and cracking resistance.

3 Modelling

A model is constituted by a computing technic and a set of physical deterioration laws describing the behaviour of the structure and its evolution under service conditions. It starts from input data describing the structure, its components, the environmental and loading conditions.

The choice of the model is critically dependent on a correct evaluation of the case. It can be profitably supported by the extensive review of the possible cases of reflective cracking given by Colombier (ref 14).
Road structures concerned by reflective cracking may be divided into the 3 following categories :

. Rigid pavements with an overlay
. Semi-rigid pavements.
. Flexible pavements with or without overlay.

Besides these classical road structures, many situations do exist in which discontinuities may have been introduced for different reasons (road widening, local repairs, transversal or longitudinal trenches, local weaknesses of the subgrade, etc).
An example given in figure 2 illustrates the stress created by a wheel travelling over an overlay (or anti-

reflective system) laying on to an old cracked surface course which is now considered as the actual base layer of the system.
The development of a crack in the overlay may generally be decomposed in three steps involving different kinds of physical processes :

- initiation of the crack
- propagation of the crack
- pavement failure when the crack reaches the surface .

At each step different physical laws can be applied according to the type of structure concerned and the predominant loading conditions applied to it.
The driving forces for crack initiation and propagation are:

. traffic
. temperature variations
. the hydric variations of the soil

These different conditions may appear independently or jointly as the main causes of the distress.

This will result in complex situations which may be treated separately in a first approach and superposed to get a final picture of the reality.

1. Traffic loads travelling over a crack in a lower base layer generate three successive stress pulses, two in the shearing mode and one in the opening mode when the load is over the crack (fig.2).

2. Temperature changes cause the base layer and overlay to try to expand and contract, generating the mode 1 opening of the crack.

3. Some observations are reporting cracking due to the thermal warping of overlaid cement slabs (ref 15). In such cases cracks are initiated at the surface and progress downwards.

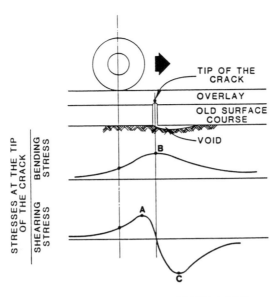

Figure 2 : Stresses induced at the cracked section of an overlay due to a moving wheel load (ref.6)

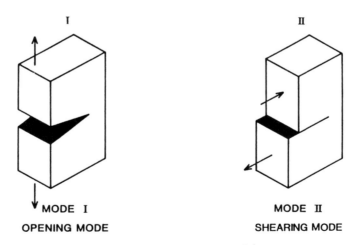

Fig 3 : Two modes of crack propagation

Table 3 : Models for reflective cracking

Model	Short description	References
A	Multi-layer linear elastic	16,17,18,19
B	Extended Multi-layer	19,20,21
C	Equilibrium equations	22
D	ODE Overlay design model	6
E	FE/FD analysis + fracture mechanics	23,24,25,26
F	Blunt crack band theory	27,28

	Model	A	B	C	D	E	F
Function	Stress/strain analysis	X	X			X	X
	Service Life prediction	X	X	X	X	X	
	Overlay design	X	X	X	X	(X)	
	Performance based comparison	X	X	X		(X)	(X)
Computing tool	Analytical	X	X	X	X		
	Finite Elements				X	X	(X)
	Finite differences					(X)	(X)
Dimensions	2 or 3 D	3	3	2	2	2	2
Loading conditions	Traffic	X	X	X	X	X	
	T.shrinkage base layer			X	X	X	X
	T.shrinkage upper layer			X	X		
	Warping				X	(X)	X
Damage law or criterium	Fatigue	X	X	X		(X)	
	Crack propagation		X		X	X	
	Thermal cracking			X			
Structure	Multilayer	X	X			X	
	Beam on foundation			X	X		
	With interlayer	(X)	(X)	X		(X)	X

(X) : only in some cases

3.1 Review of some models for analysis and design
This review will give a short description of a number of models which were proposed and published these last years (see table 3). They are the result of strong research efforts in order to develop valuable tools for structural evaluation and design.

3.1.1 Multilayer linear elastic model (A in table 3)
The multilayer linear elastic model has gained the stage of implementation since many years.
   The stress analysis tool of these models is based on the generalized elastic layer theory of Burmister (ref 16). This method works under the assumption that the structure is continuous through all of its layers and that it obeys to the conditions of application of such models i.e.

. Axi-symmetrical geometry,
. homogeneous, isotropic linear elastic materials.
. All layers extend to infinity in the horizontal plane.
. the friction between layers is either slippery or rough.

The parameters necessary to describe the structure are
. number of layers;
. Thickness of each layer;
. interface conditions between the successive layers;
. elastic properties of the individual layers (stiffness modulus and Poisson's ratio).

The fatigue phenomenon is currently used as the fundamental criterium in the design of new road structures or the calculation of overlay thicknesses (ref 16,17,18,19).
   This model has been successfully used since many years as the basic element of many structural design methods. It is now available under the form of software programs (ref 19,20).
   It addresses exclusively the initiation phase of the cracking process in the case of an initially homogeneous and continuous pavement structure.
The adjustment, for practical purposes, to the actual performance of roads was first made by calibration factors.
 One must in fact include

. the time necessary for the crack to grow up to the surface;
. the healing effect of rest periods as far as traffic loads are concerned.

Before dealing with models including discontinuities it

is worth to remember that the multilayer models are able to give in a first instance an evaluation of the original state of the structure to be treated (i.e. in the absence of any discontinuity in the lower layers).

3.1.2 Application of the linear elastic multilayer theory to crack propagation problems (B in table 3).

Although these theories are not suited to handle the case of localized damage features such as cracks different trials have been made to include the crack propagation in a simplified manner.

The approach proposed in the MOEBIUS software (ref.20) considers the pavement as initially sound. The structure is divided in as may layers as possible having each the initial properties of new asphalt. The first crack at the bottom is supposed to be initiated by fatigue.

After the crack initiation stage the properties of the different sublayers are progressively reduced from bottom to the top with a rate of propagation determined from the knowledge of Paris'law.
Such a procedure can obviously not claim to perfectly model the complexity of the cracking phenomena. The accuracy of the predictions are of course limited by the oversimplification and are also dependent on the input data.

Another trial to use a linear elastic multilayer program for the propagation of cracks from an old pavement through a new overlay was made by Van Gurp and Molenaar (ref. 21).
A comparison was first made between results obtained from a finite element analysis and that of the analysis made with the BISAR multi-layer program (ref 19). This led to a first calibration of the finite element program : the dimensions of the grid were set to values such that in the case of an uncracked structure the critical asphalt strain would be equal to the ones computed by the BISAR program.
A study of cracked structure was then made in order to set up effective moduli values for the BISAR program.
It was concluded that providing that reliable effective modulus values are introduced, the linear elastic models could be applied for overlay design purposes.

It is clear however that this type of extension of the elastic multilayer theory is merely a way to use an existing tool in a field for which it was not initially developed.

3.1.3 Models based on equilibrium equations
(C in table 3).
A procedure for the design of asphalt concrete overlays on existing portland cement concrete (PCC) has been

developed for the Arkansas State Highway transportation department (ref. 22).The procedure is based on a simple mechanistic approach in which the main features considered are the movements of the concrete slabs close to joints or cracks and the thermal movements. Equilibrium equations were used for estimating the stresses which were then used in a fatigue type of approach to estimate the life time of an overlay with or without a functional layer.
This procedure has been implemented under the form of a computer program and charts for practical overlay design.

3.1.4 ODE Mechanistic empirical method (D in table 3)
In another approach Jayawickrama and Lytton (ref 6) have developed a set of mechanistic empirical overlay design procedures to address reflective cracking in asphalt concrete overlays of existing asphalt or portland cement concrete pavements.
The basic design equations of this procedure are based on fracture mechanics and beam on elastic foundation concepts.
They address fracture in the slab due to bending and shear caused by moving wheel loads and due to opening caused by thermal movements of the cracked existing pavements. The practical overlay design procedure was finally obtained by calibration of the mechanistic equations with performance data from in-service pavements.

This very powerful and practical tool allows the design of classical overlay solutions. It can now be applied for overlays on damaged asphalt pavements as well but it does not so far cover the case of interlayer products. Anyway it can be used as a first step in the overlay design for the determination of the minimum reference thickness of a conventional type of overlay.
The design program ODE (standing for Overlay Design Equations) which was developed for the implementation of the procedure is restricted to six climatic regions of the United States. Its extension to other climatic regions requires the development of regression equations based on field data collections.

3.1.5 Finite element analysis including functional interlayers and crack propagation laws (E in table 3).
A large contribution to the introduction of the fracture mechanics concepts in road structures is due to Majidsadeh (ref. 23) in an attempt to explain the fatigue phenomenon. Further application of fracture mechanics to overlaid structures has been made since then by him and many researchers.
The use of finite element calculations in conjunction with the fracture mechanics approach needs the use of special crack tip elements in order to get the correct values of

the stress intensity factors of the Paris law.

The first extensive trials in this field concern studies carried out by Monismith et al (ref. 25) for the case of stress absorbing interlayers.

The effect of soft SAMI type interlayers were modelled in this way and their positive contribution in the case of thermally induced shrinkage of the lower base layers could be demonstrated (ref. 28).

The modelling of the reflection of transverse shrinkage cracks made by Marchand and Goacolou for bituminous overlays on cement stabilised base layers under the effect of traffic and thermal stresses has shown that the path followed by a crack can be influenced in different ways by the geometry of the structure and the component properties (ref. 26). This study also showed that the debonding mechanism leads to the development of horizontal cracks while strong bonds lead to a vertical propagation mechanism.

3.1.4 The blunt crack band theory (F in table 3)

Another way to model the behaviour of cracked structures has been used by Haas and P.E. Joseph (ref. 27).

Unlike theories based on the fracture mechanics approach these models simulate a crack as a broad zone of weak material. It can be considered indeed that an actual crack in an heterogeneous material such as a bituminous mix does not correspond with the theoretical representation which is assumed in the fracture mechanics approach.

Moreover this crack morphology can be widely variable with the grading of the mix, the mechanism that generated the crack and the temperature at which fracture occurred.

To take account of the three dimensional nature of a damaged zone, the blunt crack theory represents an effective crack as a vertical band having a width close to the maximum grading size and in which (in a way similar to the extended elastic layered approach), a weak modulus values is introduced. Such structures can be modelled either by finite elements or by finite difference computer programs (ref. 28). The approach leads to a realistic evaluation of the stress distribution in the cross section of a cracked structure containing different types of interlayers. The procedure is valuable in screening alternative treatments but it is unable so far to describe the crack propagation mechanism.

3.2 Some remarks about finite element modelling

The experience gained in the field of modelling has led to adopt finite element for finite difference programs as the stress-strain computation tools for the study of the reflective cracking phenomenon.

It is clear that we have to deal with a very complex

problem and we should be aware of the fact that we are still doing simplifications to keep it liable for practical purposes. But over simplification in models and assumptions must not lead to unreliable answers. Hence some aspects of these methods need still care and attention such as :

3.2.1 The influence of the geometry of the mesh system

High stress concentrations close to the discontinuities can be the source of important errors if special elements are not used. On the other hand the grid distribution and size of the elements can have a non negligible effect on the calculated stress magnitudes even far from cracks or discontinuities (ref. 25).

3.2.2 Limitations of two dimensional modelling

As was already mentioned by Monismith (ref. 25), ideal solutions would consider a three dimensional pavement system. It appears that in the case of traffic loading the plane strain assumption generally implicitly adopted in 2 D modelling does not correspond to reality and may even lead to wrong statements (ref. 28).

The linear elastic theory has at this point of view the advantage to give a correct three dimensional solution for continuous structures under traffic loads.

3.2.3 The boundary conditions

Finite element programs concern objects with limited spacial dimensions, they are therefore not directly usable for continuous pavement layers unless boundary conditions are introduced to simulate the reaction forces of the remaining adjacent parts of the structure.
Far from the crack these conditions can be evaluated more precisely by a preliminary study using a linear elastic model of the same structure without any crack.

3.2.4 Unconventional mechanical properties

Much needs to be done in order to improve the laws describing the behaviour of functional interlayers because they are - even more than the other components - non linear, non elastic and non isotropic.

4 Simulation testing of crack propagation

The final soundness of any computed simulation must be assessed on the basis of laboratory tests in a first instance and then on full scale projects for the final adjustments. Experimental simulation can be considered as an alternative way for assessing the performance of an anti reflective cracking system.

4.1 General features about testing procedures

Laboratory testing techniques are generally developed to examine different aspects of a system for the situations where reflective cracking may normally develop.

Although many different facilities are described in the literature, it is possible to classify them in a limited number of types according to the particular situation they are supposed to address.

Before entering into a brief description of these testing types some general remarks must be made :

4.1.1 The manufacture of test samples

Sample manufacture must reproduce the actual conditions used in full scale implementation. The most representative samples are of course those that can be taken out of a road section.

4.1.2 Size of the samples

The test samples are always limited in their geometrical dimensions, so that in some cases the loading conditions do not really correspond with any real situation in the road. This problem is similar to what is met in analytical simulation with FE methods when the boundary conditions are irrelevant.

4.1.3 Testing temperature

Owing to the high temperature susceptibility of bituminous materials it is of primary importance to control the temperature with high accuracy.

4.2 The different types of testing procedures (table 4)

In all the tests the onset of a crack and its progression in function of the time or the number of loading cycles is monitored. The tests are generally carried out for comparative purposes. In some cases however they are used in order to calibrate and verify analytical models.

4.2.1 Traffic loading

These equipments are intended to study the development of cracks under transient loading conditions. The contact pressure and loading frequencies need here to correspond with traffic loads.

4.2.1.1 Beam testing facility (ref. 29,30)

The anti- reflective overlay system is placed over a cracked subbase. This structure is cut under the form of a beam and placed over an elastic supporting layer (generally rubber with a well known reaction modulus). Repeated loads are applied at the surface of the asphaltic layer in order to simulate either mode 1 or mode 2 loading conditions. The most critical problem of this type of test

Table 4 : The different testing procedures

Loading	Testing procedure	References
Traffic loads	Beam testing wheel tracking	29,30
Thermal shrinkage of base layer	Controlled crack opening repeated loading	30,31,32
Thermal shrinkage of upper layer	Restricted beam under constant rate of temperature change	4,31,35,36
Traffic + thermal shrinkage of base layer	Combination of horizontal opening of a crack with vertical repeated loads.	37

appears at the extremities of the samples where the reaction forces of the rest of the road structures have to be introduced by one way or another.

4.2.2.2 Wheel tracking testing equipment
A slab or beam representing the structure to be tested is submitted to the action of a moving wheel. It allows investigation of reflective cracking on larger samples than permitted by the beam testing method, and under more representative loading conditions.

4.2.3 Thermal loading
Large strains can develop in base layers as a result of thermal effect causing joints or cracks in cemented materials to open and close over much longer time scales.

4.2.3.1 Thermal shrinkage of the base layer
Thermal movements of opening and closing of a crack in the base layer can be simulated in different ways. The sample is alteratively submitted to tension and compression loads which concentrate in the vicinity of the discontinuity. In some equipments the crack in the base layer is opened continuously at a constant rate.

The information provided by these tests can vary from simple visual observation of the cracking process in function of loading cycles or time, to measurement of more fundamental properties such as shear modulus of the interface and stress-strain relationships (ref.30, 31,32,33). These tests are generally carried out at low temperature where the risk of unstable cracking is the highest (estimated to be below -5°C in most cases) (ref. 28).

4.2.3.2 Thermal shrinkage of the overlay
These tests are intended for the simulation of cracks generated in the overlay by high temperature variation rates. Although this type of test is not directly representative for the reflective cracking the evaluation of the resistance of the overlay to this type of cracking is an important information. The test can be carried out on a beam of the overlay material which is maintained at a constant length in a rigid frame while the surrounding temperature is made to change at a constant rate (ref. 4,31,35,36).

4.2.4 Compound testing procedures
In order to stick closer with actual conditions where we have to deal with the combined effects of thermal shrinkage and traffic loads the Autun laboratory has developed a testing facility combining the beam bending test with thermal shrinkage at constant rate of the base layer (ref 37). This procedure allows all types of overlay systems to be tested. It is currently used for comparing different solutions and their ranking based on 2 criteria:
- the crack initiation time, the rate of crack propagation.

5 General conclusions and future prospects

There have been during the last years many advances and developments in the understanding of the cracking phenomenon. It is particularly true that the power and availability of computer facilities and software tools allow much more accurate modelling of the process and this causes some concern over whether our basic knowledge of materials characterization is sufficient to fully exploit this situation.

5.1 Future trends for explaining and characterizing the bituminous overlay properties
A better knowledge of the mechanical behaviour of these materials will improve our predictions.
The mechanical characteristics describing bituminous materials are generally dealt as if they were independent of each other. There is clear evidence now that these properties are linked.

It has been demonstrated indeed that :
- the parameters of the Paris law are dependent on the shape of the modulus master curve (ref 8)
- the end phase of the fatigue process can be explained on the basis of the crack propagation law (ref 23)
- the temperature dependence of the modulus, the fatigue law and the crack propagation law can all three be explained with the absolute rate theory developed by Eyring (ref 2.9).
- the use of the energy dissipation as the driving force of the damage instead of the classical stress-strain criteria has the potential to clarify some phenomenon such as the healing process (ref 38).

Recent advances of fundamental research are showing that we can hope in a near future to get a more satisfactory and consistent explanation of these phenomenon under the form of a unified theory and consequently more accurate evaluations will be possible.

5.2 Characteristics of interlayers
The use of interlayers is a way to improve the cracking issue. Many products have proven their effectiveness in full scale projects but negative experiences have also been observed. We should take these problems objectively into account in order to avoid misuse and problems for the future. A better knowledge of their functional possibilities and limitations can be achieved on the basis of modelling and simulation testing methods. More investigations will be necessary to characterize their in situ properties and their evolution in the long term. In this respect, very little information is available yet on their aging in presence of environmental conditions

5.3 Models
The most performant model will still be irrelevant and useless if the input data are incorrect or inaccurate.
The experience gained by the already existing programs for the design of traditional overlay procedures should serve as example for future implementations of procedures intended for overlay system design including functional interlayers.

5.4 Laboratory simulations
Laboratory simulation procedures are now existing . They are very useful in revealing and proving the effectiveness of overlay systems under different conditions of loading and environment.
However we must bear in mind that the scale factor is their most important limitation. Experimental simulation

is complementary to evaluation by modelling.
But the last step remains the validation on full scale monitored projects.

Bibliography

1. F.Bonnaure, G.Gest, A.Gravois and P.Ugé : A new method of predicting the stiffness modulus of asphalt paving mixtures.Proceedings A.A.P.T. 1977
2. L.Francken and C.Clauwaert : Characterization and structural assessment of bound materials for flexible road structures.
   Proceedings of the VIth Int.Conf on the Structural Design of Asphalt Pavements.Ann Arbor Michigan July 1987, pp. 130-144.
3. J.Verstraeten and RILEM TC 101 BAT: Results of an interlaboratory experience. Materials and structure.
   RILEM - Materials and structures to be published.
4. W.Arand : Behaviour of asphalt aggregate mixes at low temperatures. Proceedings of the fourth RILEM symposium on Mechanical tests for bituminous mixes. Budapest 1990, pp. 68-84.
5. P.C.Paris and F.Erdogan : A critical analysis of crack propagation laws. Journal of basic engineering , Transaction of the American society of mechanical engineering,Series D, vol.85 (1963),pp.528-553.
6. P.W.Jayawickrama, R.E.Smith, R.L.Lytton and M.R.Tirado: Development of asphalt concrete overlay design program for reflective cracking. RILEM Conference on Reflective cracking in pavements. Liège 1989, pp. 164-170.
7. A.A.A.Molenaar : Structural performance and design of flexible road constructions and asphalt concrete overlays.Thesis Technische Hogeschool Delft 1983.
8. R.A.Schapery : A method for predicting crack growth in non homogeneous viscoelastic media. Int.Journal of fracture mechanics.14 (1978) 293-309.
9. M.Jacobs : Determination of crack growth parameters of asphalt concrete, based on uniaxial dynamic tensile tests. Proceedings of the fourth RILEM symposium on Mechanical tests for bituminous mixes. Budapest 1990, pp.483-496.
10. Tobolski and H.Eyring : Mechanical properties of polymeric materials . Journal of Chemical Physics 11 (1943), pp.125-134.
11. J.M.Rigo et al.: Laboratory testing and design methods for reflective cracking interlayers. RILEM Conference on Reflective cracking in pavements Liège 1989, pp. 79-87.

12. R.D.Barksdale : Fabrics in asphalt overlays and pavement maintenance. N.C.H.R.P. Synthesis of highway practice 171.
    Transportation Research Board Washington D.C. July 1991.
13. Several papers in : Dossier lutte anti-fissure
    Revue générale des routes et aérodromes n°685, Paris, May 1991.
14. G.Colombier : Fissuration des chaussées, nature et origine des fissures, moyens pour maîtriser leur remontée. RILEM Conference on Reflective cracking in pavements. Liège 1989, pp. 3-22.
15. M.E.Nunn : An investigation of reflective cracking in composite pavements in the United Kingdom. RILEM Conference on Reflective cracking in pavements. Liège 1989, pp. 146-161.
16. J.Verstraeten : Stresses and displacements in elastic layered systems. Proceedings of the 2nd Int.Conf on the Structural Design of Asphalt Pavements. Ann Arbor, August 1967.
17. A.I.M.Claessen, J.M.Edwards, P.Sommer and P.Ugé : Asphalt Pavement Design. the Shell method. Proceedings of the IVth Int.Conf on the Structural Design of Asphalt Pavements. Delft, July 1977, pp. 39-74.
18. B de la Taille, P.Schneck and F.Boudeweel : ESSO Overlay design system. Proceedings of the Vth Int.Conf on the Structural Design of Asphalt Pavements. Delft, July 1982, pp. 682-694.
19. C.P. Valkering and D.R.Stapel : The Shell pavement design method on a personal computer. Proceedings of the 7th International Conference on Asphalt Pavements. Nottingham 1992, Vol. 1, pp. 351-374.
20. B. Eckman : ESSO MOEBIUS Computer software for pavement design calculations. User's manual. Centre de Recherche ESSO. Mont Saint Aignan, France, June 1990.
21. C.A.P.M.Van Gurp and A.A.A.Molenaar : Simplified method to predict reflective cracking in asphalt overlays. RILEM Conference on Reflective cracking in pavements. Liège 1989, pp. 190-198.
22. S.B.Seeds, B.F.Mc Cullough and F.Carmichael : Asphalt concrete overlay design procedure for portland cement concrete pavements Transportation Research Record 1007 Washington D.C.1985, pp. 26-36.
23. K.Majidzadeh,E.M.Kaufmann and D.V.Ramsamooj : Application of fracture mechanics in the analysis of pavement fatigue. Proceedings A.A.P.T., Vol. 40, 1970, pp. 227-246.

24. K.Majidzadeh, L.O.Talbert and M.Karakouzian : Development and field verification of a mechanistic structural design system in Ohio. Proceedings of the IVth Int.Conf on the Structural Design of Asphalt Pavements. Delft, July 1977, pp. 402-408.
25. C.L.Monismith : Reflection cracking. Analyses, laboratory studies, and design considerations. Proceedings A.A.P.T., vol. 49, (1980), pp. 268-313.
26. J.P.Marchand and H.Goacolou : Cracking in wearing courses. Proceedings of the Vth Int.Conf on the Structural Design of Asphalt Pavements. Delft July 1982, pp. 741-757.
27. Haas and P.E Joseph : Design oriented evaluation of alternatives for pavement overlays. Conference on Reflective cracking in pavements. Liège 1989, pp. 23-46.
28. L.Francken and A.Vanelstraete : Interface systems to prevent reflective cracking. Modelling and experimental testing methods. 7th ISAP conference on asphalt pavement. Nottingham, 1992 vol.1, pp. 45-60.
29. S.F.Brown, J.M Brunton and R.J.Armitage : Grid reinforced overlays. RILEM Conference on Reflective cracking in pavements Liège 1989, pp. 63-70.
30. D.Sicard : remontée des fissures dans les chaussées. Essais de comportement en laboratoire par flexion sur barreaux. RILEM Conference on Reflective cracking in pavements Liège 1989, pp. 71-78.
31. R.A.Jimenez, G.R.Morris and D.A.Dadeppo : Tests for strain-attenuating asphaltic materials. Proceedings A.A.P.T., vol. 48, pp. 163-191.
32. J.P Antoine : Matériel et méthode d'essai pour l'étude en laboratoire de la résistance à la remontée des fissures au travers de revêtements routiers.
RILEM Conference on Reflective cracking in pavements Liège 1989, pp. 88-89.
33. L.Francken : Fissuration de structures semi-rigides Essais de simulation. Proceedings of the fourth RILEM symposium on Mechanical tests for bituminous mixes. Budapest 1990, pp. 419-431.
34. T.Brooker, M.D.Foulkes and C.K.Kennedy : Influence of mix design on reflection cracking growth rates through asphalt surfacing. Proceedings of the VIth Int.Conf on the Structural Design of Asphalt Pavements. Ann Arbor Michigan July 1987, pp.107-120.
35. J. Eisenman, U. Lampe, U. Neumann: Effect of polymer modified bitumen on rutting and cold cracking performance. Proceedings of the VIIth Conference on Asphalt Pavements, Vol. 2, pp. 83-94 - Nottingham m 1992.

36. H. Kanerva : Effect of asphalt properties on low temperature cracking of asphalt mixtures. Proceedings of the VII Conference on Asphalt Pavements , Vol. 2, pp. 95-107, Nottingham 1992.
37. J.H.Vecoven : Méthode d'étude de systèmes limitant la remontée de fissures dans les chaussées. RILEM Conference on Reflective cracking in pavements, Liège 1989, pp. 57-70.
38. A.C..Pronk and P.C.Hopman : Energy dissipation : the leading factor of fatigue . Proceedings of the SHRP conference Highway research: Sharing the benefits. London 1990, pp. 255-168.

5 GEOTEXTILE USE IN ASHPHALT OVERLAYS – DESIGN AND INSTALLATION TECHNIQUES FOR SUCCESSFUL APPLICATIONS

F.P. JAECKLIN
Geotechnik Consulting Engineers, Ennetbaden, Switzerland

Abstract
The paper summarizes the current practical installation techniques and sources of error in installation as well as design concepts. This highlights applications that are prone for success and it stresses situations in which geotextile use might be a pitfall for reasons of inadequate subbase or asphalt conditions. Thus the selection and design of retrofitting depends on a number of aspects including detailed analysis of the evaluation of the asphalt technology on the individual site.
Another subject to evaluate is the selection of geotextile nonwoven fabric versus geogrid. This aspect is normally related to the concept of 'reinforcing' with a high strength grid, versus 'SAMI – stress absorbing membrane interlayer' using non woven fabric. The more detailed study indicates that this is not always the case and geogrids may well act as SAMI as well, whereas some nonwoven may act as a reinforcing.
The paper concludes that sound engineering concepts and comprehensiv understanding of the individual site conditions combined with asphalt technology are necessary for good results of geotextile use in asphalt overlays.
The conclusions base on the Swiss geotextile manual, section geotextiles in asphalt overlays as prepared by the author and on tables and papers of Mathias Blumer. The various logical steps to analyze problems of existing pavements are outlined for conceiving alternatives for small repair or total retrofitting of roads for minimum expenditure for maximum efficiency of limited road construction budgets.

1. ANALYZE DAMAGING REASONS FIRST

For reconstruction and retrofitting of existing roads the following procedures are efficient and adequate. Proceed as follows:
   1- first systematically observe the types damages,
   2- observe the development of the damages over the years
   3- take samples for testing (check penetration, softening point)
   4- take load readings on the pavement (Benkelman beam, etc)
   5- investigate the local subground and groundwater conditions
   6- find combinations of various reasons that might be involved
   7- ask a specialist for damage evaluation if there is some doubt.
Concept for engineering damage evaluation : first find the technical reasons that caused the damage, then conclude for remedial measures.

Reflective Cracking in Pavements. Edited by J.M. Rigo, R. Degeimbre and L. Francken.
© 1993 RILEM. Published by E & FN Spon, 2–6 Boundary Row, London SE1 8HN. ISBN 0 419 18220 9.

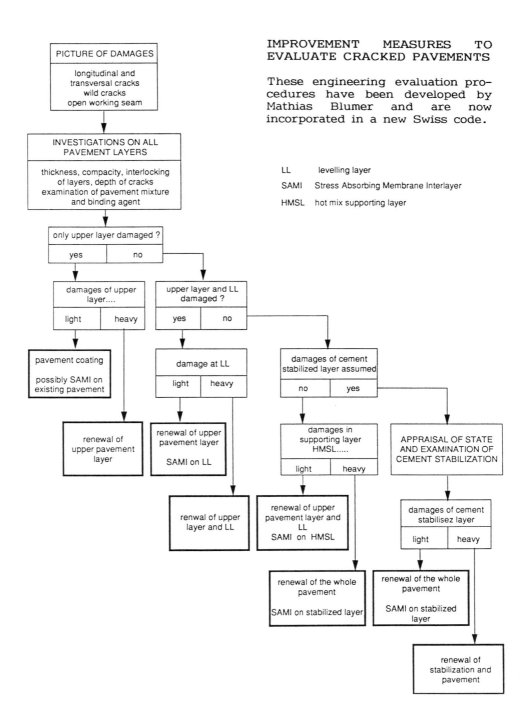

2. ENGINEERING SYSTEMATICS FOR ROAD RETROFITTING

CASE 1: LOCAL DAMAGES ONLY

- Types of damages:
  - locally crushed areas
  - elephant skin fissures
  - pot holes
  - porous areas with fissures
  - open seams
  - slippery surface
- Possible causes of damage:
  - insufficient bearing capacity
  - insufficient frost resistance
  - material/ construction deficiency
  - aging or fatigue
- Additional on site testing:
  - possibly deflection testing
- Additional laboratory testing:
  - penetration and softening point

TYPES OF REPAIR - Local repair:
  - fill pot holes
  - grind and fill ruts and holes
- Overall repair:
  - surface dressing using tack coat and chips
  - use special binders
  - repair old surface first
  - GEOTEXTILE (SEAL AND SAMI)
  - GEOGRID FOR STRENGTH
  - asphalt pavement overlay
- Increase road strength capacity:
  - depends on traffic and tests
- Totally renew the road:
  - only if tests show bad results

CASE 2: EXTENDED SURFACE DAMAGES

- Type of damages:
  - lean surface and sanding
  - intense wear
  - reflection cracks
  - wild, non systematic cracks
- Possible causes of damage:
  - advanced aging of binder
  - material/ construction deficiency
- Additional on site testing:
  - possibly deflection testing
- Additional laboratory testing:
  - drill cores: layer thickness
  - measure adhesion of layers
  - binder, gradation, components

TYPE OF REPAIR: no local repair
- Overall repair:
  - surface dressing using tack coat and chips
  - GEOTEXTILE (SEAL AND SAMI)
  - GEOGRID FOR REINFORCING
  - one or two layers overlay

    or:
    anyway as needed:
  - reforming or repaving
  - grind old pavement first
  - add leveling layer as needed
  - use special binders
  - repair old surface first or grade
- Increase road strength capacity:
  - depends on traffic and tests
- Totally renew the road:
  - only if tests show bad results

CASE 3: EXTENDED SURFACE DAMAGES
- Type of damages: - deep ruts and waves
                                            - slippery surface
- Possible causes of damage: - insufficient asphalt resistance
                                              against permanent deformations
                                            - material/ construction deficiency
- Additional on site testing: - measure evenness
                                            - tire friction
- Additional laboratory testing: - drill cores: layer thickness
                                            - measure adhesion of layers
                                            - binder, gradation, components
TYPE OF REPAIR: no local repair
- Overall repair: - surface dressing using
                                            - coat binder and crushed rock
                                            - GEOTEXTILE (SEAL AND SAMI)
                                            - GEOGRID FOR REINFORCING
                                            - one layer asphalt pavement
                                              overlay
              or: - two layer asphalt overlay
              or: - reforming or repaving
         anyway as needed: - grind old pavement first
                                            - add leveling layer as needed
                                            - use special binders
                                            - repair old surface first or grade
- Increase road strength capacity: - depends on traffic and tests
- Totally renew the road: - only if tests show bad results

CASE 4: EXTENDED STRUCTURAL DAMAGES
- Type of damages: - very uneven surface in x,y
                                            - large single cracks
- Possible causes of damage: - uneven settlement
                                            - uneven frost heave
                                            - aggregates not frost resistant
                                            - subbase aggregates too thin
                                            - insufficient subground drainage
- Additional on site testing: - deflection readings
                                            - drainage facilities and
                                              performance
- Additional laboratory testing: - drill cores: layer thickness
                                            - measure adhesion of layers
                                            - binder, gradation, components
                                            - subbase frost resistance
                                            - underground frost resistance
TYPE OF REPAIR: no local repair
- Overall repair: - GEOTEXTILE (SEAL AND SAMI)
                                            - GEOGRID FOR REINFORCING
                                            - one THICK or two overlays
         anyway as needed: - grind old pavement first
                                            - add leveling layer as needed
                                            - use special binders
- Increase road strength capacity: - add overlay for thickness
                                            - add strength by partial renewal
- Totally renew the road: - reconstruct the whole road

CASE 5: EXTENDED STRUCTURAL DAMAGES
- Type of damages: - extended deformations
                                      - visible relocations
                                      - intense elephant skin cracks
- Possible causes of damage: - insufficient bearing capacity
                                      - no frost resistance of aggregate
                                      - insufficient drainage
- Additional on site testing: - deflection measurements
                                      - drainage performance
- Additional laboratory testing: - drill cores: layer thickness
                                      - measure adhesion of layers
                                      - binder, gradation, components
                                      - frost resistance of subbase
                                      - underground frost resistance
TYPE OF REPAIR: no local repair
- Overall repair: - yes
- Increase road strength capacity: - add overlay for thickness
                                      - add strength: partial renewal
- Totally renew the road: - reconstruct the whole road

3. RETROFITTING CONSTRUCTION WORKS

Based on the extension of damages and the traffic volume and the damage reasons a first concept decision must be made:
    A : local repair
    B : grind, repair cracks, place asphalt overlay on original level
    C : repair cracks and place asphalt overlay on top
    D : complete renewal
Each of these alternatives is possible with or without geotextile (geogrid or geofabric) and the related technology for extended live span of repair work. Thus the systematics and design and repair approach are of basic interest applicable for all cases.

4. DESIGN PROCEDURES FOR PAVEMENT RETROFITTING

1. Analyze the damage causes
2. Determine minimum requirements for repair and test section such as traffic volume, type of pavement, thickness of pavement, temperature for laying
3. Determine concept of retrofitting including type of geotextile, purpose and intended function
    a. geofabric for water sealing and SAMI (stress absorbing membrane interlayer) or
    b. geogrid for reinforcing interlayer
4. Specify the geotextile according to the function in mind.
5. Then specify the type of binder
    a. asphaltic emulsion seal coat or
    b. hot spray asphaltic seal coat
    c. with or without polymeric modification
6. Determine other design details such as
    - depth of grinding if any
    - thickness of overlay asphalt pavement

- possibly two overlays
- edit work description, specifications and contract documents
- list special site conditions and special local requirements
- draw time schedule and work sequence and dead line for completion of various work steps
- add special description, explanations, and specifications for working with geotextile fabrics or grids. List special errors to be omitted with these new materials.

Obviously proper performance of these engineering tasks for repair outlined above require an understanding of the causes and mechanics for the reasons for damages.

5. CRACKING : CAUSES, DEVELOPMENT AND IMPORTANCE

The worldwide success of asphalt pavement lies in the ability of bitumen material to withstand traffic loads and temperature induced loads without seams and for an extended period of time due to their ability to absorb slow acting deformations by plastic creep and still show a quite elastic behavior for short term traffic loads.

This elasto-plastic material behavior is the key. However the good qualities disappear over the years by alteration, aging, and fatigue and cracking in various patterns and characteristics are the result requiring repair or retrofitting or total renewal.

The development in asphalt technology therefore concentrates on how to reduce the cracking potential by reduced aging by better binders using polymer modification and special means to reduce some extreme effect of temperature changes near the top surface.

For this last goal various techniques have been developed: use additional materials in interlayers to either absorb the stresses or reinforce against the stresses for reducing tension in asphalt and thus extend the useful life span of retrofitting construction work. Such methods are:

A : SAMI - Stress Absorbing Membrane Interlayer - such as Geotextile fabric embedded in a rather thick seal coat for binding the top layer firmly to the existing asphalt, yet allowing some slow creep as needed by temperature changes.

B : The very same Geotextile fabric interlayer with thick tack coat surface dressing and chips to seal off water penetration to the lower asphalt layers and to the subbase and subground, thus reducing adverse effects of freezing and thawing and improving the structural strength.

C : Geogrid reinforcing interlayer placed under a new top pavement accept some near surface stress due to local settlement along a trench, a widened road or due to thermal stresses.

Such effects are described by the following figures. The purpose is to understand the mechanics for best putting the concept to work in each individual case.

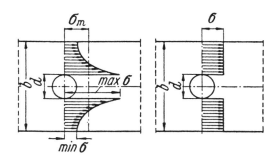

FIG 1 : TENSION PEAK AT END OF GROOVE OR CRACK

1. Presumably tension in a homogenious material is distributed evenly in any cross section, as seen on the right.
2. This asumption is not true closed to a hole, because of stress concentrations near the hole: locally the tension is three times the average in case of a circular hole and many times more if the void is flatter.
3. Tensions develop to extreme peaks near ends of cracks and thus tension may exceed tensional strength and cracks extend even further, aggravating the effect: The end of a crack is the most unfavorable spot where stresses accumulate and tend to lenghten the crack.
4. In a asphalt pavement tensions develop due to shrinkage from temperature decrease or due to shrinkage of the cement stabilized subbase.

Conclusion : If cracks start to develop there is great probability they develop even further and widen more.

A cement stabilized subbase shrinks and it shrinks even more the more cement has been admixed. The 'better' dosage results in a harder stabilization, but it produces cracks that might be further apart, yet much wider because each crack has to absorb a longer shrinkage length and it has to absorb more shrinkage because of more cement dosage.

Reflection cracking above cement stabilized subbase are therefore very probable unless special means are taken, such as an additional gravel layer or a SAMI - geotextile fabric directly above it.

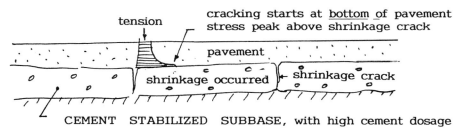

FIG 2 : PEAK STRESS IN PAVEMENT ABOVE SHRINKAGE CRACK

The peak stress induced by the cement stabilized subbase and its shrinkage effect is greatly reduced by the stress absorbing effect of the creep within the thick seal coat and the GEOTEXTILE (SAMI).

A GEOGRID may produce a similar effect in absorbing some stress which in turn reduces distress in the neighboring asphalt.

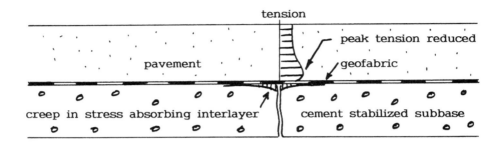

FIG 3 : GEOTEXTILE (FABRIC OR GRID) REDUCES PEAK STRESS

Low temperatures in cold winter nights affect the very top surface of the pavement the most, thus there is a temperature induced peak stress near surface. This effect is visible at the edges of concrete pavement sections or even asphalt pavement that tend to raise.

As a further consequence such pavement tends to crack in wide open singular cracks, sometimes along the center line of roads.

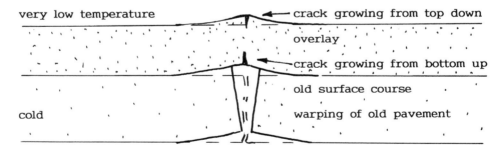

FIG 4 : LOW TEMPERATURES INDUCE PEAK TENSION STRESSES NEAR SURFACE

The use of geofabric directly on or on a leveling layer on cement stabilized subbase should be mandatory for new constructions in areas where road salts are being used:
1. The geofabric and tack coat absorbs shrinkage cracks from below and protects the bituminous concrete from initiating reflective cracking
2. The tack coating seals the subbase off from road salts which are very detrimental to many cement stabilized layers, causing substantial loss of strength

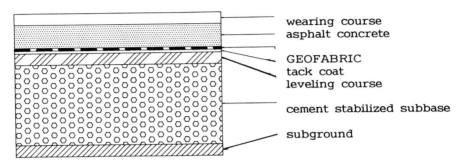

FIG 5 : USE OF GEOFABRIC ON CEMENT STABILIZED SUBBASE FROM THE BEGINNING

Number of load cycles versus crack length. Obviously the non reinforced samples have much longer and more pronounced cracks, whereas the reinforced asphalt test samples have shorter cracks. Note that samples with geogrids placed right at the very bottom of the sample show no cracking at all. A structural engineer would expect somewhat similar effects in reinforced concrete.

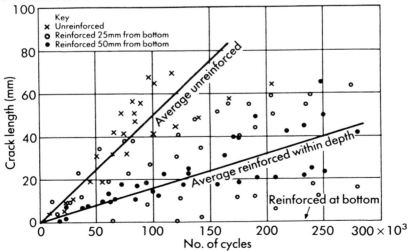

FIG 6: RESULTS OF REFLECTION CRACKS

6. EFFECTS OF GEOFABRICS

SEALING : The geofabrics (and some geogrids) allow for spraying an extended amount of seal coat (more than 1 kg/m2), some 3 to 4 times as much as used for just binding an overlay. Such quantity would not be possible without the geofabric, because excessive 'bleeding' would take place.
The waterproof sealing greatly improves the long term behavior because several effects are eased: less freezing and thawing and softening due to excessive water content and no detrimental effects from road salts in cement.
Obviously such effect are at best if combined with enhanced drainage facilities.

MORE REGULAR BONDING TO THE LOWER ASPHALT LAYER: The intense seal coat spraying means an intense and more regular bonding. This improves the overall behavior of the top pavement. However to bonding may not be more than in cases without geofabric, it is the more even distribution that counts.

REDUCE PEAK STRESSES IN THE SURFACE PAVEMENT: The creep capability of the rather thick seal coat allows for reduction of local peak stresses which in turn reduces the tendency for cracking.

GREATLY REDUCE AND OR PREVENT REFLECTION CRACKING : Probably the most important effect of this creep capability helps reducing reflection cracks and their negative effect on further deterioration of the pavement that proceeds much faster once some racks have started. This is why the understanding of long term pavement preservation of a good maintenance crew can be seen at the quality and responsive follow up on sealing cracks.

7. EFFECTS OF GEOGRIDS

1. Geogrids take some of the temperature and long term tensions and thus reduce stresses in the asphalt pavement
2. Such reinforcing effect is positive in areas closed to potential cracks , especially along construction seams, over transverse drainage ditches, over culverts crossing underneath, at road widenings, and at spots of irregular bearing capacity.

Tests at room temperature have shown the following effects:

1. Geogrid reinforced samples show higher bending resistance than those without
2. At or near breaking load reinforced samples show much larger deformation than those without
3. Geogrid reinforced samples show a much earlier stress increase than those without, the deformation modulus is increased
4. Non reinforced samples result in a single crack, whereas reinforced samples distribute deformation into various cracks of much finer width

The diagram displays how a ten year interval maintenance program affects the long term road. It essentially demonstrates the slow road decay over the years unless substantial repair is made using long term objectives for maintaining their value.

The concept requires to recognize the status of any given section and determine whether a small job like an overlay is adequate or thorough reconstruction is needed at a certain point.

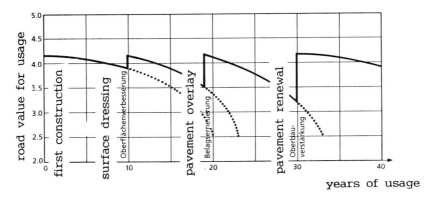

FIG 7 :   ROAD VALUE HISTORY WITH 10 YEARS REPAIR INTERVALS

For this road section extensive core drilling was made to find the thickness of the individual asphalt layers. Evidently the surface deformations are not from insufficient bearing capacity as originally thought of, but from plastic squeezing of top layer to the sides. Thus grinding and an overlay was the adequate repair method.

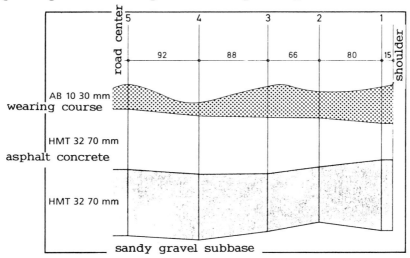

FIG 8 :   CORE DRILLING TO FIND REASON FOR UNEVEN SURFACE

FIG 9 : Deflection measurement using the Benkelman beam for individual readings at single spots.

FIG 10 : Deflection measurement using the Deflectograph Lacroix for repetitive readings at a large number of spots.

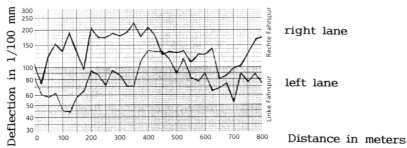

FIG 11 : Deflection measurement result of a road with pronounced difference of readings on left side versus right side. Consequently there was thicker overlay placed on the downhill side for compensating lower bearing capacity.

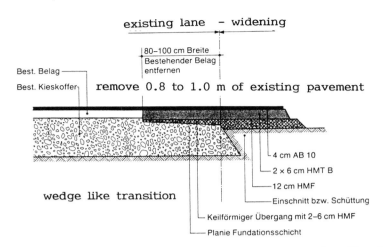

FIG 12 : Example of a road widening using a wedge like transition area to increase the strength at the weak point of joint to the existing subbase, preferably enhanced with geogrid reinforcing.

8. USE OF DEFLECTION READINGS FOR PAVEMENT DESIGN

Extensive research was done in Switzerland compiling experience data for finding the relationship between deflection readings and necessary overlay thickness, depending on expected traffic volume. These data are represented in the design graph figure 13 using the variables as defined below:

TRAFFIC VOLUME		TF	heavy trucks
T1	very light traffic	10 to 30	less than 25
T2	light traffic	30 to 100	25 to 75
T3	medium traffic	100 to 300	75 to 250
T4	heavy traffic	300 to 1000	250 to 750
T5	very heavy traffic	1000 to 3000	over 750

TF = number of truck equivalents per day

d_v = controlling deflection value $d_v = d_{av} + 2s$ in 1/100 mm

d_{av} = average deflection

s = standard deviation from average calculation on a selected road length of similar damages

Deflection readings to be taken away from frost and thawing periods and at pavement temperature above 25 degrees C

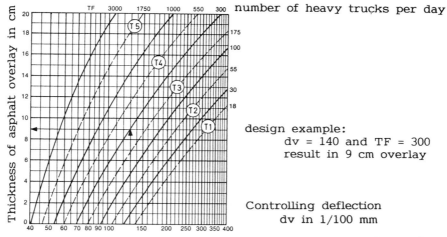

design example:
dv = 140 and TF = 300
result in 9 cm overlay

Controlling deflection
dv in 1/100 mm

FIG 13 : DESIGN GRAPHIC to determine additional overlay and pavement thickness for 20 year life span from controlling deflection value and from anticipated traffic volume.

This design chart is useful for standard cases. Data is not available yet to what extent geotextiles and geogrid would reduce deflection and thus reduce overlay thickness and cost. Possibly a reduction of deflection with geotextiles is not feasible and therefore savings of pavement thickness are not justified for traffic loads. The geotextile and geogrid effect is rather related to settlement and temperature effects and results in a longer life span by retarding reflective cracking.

9. BITUMENOUS BINDER QUALITY AND REPAIR POTENTIAL

Data compiled at the Canton of Zurich road laboratory from asphalt samples with reference to observed damages resulted in an optimum range for highways with heavy traffic. This range is defined with quality tolerances for the bituminous binder to be:

- R + K = 52 to 62, softening point, and
- P = 30 to 50 penetration

These findings are extremely significant. The consequence is that road sections ready for repair must first be tested in all layers. If some of the asphalt concrete layers show high softening or low penetration, mostly due to aging, the addition of a geotextile enhanced overlay is useless. Again the decision for repair depends on more factors than just geotextile.

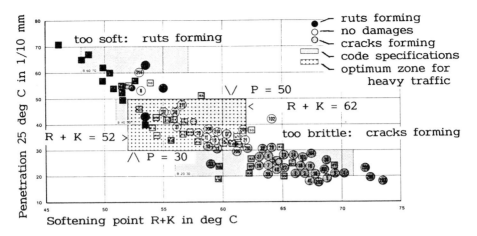

FIG 14 : BINDER QUALITY AND DAMAGE POTENTIAL

Soft binders with a softening point below 52 and or a penetration higher than 50 are too soft and tend to form deep ruts. Hard binders with a softening point above 62 and or penetration below 30 are too hard and tend for brittle cracking.

10. SPECIFICATIONS AND INSTALLATION PROCEDURES

TASK FORCE 25 - The American 'Task Force No. 25', a code committee has compiled technical data and experiences to summarize and specify proven techniques.

SWISS GEOTEXTILE MANUAL - Their result is a 6 page specification and description do be used for tenders. This task force 25 document was used as one of the basics for the Swiss Geotextile Manual, section on geotextiles and geogrids in bituminous pavements, SVG 1990. Since this manual is more recent and comprises 68 pages in German and French it covers the subject in more detail and for most practical purposes. Even though the writer was the author of this manual, very much credit is owed to the Technical Committee of SVG consisting of a knowledgeable and truly international team of specialists from government agencies, universities, consultants, and very important of many geotextile and geogrid manufacturers, who all collect data and information. For this reason this manual has become a thoroughly checked and valuable tool for the practitioners.

SIMILAR GEOFABRIC INSTALLATION - For geofabrics in overlays the task force 25 document and the Swiss manual essentially require the very same installation procedures and materials, which was to be expected. They are not listed here, since the documents are easily available. They both ask for similar materials, for similar equipment for automatically spraying and good temperature control of tack and asphalt concrete, for the same surface cleaning, wider spraying area, overlaps, and cuts on wrinkles, similar seal quantities, essentially for the whole operation.

SOMEWHAT DIFFERENT GEOFABRICS - There are some slight discrepancies, such as the minimum 80 lbs/ft = 1.2 kN/m tensile strength versus the European minimum of 4 kN/m. This higher minimum stems from the European manufacturers producing somewhat stiffer material for easier installation. Similarly the American elongation at break is 50%, whereas the European is 30% only, because of slightly bonded geofabrics for better handling.

The Swiss manual specifically requests polymer modified hot tack or 70% polymer modified emulsion, whereas the American specs refer to 60% emulsion and no polymer modification is mentioned. The polymer significantly increases plasticity at a wider temperature range in the pavement and thus increases long term performance. Possibly this difference also shows the changing technology in tack coat over the years.

11. CONCLUSIONS FOR ENGINEERING PRACTICE

These two basic documents on geosynthetics in asphalt overlays are supplemented by numerous brochures and technical documents as handed out by manufacturers. Obviously the question arises how these documents compare:

1. BROCHURES AND DOCUMENTS from manufacturers are good introductions to get to know technology and procedures. They cannot be as complete as the specifications and they rarely details on the items to avoid.

2. TASK FORCE 25 SPECIFICATIONS are prepared for the American bidding procedures and they are certainly very handy and necessary for it. They cover the material and installation of geofabrics in clear and comprehensive technical detail.

    However the task force document is for contractors and not for engineering. It does not explain design and typical sections, nor does it provide information which type of retrofitting is adequate in what case.

3. GEOGRIDS IS A SEPARATE SUBJECT - As mentioned the task force covers geofabrics use, but no geogrids yet. Apparently this technology is newer and originated in Germany for woven polyester grids, in England for extruded polypropylene grids and in Canada for glass fiber grids.

3. ENGINEERING MANUAL - The Swiss manual covers these products including design features, guidelines for specifications and very detailed installation instructions. However it is not a specification to be used as a standard for all cases, since there is room for adapting to products and special site conditions.

specification to be used as a standard for all cases, since there is room for adapting to products and special site conditions.

4. ASPHALT TECHNOLOGY – The geotextile and even the reflective cracking literature refers to many details of installation and geosynthetic technology, yet the asphalt specifications and typical parameters are hardly even mentioned. From this geotextile specialists and manufacturers may conclude asphalt properties are of secondary importance as long as the installation and the overlay are correct. This paper demonstrates, that this is absolutely not the case. There is no benefit in geotextile overlay on a weak subbase or even on an aged asphalt concrete, which invariably will crack severely and spoil the effort.

5. THINNER OVERLAY TO SET OFF COST FOR GEOTEXTILE – It is natural to search for a trade off in economizing the overlay thickness to set off the extra cost for geotextile and tack coat. However design procedures to justify such skinny design are not known to date and they are not recommended. An extended research on Swiss highways to check long term performance of overlays on geotextile is near completion and shows excellent performance of medium to thick overlays. On a 10 year old project the only places with cracks is where the asphalt course tapers off. A recent trend in Canton of Zurich is using 20 to 30 mm of overlay on geofabrics with good results however. The key to such performance is top precision for quality asphalt concrete, exact temperatures for compaction etc. To further illustrate they do not overlap geofabrics, they join them precisely side by side. Another concept uses more material and less labor, thus some more material thickness allows for little more construction tolerances and the longer performance life span.

6. GEOFABRIC VERSUS GEOGRIDS – For many geofabric or geogrid is a matter of philosophy or even religion. In reality the two technologies are for different purposes and they function totally different. They are for different engineering problems and for different applications in most cases.

For others the ease of installation is a determining factor. Again there are geofabrics easy to install and there are geogrids that are installed by pressing with a pneumatic roller only.

Generally geofabrics are for water seal and SAMI, geogrids for reinforcing. However extruded geogrids use 2 kg/m2 of tack coat and chips, thus they are more seal and SAMI than most geofabrics. Some geofabrics are made of very stiff glass fibre and may well act much earlier in stress than some grids in polypropylene with natural creep characteristics. This remark means the world is not that simple and it will be even more complex looking at some geocomposits using geogrid and geotextile fastened together.

The final conclusion from this wide scope of applications simply means the design engineer should first identify the task of his project, investigate the technical problem and make decisions step by step to fin reasonable solutions that fit the purpose and the budget.

REFERENCES

[1] Dr. Felix P. Jaecklin: Geotextileinlagen in bituminösen Belägen, neues Kapitel 11 des Geotextilhandbuches des SVG, Schweizerische Gesellschaft der Geotextilfachleute, EMPA, Postfach, CH-8001 St. Gallen, 1990 (Swiss Geotextile Manual SVG, Section on Overlays on Geotextiles)
[2] Dr. Felix P. Jaecklin: Erfahrungen mit Geotextileinlagen und Einfluss der Belagstechnologie, Proceedings K-Geo 92, 2. Internationaler Kongress Kunststoffe in der Geotechnik, May 1992, Luzern, Switzerland, published by SVG, Empa, St. Gallen
[3] Mathias Blumer, Dipl. Ing.: Strassenbau and Strassenerhaltung mit Asphaltmischgut, SMI, Belag and Beton, CH-6023 Rothenburg
[4] Mathias Blumer, Dipl. Ing.: Massnahmen zur Verhinderung von Rissbildungen, Strasse und Verkehr, VSS, Heft 5/91, p. 251
[5] K. Gossow: Der Einsatz von Vlies im Oberbau von Asphaltstrassen, Erfahrungen aus der Praxis, Das stationäre Mischwerk, 1/1980
[6] B. R. Graf: Auscultation et entretien lourd de l'autoroute N1 Genève-Lausanne, Strasse und Verkehr, 2/1987
[7] Prof. W. Arand: Kälteverhalten von Asphalt, Teil 1, Bewertungshintergrund zur Beurteilung des Walzasphaltes bei Kälte, Die Asphaltstrasse, 3/1987
[8] M. Kronig, B. Kuhn: Untersuchungen an Asphaltbelägen im Kanton Zürich, Strasse und Verkehr, 10/1990
[9] Dr. Ing. K. Born: Konstruktionselemente des Stahlbaues, Stahlbau Band 2, 2. Auflage, Okt 64, Seite 6 (Kerbspannungen).

PART TWO
DESIGN MODELS FOR REFLECTIVE CRACKING IN PAVEMENTS

6 CAPA: A MODERN TOOL FOR THE ANALYSIS AND DESIGN OF PAVEMENTS

A. SCARPAS and J. BLAAUWENDRAAD
Structural Mechanics Division, Delft University of Technology, Netherlands
A.H. de BONDT and A.A.A. MOLENAAR
Road and Railroad Research Laboratory, Delft University of Techology, Netherlands

Abstract
In this contribution CAPA, a user-friendly, PC based, finite elements system, capable to analyze and dimension pavement overlays is presented. Because of the generality of the finite elements method, arbitrary geometry, boundary conditions, loading and interlayer bonding can be specified. A variety of options enables the simulation of discrete cracks in the pavement and their interaction with the surrounding materials. Starting from an initial crack length in the pavement, the system can automatically propagate the crack all the way to the surface computing concurrently the relevant fracture mechanics parameters at successive crack tip positions. A user friendly graphical screen input facility, together with a fully fledged mesh generator and extensive pre- and post-processing graphic facilities allow the designer to quickly evaluate the efficiency of various overlay techniques.
Keywords : Crack Interface Shear Transfer, Finite Element Method, Fracture Mechanics, Overlay Design, Overlay Reinforcement.

1 Introduction

The rehabilitation of cracked roads by overlaying is rarely a durable solution since, after a while, the cracks propagate rapidly through the new layer, Fig. 1. This phenomenon is commonly known as reflective cracking, it occurs in almost all types of pavements and imposes heavy financial strains on national and local pavement maintenance authorities.

For overlaying to become cost effective the speed of propagation of the existing crack to the surface has to be reduced. Various solutions are presently available such as :
* placing reinforcement between the cracked top layer and the overlay,
* placing a stress absorbing membrane between the cracked top layer and the overlay,
* modifying the mechanical characteristics of the overlay

all of which are currently the subject of intense international debate and investigation.

Since it appears that, in the foreseeable future, no common concesus can be achieved as to the choice of a unique optimum among the above three alternatives, pavement designers will have to consider more than one options in their calculations and implement the one that optimizes their design goals.

Because of its generality and its sound theoretical base, the finite element method consists a very powerful computational tool capable to assist the designer in his evaluation. A few years ago, utilization of finite elements for ordinary design calculations would be impractical. The success of the finite element method depends on the availability of computers powerful enough to solve the resulting large systems of equations. For this reason it was considered until recently the sole domain of researchers and academics.

Fortunately, the modern developments in micro-computer engineering enable, at present, implementation of the method to very modest sized personal computers commonly used by engineers for routine design calculations. In this contribution CAPA, a user-friendly, PC based, finite elements system, capable to not only analyse but also dimension unreinforced or reinforced overlays will be presented.

2 General characteristics of CAPA

For an efficient modelling of the in-situ conditions of a cracked pavement, in addition to ordinary quadrilateral finite elements, various other types of elements are necessary, Fig. 1. It is the purpose of this section to briefly review, correlate and present a summary of the features of these elements as implemented in CAPA.

2.1 Pavement idealization

Numerically integrated eight-noded isoparametric quadrilateral elements are used for modelling the geometry of the pavement layers. Extensive graphics oriented input facilities allow rapid geometric modelling by means of a powerful mesh generation module.

Constitutive modelling is within the framework of the theory of elasticity. An incorporated material library allows the choice of isotropic or orthotropic constitutive models for any of the constituent materials of the pavement.

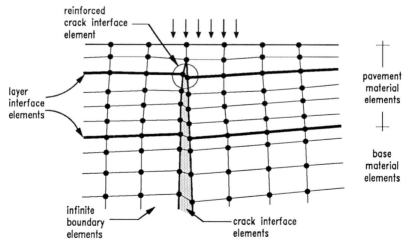

Fig. 1 Finite elements simulation of reflective cracking

2.2 Reinforcement idealization

Discrete reinforcing bars can be modelled by the use of three noded numerically integrated isoparametric truss elements.

2.3 Reinforced soil idealization

Discrete modelling of all reinforcing bars in reinforced soil applications is neither necessary nor feasible. In the uncracked regions of the soil mass, reinforcement contributes to the stiffness of the surrounding soil and shares a portion of the imposed load. For this type of situations, a special "reinforced soil" element is included in CAPA.

This is a numerically integrated, eight noded isoparametric quadrilateral element whose stiffness is formed automatically by CAPA by the superposition of the stiffnesses of an ordinary quadrilateral soil finite element to that of one or more "smeared reinforcement" elements, Fig. 2.

A smeared reinforcement element corresponds to a set of parallel reinforcing bars having inclination θ with the horizontal. Even though each set of distributed bars is treated as an individual finite element, it has no shear rigidity and can only carry forces along the specified bar axis direction.

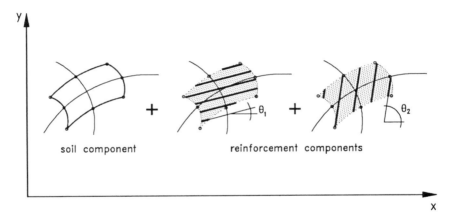

Fig. 2 Smeared reinforced soil element

The thickness of each smeared reinforcement element is calculated by CAPA on the basis of the reinforcement percentage ratio of each set. A maximum of two arbitrary oriented, independent sets of uniformly distributed reinforcing bars can be specified.

2.4 Layer interface idealization

The degree of bonding between the successive layers can influence significantly the overall structural response of the pavement. The interface regions between the layers constitute discrete discontinuities in the otherwise homogeneous -for engineering analyses purposes-body of the pavement. For this reason their simulation within the context of the finite element method requires the use of specially dedicated elements.

In CAPA, bonding between pavement layers can be simulated by means of interface finite elements connecting the nodes of the upper material layer with those of the lower layer, Fig. 1. A typical 2-D quadratic interface element is shown in Fig. 3. It consists of three pairs of nodes. The thickness of the element w in its undeformed configuration can be specified to be very small or even zero.

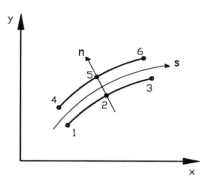

Fig. 3 Quadratic interface element

The constitutive relation associating local tractions to relative local displacements can be expressed as :

$$\begin{Bmatrix} dV \\ dN \end{Bmatrix} = \begin{bmatrix} D_s & 0 \\ 0 & D_n \end{bmatrix} \cdot \begin{Bmatrix} ds \\ dw \end{Bmatrix} \qquad (1)$$

in which s is the interlayer slip. By varying the stiffness characteristics of the element, various degrees of interlayer bonding can be simulated.

An extensive program of investigation is currently undertaken at Delft University of Technology with the aim of identifying the factors influencing interlayer bonding.

2.5 Modelling of boundaries

The nature of most pavement engineering problems is such that, on one hand, realistic modelling of far field boundary conditions is necessary if accurate results are to be obtained while, on the other hand, the area of interest is dwarfed by the extent of the surrounding medium. As a result, an excessively large number of finite elements is required for the geometric modelling of areas of the structure which are otherwise of little interest to the analyst.

Infinite elements present an attractive alternative. By being able to map the infinite domain surrounding the area of interest to a finite space, they enable a more accurate and at the same time a more economical -in terms of mesh size and execution time requirements- analysis.

3 Finite element modelling of cracking

Cracks also constitute regions of discrete discontinuity in the body of the pavement. For this reason their simulation within the context of the finite element method requires the use of specialized techniques.

Depending on crack location and the nature of the imposed loading (e.g. traffic, thermal etc.) the relative displacements of the crack faces differ, de Bondt [1992]. In purely flexural modes of deformation -as those under the centre line of a wheel load, or due to uniform thermal loads- only opening displacements occur, while, in shear-flexural cracks - as those due to off-center wheel loads- both, opening and sliding displacements occur. In the following various aspects of crack simulation and their implementation in CAPA will be briefly reviewed.

3.1 Crack interface idealization

Load transfer measurements at cracked pavements, de Bondt and Saathof [1993], indicate that aggregate interlock may play an important role in determining the structural response of the pavement.

The degree of aggregate interlock depends on several factors the most notable among which is the magnitude of the compressive force normal to the crack face. It is this normal force which prevents unrestrained crack dilatancy, Fig. 4(a), and enables the development of the friction mechanism at the interface, Tassios and Scarpas [1987], Fig. 4(b).

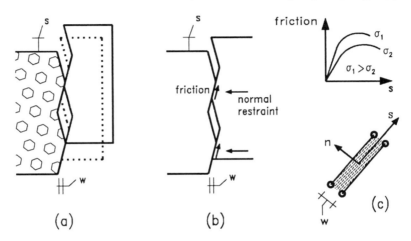

Fig. 4 Crack interface idealization

The phenomenon of crack interface shear transfer can be simulated in CAPA by means of interface finite elements connecting the nodes on either side of the crack line. Their constitutive formulation is similar to that presented in section 2.4.

An extensive program of investigation is currently undertaken at Delft University of Technology with the aim of identifying the appropriate stiffness coefficients for use in Eq. 1.

3.2 Reinforced crack interface idealization

Externally applied normal compressive forces are not necessary for the development of the friction mechanism in case of reinforced cracks. As shown in Fig. 5, because of the wedging action of the crack faces the reinforcement is tensioned. For equilibrium, an equal and opposite force is developed at the interface. This plying action of the reinforcement not only restrains crack dilatancy but also, enables the development of the friction mechanism.

In CAPA, modelling of the restraining action of the reinforcement crossing a crack and of its contribution to the phenomenon of shear transfer at the crack interface is achieved by means of a specially developed "reinforced crack interface" element. The overall behaviour of the element is determined as the sum of the individual responses of the basic force transfer mechanisms as identified in Fig. 5.

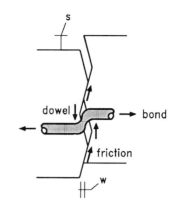

Fig. 5 Force transfer mechanisms at a reinforced crack interface

3.3 Singularity formulation

Elastic materials exhibit an $1/\sqrt{r}$ singularity in the distribution of strains in front of the tip of a crack. Utilization of conventional finite elements for simulation of this type of singularity is extremely inefficient as it requires a prohibitively large number of elements around the tip region.

Instead, the singularity can be accurately and elegantly modelled if the elements surrounding the crack tip node are substituted by specially developed "crack tip" elements, Barsoum [1976], Fig. 6, capable of representing the required discontinuity within themselves.

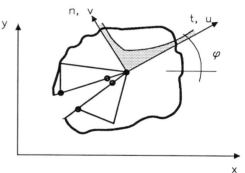

Fig. 6 Crack tip simulation element

4 Overlay design methodology

From the theory of fracture mechanics, if the stress intensity factor in front of a crack in a solid is known for different crack lengths then the number of load repetetions required for the crack to propagate from its original length c_0 to a length c_1 can be estimated by means of Paris' law :

$$N_{tot} = \frac{1}{A} \cdot \int_{c_0}^{c_1} \frac{1}{\left(K_{I,eq}(c)\right)^n} \cdot dc \qquad (2)$$

in which : $K_{I,eq}(c)$ = the equivalent stress intensity factor at the crack tip encompassing the effects of both modes of deformation

A, n = experimentally determined material fracture parameters

From the finite elements point of view this procedure ordinarily implies a series of successive analyses, in each one of which the mesh is manually modified to allow for crack propagation, one element row at a time. At the end of this series of analyses the designer would normally have to compute the integral of Eq. 2 by hand.

In CAPA, an incorporated powerful remeshing technique completely automates the above task and hence greatly speeds up the process of evaluation of alternative design solutions. Details can be found elsewhere, Scarpas [1991]. Only a brief outline will be presented here. Starting from the initial cracked pavement configuration as input by the designer, the following steps are performed by the system :

(i) the ordinary quadrilateral elements surrounding the crack tip are substituted by singularity elements.

(ii) an analysis is performed for the specified load combination and material properties. The relevant fracture mechanics parameters at the crack tip as well as the displacements, the stresses and the strains are computed and stored on file.

(iii) the singularity elements surrounding the current crack tip are automatically replaced with ordinary elements and the new crack tip node is determined.

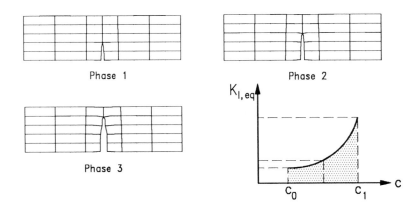

Fig. 7 Automatic crack propagation procedure in CAPA

As illustrated in Fig. 7, the above procedure is repeated for the new crack tip until crack propagation reaches the top of the overlay. In situations where it is desired to include in the analysis the force transfer characteristics of the crack faces, the newly created nodes on either side of the crack line can be joined by means of crack interface elements. Their characteristics are determined on the basis of the material properties of the surrounding materials.

Once a set of pairs ($K_{I,\,eq}$, c) has been obtained, CAPA evaluates the number of traffic passages to failure by numerical integration of Eq. 2.

The voluminous amount of information capable to be generated by the finite elements method consists both, its strong advantage over other less general methods but, also, its main weakness since, reduction of large amounts of data is a laborious, time consuming operation.

To enable the designer to concentrate on the design aspects of his project rather than on data handling and reduction, extensive graphically controled presentation facilities are included in CAPA. Through these, multicolour screen and hard copy plots of deformations and stresses can be obtained.

5 Conclusions

The generality of the finite element method allows the designer to realistically incorporate in his analyses various aspects of pavement behaviour which would be impossible to include otherwise.

By utilizing the latest developments in micro-computer engineering, CAPA, a user-friendly, PC based, finite elements system, can significantly contribute in both, the analysis and the design stages of pavement engineering.

6 Acknowledgement

The financial assistance provided by the Netherlands Technology Foundation (STW) through a research grant to Delft University is greatfully acknowledged.

7 References

Barsoum, R.S. (1976) On the Use of Isoparametric Elements in Linear Fracture Mechanics, **Int. Journal of Numerical Methods in Engineering**, 10, 25-37.

de Bondt, A.H. (1992) **Cracking of Asphalt Concrete Overlays.** Road and Railroad Research Laboratory, Delft University of Technology, The Netherlands.

de Bondt, A.H. and Saathof, L.E.B. (1993) Movements of a Cracked Semi-Rigid Pavement Structure, in **RILEM Conference on Reflective Cracking in Pavements**, Liege.

Scarpas, A. (1991) Simulation of Fracture Zone Development in Reinforced Concrete Members, in **RILEM Conference on Fracture Processes in Concrete, Rock and Ceramics**, Noordwijk.

Tassios, T.P. and Scarpas A. (1987) A Model for Local Crack Behaviour, in **IABSE Colloquium on Computational Mechanics of Reinforced Concrete**, Delft.

7 SEMIRIGID PAVEMENT DESIGN WITH RESPECT TO REFLECTIVE CRACKING

I. GSCHWENDT
Slovak Technical University, Bratislava, Czechoslovakia
I. POLIAČEK
Viaconsult, Bratislava, Czechoslovakia
V. RIKOVSKÝ
Research Institute of Civil Engineering, Bratislava, Czechoslovakia

Abstract
This paper deals with design of asphalt pavements with cement bound base layers. In standard design method we use the mathematical solution of pavement model, but the deformation characteristics of base materials are design and not real values. In the new design procedure for semirigid pavement with analytical evaluation we compare the radial stress induced in the critical layer by load with the bending strength. In this paper the calculation of the needed thickness of asphalt layers for protection of the surfacing against the reflective cracks is described. There is the description of road test section (built in 1987) with stabilization in base course with geotextile membrane and/ or additional compaction as a protection against cracking.
<u>Keywords:</u> Reflection cracking, Mechanistic-empirical design method, Geotextiles, Field testing

1 Introduction

The design procedure for semirigid and flexible pavements is in principle the same: for calculation of stresses and strains in structure layers we use the mathematical solution for model of the pavement which is the multilayer elastic system. Each of the layers is determined by thickness and characterized by modulus of elasticity and Poisson`s ratio of material. These deformation characteristics are not real, but design values. They were derived on the basis of mechanical properties of road construction materials and estimated from laboratory tests, taking into account behaviour of materials in these pavement stucture during its lifetime. For very often used cement stabilization and cement bound materials of the design values of characteristics are as follows:
. cement stabilization
  $I.^{st}$ quality (determined by laboratory test of compressive strength),
  C_7 = 2,5 - 3,5 MPa, E = 1200 MPa, μ = 0,25
  $II.^{nd}$ quality

$C_7 = 1,8 - 2,5$ MPa, $E = 1000$ MPa, $\mu = 0,25$

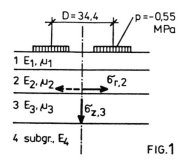

FIG.1

- cement bound granular material
  - I^{st} quality (determined by laboratory test of compressive strength)
    $C_{28} = 7 - 12$ MPa, $E = 2500$ MPa, $\mu = 0,22$
  - IInd quality
    $C_{28} = 6 - 11$ MPa, $E = 2000$ MPa, $\mu = 0,22$

The compressive strength of materials is measured by standard laboratory method using cylindrical or cubic test specimens.

According to general rules for evaluation of pavement structures, is necessary to control the frost protection and mechanical efficiency of pavement structure, which must be during the whole lifetime greater than the requested minimum value /1/.

At the beginning of the control is calculat the radial stresses in the bottom of each layer (FIG.1). After then we need to compare the calculated stress with the bending strength of bound (stabilized) material.

The design values of bending strength R are:
- for the cement stabilization
  I^{st} quality $R = 0,5$ MPa, IInd quality $R = 0,4$ MPa
- for the cement bound granular material
  I^{st} quality $R = 1,0$ MPa, IInd quality $R = 0,8$ MPa

The permissible stress is in relation to number of standard axle load repetitions. In this way we take into account the fatigue of material. The fatigue strength Ri,f we can express with the equation

$$R_{i,f} = S_N R_i \qquad /1/$$

where $\qquad S_N = a_i - b_i \log N_i \qquad /2/$

is fatigue coefficient and

a_i, b_i are qualitative parameters of the fatigue relation for material,
N_i - number of the standard axle load repetitions,
R_i - the design value of the bending strength

For soil cement stabilization (or any cement bound material) with compressive strength C_{28} higher than 6,0 MPa

$$S_N = 1,0 - 0,07 \log N \qquad /3/$$

for soil cement stabilization (or any cement bound base material) with compressive strength C_{28} less than 6,0 MPa

$$S_N = 0,9 - 0,08 \log N \qquad /4/$$

The designed pavement structure fulfills the requirements when the equation /5/ is satisfied:

$$\sum_i q_i \frac{\sigma_{r,i}}{S_N \cdot R_i} \leq 1 \qquad /5/$$

where $\sigma_{r,i}$ is radial stress at the bottom of the critical layer (MPa),
q_i - relative duration (time) of constant conditions for stress calculation.

The changes of external pavement "working" conditions - the temperature regime and changes in traffic load (the growth of heavy vehicles number) we can take into account and consider in pavement structure evaluation by integration method summarizing the partial damages ΔSV (damage portions) of the material in the critical layer /2/:

$$\sum \Delta SV \leq 1,0 \qquad /6/$$

while

$$\Delta SV = \frac{\tau \cdot \sigma_{r,i}}{/a_i - b_i \log(\delta_i N_C)/ \, R_i} \qquad /7/$$

where

τ - relative duration of the period i,
$\sigma_{r,i}$ - radial stress at the bottom of the critical layer (MPa)
a_i, b_i - qualitative parameters of the fatigue relation, also /2/,
N_C - number of the design axle load repetition,
δ_i^C - relative length (time) of the pavement exploitation,
R_i - design value of the bending strength (MPa).

2 Crack control method

Evaluating the design of semirigid pavements, we are to apply two independent criteria. The designed pavement is, as we have seen, evaluated as concerns the fatigue bending strength of the material of the critical pavement layer. In this evaluation we suppose, that the cement bound layer is cracked. The design value of the elasticity modulus of the cracked bound layer we take as one half of the real modulus value. Critical layer of the pavement is normally the last asphalt bound layer over the cement bound layer in the pavement. The service efficiency $/N_c/$ of this layer should be greater than the requested value /the calculation scheme is on the FIG. 4a/.

If the pavement fulfills the first criterion, we have to find out whether the thickness of the asphalt layers is greater than the required thickness, which gives guarantee that the shrinkage cracks in the cement bound base layer will not reflect through the asphalt layers to the top of the pavement during its lifetime.

In this - second - part of assesment we
- estimate the distance of the cracks in the cement bound base course and the penetration of asphalt binder at the end of requested lifetime of the pavement,
- find out from the requested pavement service efficiency /which corresponds to the requested lifetime and traffic volume/, estimated distance of the cracks in the cement bound base course and the penetration of the asphalt binder, with the help of. FIG. 2, the thickness of the asphalt layers to which the reflective cracks reach.

When we do not use other means of crack regulation, we suppose that the distance of the shrinkage cracks reaches 10,4 m. The penetration of the asphalt binder can be determined from FIG. 3.

The rest service efficiency /restlife/ of pavement, which is damaged by the shrinkage and reflective cracking can be calculated according to the scheme in the FIG. 4b. The moduli of the asphalt layers, which are damaged by the reflective cracks, we consider in the calculation also with the half values of the real moduli values.

The pavements rest service efficiency should be greater than

$$N_{c,II} > 0{,}25 \text{ to } 0{,}35 \, N_{c,I} \qquad /8/$$

In this equation

$N_{c,I}$ - the service efficiency (service life) of the pavement calculated for the first criterion (fatigue bending stress concept);

N_c,II - the service efficiency (service life) calculated

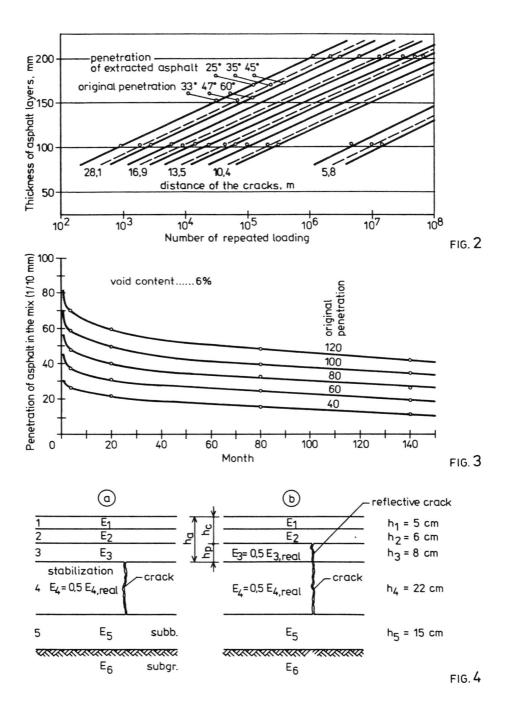

FIG. 2

FIG. 3

FIG. 4

for the pavement damaged with the reflective cracks.

We can substitute the assessment of the thickness of asphalt layers in the pavement with proper design of the pavement structure by the help of the graphs in FIG. 2 and 3. The requested thickness of the asphalt layers in the semirigid pavement, which secures that the cracks do not reach the surface of pavement, can be determined. In this case the distance of the shrinkage cracks we consider with the value at least 13,5 m (for greater security). This approach is, of course, less economical as the standard way of assessment.

3 Full scale experiments

In order to test (verify) and compare the theoretical assumptions and real behavior of a pavement structure with and without a geotextile membrane as a protection against reflective cracking , a full scale experiment have been organized. The test was realized on a circular test road /3/ and on motorway construction. The original pavement structure design on motorway with 20 cm thickness of asphalt layers on cement bound granular material: 4 cm mastic asphalt, 8 cm asphalt concrete, 8 cm dense bitumen macadam, 20 cm cement bound granular material and 20 cm sand and gravel subbase. The design of the new structure with geotextile:
- 4 cm mastic asphalt
- 8 cm asphalt concrete
- geotextile (non woven) interlayer
- 24 cm cement bound granular material (5 % cement content, R_{28} = 13,9 MPa)
- 24 cm sand and gravel subbase

For construction of pavement the contractor used standard road building methods. In part of test section-before laying the geotextile on the surface of cement bound material, the modified asphalt was applied-tack coat 1,0, max. 1,25 kg on m^2. Non-woven geotextile (TATRATEX) of mass 300 g.m^{-2} was laid in sheets 3,5 m width. The road was constructed during 1985. The traffic load is expressed by 1860 mean daily number of commercial vehicles.

For new pavement structure we calculate the theoretical service efficiency (service life) $N = 3,87 \cdot 10^7$ standard axles by the equation

$$\log N = \frac{a_i - \sum q_i \frac{\sigma_{r,i}}{R_i}}{b_i} \qquad /9/$$

(see formulae 5 and 2).

The state and the behavior of the pavement was assessed on the bas of the deflection meassurements with FWD. The annual visual condition survey has been realized too. Until now, the realized traffic load is 4,78 mil. heavy vehicles, which represents about 25 % of design traffic load. On the section with geotextiles in structure there are no reflective cracks and the bearing capacity is satisfactory. There are some places with permanent deformations of surfacing; ruts are 8 mm deep. The ruts depth for the end of design period was calculated to be 24 mm.

4 Conclusion

As we could see, there exist possibilities to asses the designed semirigid pavement from the point of view of reflection cracking.
  We are able to
- find out wheter the designed thickness of asphalt layers in the pavement is sufficient to ensure that the shrinkage cracks in the cement bound base layer do not reflect to the top of pavement in the given time, or
- to design the structure of the semirigid pavement so that the reflective cracks can not reach the top of the pavement.

5 References

1. GSCHWENDT, I.- POLIAČEK, I.: The design and assessment of pavement structures. Publ. of Research Institut of Civil Engineering, No 223/1987, Bratislava

2. NOVOTNÝ, B.: The fatigue criterion for evaluation of the flexible pavements. Silniční obzor 7/1991, Praha

3. RIKOVSKÝ, V.: Pavement performance test on circular test track. Silniční obzor 45, No 2/1986, pp.35-38

8 NUMERICAL MODELLING OF CRACK INITIATION UNDER THERMAL STRESSES AND TRAFFIC LOADS

A. VANELSTRAETE and L. FRANCKEN
Belgian Road Research Centre, Brussels, Belgium

Abstract

This paper deals with the study of interface materials to prevent reflective cracking and more particularly discusses the effect of these products on the initiation of cracks in the asphaltic overlay. A functional classification of interface systems is given, intended to reduce the wide variety of products commercially available to a limited number of families with comparable characteristics.

The results obtained from numerical modelling of interface systems subjected to traffic and thermal loads are discussed. Several classes of products are studied, from flexible systems such as SAMI's to reinforcing ones (e.g. metal grids). The magnitude of the induced strains ad the effect on the crack initiation time are compared to those of the reference structure without interlayer. For the case of traffic loads, a comparison is made between the results obtained from two-dimensional calculations and three-dimensional simulations.

Keywords:Reflective Cracking, Interface Systems, Structural Evaluation, Numerical Modelling.

1 Introduction

This paper deals with the study of interface materials to prevent reflective cracking. Unlike many research studies based on fracture mechanics, which are generally focused on the propagation phase of a crack, the studies carried out in this field at the Belgian Road Research Centre, are taking more particularly account of the behaviour of the system during the initiation phase. The assessment of the effect of an interlayer must be made on the basis of its behaviour as a component of the complete road structure. The final result depends not only on the interface but rather on the combination of materials of the different layers, from the subgrade to the wearing course.

Reflective Cracking in Pavements. Edited by J.M. Rigo, R. Degeimbre and L. Francken.
© 1993 RILEM. Published by E & FN Spon, 2–6 Boundary Row, London SE1 8HN. ISBN 0 419 18220 9.

2 Functional classification of interface systems

Table 1 gives an overview of the types of interlayer systems now commercially available. This classification is intended to improve the identification of a given interlayer among the wide variety of products available. It is based on its functionality and simplifies the choice by the reduction to a limited number of families having comparable characteristics. If completed by a performance evaluation, this classification could help to decide which system would be the best solution for a particular problem.

Table 1 : Functional classification of interlayer products
SAMI : Stress absorbing membrane interlayer ; WOV : Woven
NWM : Non woven, mechanical or spun bound ; GGR : Geogrid
NWT : Non woven thermally bound ; MGR : Metallic grid

PRODUCT TYPE :	SAMI	NWM	NWT	WOV	GGR	MGR
BASE MATERIAL :						
Polyethylene		X	X			
Polypropylene		X	X	X	X	
Polyester		X	X	X	X	
Elastomeric bitumen	X					
Glass	X			X	X	
Steel						X
FUNCTION :						
Strengthening				X	X	X
Creep-sliding	X	X	X	X		X
Stress relieve	X	X	X			
Waterproofing	X	X	X			
LAYING PROCEDURE:						
Tack coat (emulsion or binder)		X	X	X	X	
Poured within	X					
Slurry seal					(X)	X
Nailing					X	X

3 Numerical modelling

The present paper discusses results of an analysis of stresses and strains induced by traffic and thermal loads and their effect on the crack initiation time for several interface systems.

3.1 Simulations of traffic loads
Description of the simulated structure

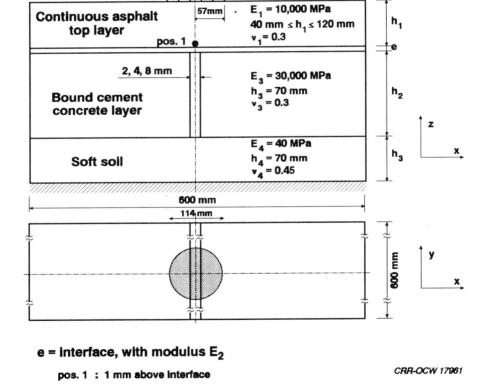

e = interface, with modulus E_2

pos. 1 : 1 mm above interface

CRR-OCW 17961

Fig.1 : Representation of the cracked structure used for the simulation of the effect of interfaces on crack initiation as a result of traffic loading.

Simulations, based on linear elastic analysis, of the effect of interface systems on the crack initiation in asphalt are performed in two and three dimensions and are compared. The structure and the simulation conditions correspond with the test conditions of the wheel tracking apparatus, currently being developed at the BRRC. The structure consists of a 70 mm thick cement concrete block of 60 cm X 60 cm with a discontinuity or crack (fig.1). A Poisson's ratio $\nu = 0.3$ and a stiffness modulus of 10000 MPa are assumed for the overlying asphalt layer. These are considered as realistic values for the frequencies involved. Complete adherence between the different layers is assumed. The case of a vertical load of 1 MPa, positioned symmetrically with respect to the crack is considered. A centered vertical loading only involves mode 1 movement of the crack. Hence, crack initiation will most probably occur just above the crack tip, since the largest stresses and strains are obtained there. The X-strain and X-stress in the asphalt layer in the prolongation of the crack and under the center of the load are therefore used for evaluation.

The three-dimensional analysis is performed by the finite element package SYSTUS (1). The centered vertical load is represented by a uniform vertical pressure acting on a circular surface with a radius of 57 mm (see fig.1). Owing to the symmetry of the problem, only one quarter of the structure is calculated. The two-dimensional analysis is performed with the finite difference program FLAC (2). Only the X- and Z-geometry of the sample are taken into account. As a result of the plane strain assumptions used in this analysis, the sample is supposed infinitely long in the Y-direction. The same holds for the load in the Y-direction; in the X-direction the pressure is applied over a distance of 57mm. We note that a 2D-analysis performed with "SYSTUS" leads to comparable results (within 5 %).

Criterion for comparing different interface systems
The effect on the crack initiation time is studied by making use of the fatigue law :

$$\epsilon_{ini} = K N^{-a} \tag{1}$$

giving the lifetime before cracking, N, of a given bituminous material for a given initial strain, ϵ_{ini}. The factor K depends on the mix composition ; a is the slope of the fatigue curve. A mean value of $a = 0.21$ was found for dense bituminous mixes studied at the BRRC.(3) This leads to the following expression for the factor of difference in crack initiation time with and without interface, Cr :

Comparison of various interface systems
Five interface systems for the case of a 4 mm crack width

$$CI = \frac{N_{interf.}}{N_{no\ interf.}} = [\frac{(e_{xx})_{interf.}}{(e_{xx})_{no\ interf.}}]^{-\frac{1}{0.21}} \qquad (2)$$

and a 72 mm thick asphaltic layer are studied :

- Reference without interface
- 2 mm thick soft interface with E_2=100 MPa (e.g. SAMI, NWM)
- 2 mm thick soft interface with E_2=1000 MPa (e.g. WOV, GGR)
- 2 mm interface of comparable stiffness with asphalt layer (E_2=10000 MPa)
- 2 mm reinforcing interface with E_2=15000 MPa (e.g. MG).

The results of the X-strain as a function of the distance above the interface (in the prolongation of the crack and under the center of the load) obtained from the two- and three-dimensional analysis are presented in fig.2(a) and fig.2(b) respectively. The study leads to the following statements for the X-strains in that discussed region :

1) The type of loading surface (circular in the 3D-model and infinitely extended strip in the Y-direction for the 2D-model) leads to considerable differences in the X-strain curves :
   - The ϵ_{xx}-values obtained from the 3D-analysis are smaller than those calculated with the 2D-model.
   - The ϵ_{xx}-curves have a different shape. From the 2D-analysis it is found that ϵ_{xx} decreases continuously as the distance above the interface increases. For the 3D-analysis a decrease in ϵ_{xx} is also found very close to the crack tip ; but then ϵ_{xx} increases to a maximum at about 5 cm above the interface to decrease very steeply again for larger distances. The increase in ϵ_{xx} implies that not only the region very close to the crack tip is important for initiation of cracks.
   - The neutral line is located closer to the top of the structure in the case of circular loading (3D-analysis). Its position is only slightly affected by the use of an interface system.
2) Concerning the effect of interface systems :
   - For regions very close to the crack tip (< 3 mm), the strains in the asphalt are reduced by the presence of an interface with comparable or higher stiffness than the asphaltic overlay. They lead to a considerable improvement in the crack initiation time (see table 2). These interface systems are effective for the prevention of crack initiation, provided the interface system itself can withstand the strains induced in it.

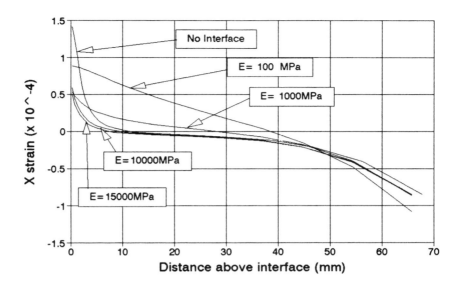

Fig. 2(a)

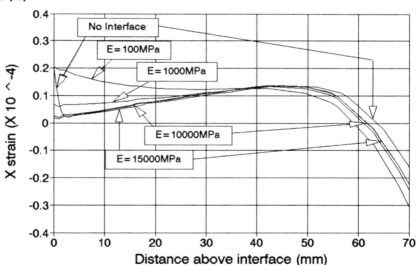

Fig. 2(b)
Fig. 2 : Calculated X-strain as a function of distance above interface (in the prolongation of the crack and under the center of the load) for different stiffnesses of the interface (for a 72 mm thick asphalt layer). (a) : from two-dimensional simulations ; (b) : from 3D-analysis. The decrease in crack initiation time for these

reinforcing systems calculated from 3D-analysis is higher than that obtained from 2D-calculations (see table 2). A decrease with a factor of 2000 is found for the reinforcing interface with E = 15000 MPa from 3D-analysis, a factor of 380 was obtained from 2D-calculations.
- For distances further from the crack tip, interface systems do not reduce the X-strain and the crack initiation time considerably. This is especially the case for circular loading (3D-analysis). The ϵ_{xx}-values at the maximum 5 cm above the interface are not drastically affected by the use of an interface system. As these strains are comparable to those very close to the crack tip, a high risk for crack initiation in this region will remain, even with the use of reinforcing interfaces. We note however that also the structures geometry should be important as to the ϵ_{xx}-value at the maximum relative to that close to the crack tip. This will be investigated in more detail in the future.
- Interfaces with smaller stiffness than the asphaltic overlay (especially the 100 MPa - interface) rather have a negative influence on the X-strain and on the crack initiation time. This holds for the circular loading (3D-analysis) as well as for the infinitely extended loading plate in y-direction (2D-calculations). The 1000 MPa stiff interface gives rise to a decrease in X-strain and crack initiation time very close to the crack tip, but provides an increase at larger distances.

Table 2 : X-strain, ϵ_{xx}, and difference in crack initiation time, Cr, in the asphalt layer in the prolongation of the crack and under the center of the load, obtained from two- and three-dimensional simulations, for different interface types. (72 mm thick asphalt layer and 4 mm crack width).

Interface type	2 Dim. Analysis 1 mm above interface		3 Dim. Analysis 1 mm above interface	
	ϵ_{xx} (X 10^{-4})	Cr	ϵ_{xx} (X 10^{-4})	Cr
Reference without interlayer	1.14	1	0.072	1
SAMI E=100 MPa	0.87	3.6	0.192	0.0092
WOV / GGR E=1000 MPa	0.50	50	0.059	2.5
E=10000 MPa	0.40	146	0.020	440
MGR E=15000 MPa	0.33	380	0.015	2030

3.2 Simulations of thermal movements of the base layer

This study is carried out with the two-dimensional finite difference program FLAC (2) and concerns a linear elastic modelling of a structure similar to that of the test sample used in the laboratory simulation apparatus of the BRRC for the experimental study of the effect of interface systems under thermal loading. It is described in more detail in (4) and concerns the simulation of a semi-rigid pavement with a joint, opened under the effect of horizontal tensile stresses applied uniformly on the sides of the concrete base material. The effect of the type of interface on the crack initiation phase was studied. Five structures with a fixed overlay material (E=1000 MPa, h = 70 mm) placed on different interlayer systems have been compared :

- Reference structure without interlayer
- 2 mm soft interlayer (E = 20 MPa)
- 2 mm stiff interlayer of metal grid (E = 10000 MPa)
- 2 mm stiff interlayer over 2 mm slurry seal (E=100 MPa)
- 2 mm stiff interlayer over 2 mm SAMI (E = 20 MPa).

A systematic study of the influence of the width of the crack, the thickness of the asphaltic overlay and its modulus was also performed. The study leads to the following conclusions. More detailed results are given in (4).

- All interlayer systems, regardless of their stiffness, are effective for preventing crack initiation in the asphalt layer very close to the crack tip, provided they can withstand the strains induced in them. Particularly at low temperatures very high strains can be obtained in the interlayers, which can exceed the critical values and hence limit the effect of such products.
- The best solution is obtained by the combination of a soft interlayer (allowing an easy horizontal movement) and a strenghtening material at the base of the overlay.
- The overlay thickness has only a limited influence on the stresses and strains just above the crack tip.

4 Conclusions

In this paper a functional classification of interlayer products is proposed. Computer simulations based on linear elastic modelling were performed to study the effect of these systems on the crack initiation phase in asphaltic overlays induced by thermal and traffic loads. As to the effect of traffic loads, a comparison is made between two- and three-dimensional simulations. The study leads to the following statements :

1) For the case of thermally induced cracks all interface systems, regardless of their stiffness, are efficient for preventing crack initiation in the asphalt layer very close to the crack tip. The nature and properties of the interlayer are of primary importance. The influence of the overlay thickness is small. The best results are obtained with systems combining a soft interlayer with a strenghtening material.

2) In the case of traffic loads, interlayer systems with comparable or higher stiffness than the asphalt overlay are efficient for preventing crack initiation very close to the crack tip, provided they can withstand the strains induced in them. The 3D-simulations show that for distances further from the crack tip, interface systems do not reduce the X-strain and the crack initiation time considerably. Since in some cases, depending on the structures geometry (e.g. the overlay thickness), very high strains can be found in this region a high risk for crack initiation remains, even with the use of reinforcing interfaces.

3) Complementary observations obtained from survey of full scale experiments indicate that a good bonding between the different layers of the system is of primary importance. It must be emphasized that the overlay thickness is predominant for what concerns the structural behaviour of the pavement as well as for the propagation fase of the crack.

5 Acknowledgments

This project was sponsored by the IRSIA (Institut pour l'encouragement de la Recherche Scientifique dans l'Industrie et l'Agriculture), which the authors wish to thank for its financial support. They are also grateful to Mr. J.Reichert, Director of the Belgian Road Research Centre, and Mr J.Verstraeten, Deputy Director.

6 References

(1) FRAMASOFT + CSI, "SYSTUS", Paris, France.
(2) ITASCA Consulting Group,INC, "FLAC Fas Lagrangian Analysis of Continua", Minneapolis, Minnesota.

(3) **Francken L. and Clauwaert C.**, "Characterization and Structural Assessment of Bound Materials for Flexible Road Structures", Proceedings of the VIth Int. Conf. on the Structural Design of Asphalt Pavements, Ann Arbor, Michigan, 1987, pp. 130 - 144.

(4) **Francken L. and Vanelstraete A.**, "Interface Systems to prevent Reflective Cracking", Proc. of the 7th Int.Conf. on Asphalt Pavements, Nottingham, 1992, pp. 45 - 60.

9 EVALUATION OF CRACK PROPAGATION IN AN OVERLAY SUBJECTED TO TRAFFIC AND THERMAL EFFECTS

J.M. RIGO, S. HENDRICK, L. COURARD and C. COSTA
Geosynthetics Research Centre, Materials Research Centre, University of Liege, Belgium
S. CESCOTTO
M.S.M., University of Liege, Belgium
P.J. KUCK
Hoechst, Bobingen, Germany

Abstract
The authors are actively involved in the development of finite element programs to study crack growth in the overlay of a structure including an anti-reflective cracking layer based on a bitumen impregnated geotextile. The program considers various positions of lorry axles and determines the total effect on the crack tip. This is combined with a temperature variation in the whole structure.
The two effects are combined in order to evaluate local damages. The F.E. program works on the following assumptions :
- the materials are elastic;
- secant moduli of the bitumen are taken into account;
- the crack has an effective width equal to the overlay maximum aggregate size;
- traffic and thermal damages can be added;
- the road structure is a semi-rigid type structure;
- the crack opening follows mode 1;
- the geotextile acts as a bitumen container and not as a reinforcement.

Keywords : reflective cracking, simulation, finite elements, model, traffic, thermal effects.

1 Introduction

The SAPLI 5 finite element programme has been developed at the M.S.M. department of Liege University for structural design. It works on mainly linear elastic mode.
  As such it has been applied to simulate the crack growth in a structure constituted of (from top to bottom) (figure 1) :
- an asphalt overlay (6 cm);
- an interlayer made of a bitumen saturated geotextile (2 mm);
- a cement concrete slab (20 cm) including a vertical crack (5 mm width);
- a 15 cm thick granular subbase;
- the soil.

Reflective Cracking in Pavements. Edited by J.M. Rigo, R. Degeimbre and L. Francken.
© 1993 RILEM. Published by E & FN Spon, 2–6 Boundary Row, London SE1 8HN. ISBN 0 419 18220 9.

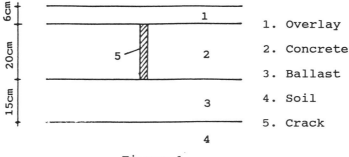

Figure 1

For this particular configuration, the objective of the work was to estimate the overlay service-life under traffic and thermal actions simultaneously.

The service life of the overlay is defined as the number of cycles of traffic and thermal loads necessary to propagate the crack across the overlay.

Fracture of a heterogeneous aggregate material such as concrete has been modelled in large finite element programs as systems of parallel cracks that are densely distributed over the finite element. This was originally proposed by Rashid (1982) and is known as the "Blunt-crack band Theory".

It has been presented by Haas (1989); Clauwaert-Francken (1989) and proved by field and laboratory observations (Rigo et al, 1989) that failures in concretes propagate with tortuosity creating a damaged zone instead of a unique crack. The cracked band width may be equal to the largest aggregates size.

The SAPLI 5 program has been used to evaluate the traffic and the thermal stresses separately and the damages induced by these stresses were evaluated and combined which permits the estimation of the overlay service life.

2 Traffic and thermal damages and their combination

2.1 Basic principle

The investigated structure has been divided into small elements by means of a 2D-finite element mesh (figure 3). The principal stresses and the resulting damages are evaluated for each element under traffic and/or thermal effects taking the stress history of each element into account. Once the damage is equal to unity for a given element, this one is removed from the mesh and the next calculation step is based on a mesh with an empty element as a simulation of the crack. This is realized step by step until the simulated crack reaches the top surface of the overlay.

2.2 Damage and crack propagation

The damage of an element is defined as the ratio between the defects surface (S_D) and the total surface (S).

$$D = \frac{S_D}{S} \qquad (1)$$

If S' represents the resisting surface,

$$S' = S - S_D \qquad (2)$$

The effective stress (σ') in an element is subjected to modifications in function of the evolution of the effective resisting surface

$$\sigma' = \sigma \cdot \frac{S}{S'} = \frac{\sigma}{1 - D} \qquad (3)$$

If a material is subjected to damage, its elastic behavior is also affected and

$$E' = (1 - D) \cdot E \qquad (4)$$

where
E = elastic modulus of the virgin material
E' = elastic modulus of the damaged material.

By assuming that the damage has a linear evolution versus the number of the applied stresses cycles,

$$D = \Sigma D_i = \Sigma \frac{n_i}{N_i} \qquad (5)$$

where
D_i = the damage caused by a given stress level σ_i
N_i = the number of stress cycles at level σ_i corresponding to the failure ($D_i = 1$).
It is given by the well known Wöhler curves
n_i = the number of stress cycles at level σ_i already applied to the element.

Schmidt (1989) presented a model of the crack propagation into a bituminous surfacing. This principle is applicable to each element of the finite element mesh presented in figure 1 :
- for the initial mesh configuration (overlay without any crack) (k = 1). The stresses are evaluated for each element. For the one that is the most loaded it is possible to evaluate the service-life by referring to the Wöhler curve

$$n_1 = N_{1,1} = (N_{i,k})_{i=1, \ k=1} \qquad (6)$$

where
n_1 = number of cycles applied to the virgin structure before removing the first element
$N_{1,1}$ = number of cycles supported by the most loaded element (N_1) before failure with the first mesh configuration.

- after failure and removal of the first most loaded element, the stresses are evaluated with the new finite element configuration. For the newly most loaded element (element 2), the service-life is evaluated by equation (7)

$$D = \frac{n_2}{N_{2,2}} + \frac{n_1}{N_{2,1}} \qquad (7)$$

where
n_1 = see equation (6)
n_2 = number of cycles applied to the second element to be removed from the mesh
$N_{2,1}$ = the maximum number of cycles to be applied for the stress on element 2 evaluated with configuration 1 of the finite element mesh
$\frac{n_1}{N_{2,1}}$ = damage supported by element 2 before element 1 failed (with configuration 1)
$N_{2,2}$ = the maximum number of cycles to be applied on element 2 for the stress evaluated with configuration 2

- for configuration 2, the number of cycles to apply before element 2 fails is

$$n_2 = N_{2,2} \cdot (1 - \frac{n_1}{N_{2,1}}) \qquad (8)$$

- for the configuration 3, the number of cycles to be applied before element 3 fails is

$$n_3 = N_{3,3} \cdot (1 - \frac{n_1}{N_{3,1}} - \frac{n_2}{N_{3,2}}) \qquad (9)$$

- the same procedure is repeated for various configurations until the crack reaches the top surface of the overlay;

- a more general version of equation (9) is given hereafter

$$n_i = N_{i.i} \cdot (1 - \sum_{j=1}^{j=i-1} \frac{n_j}{N_{i,j}}) \qquad (10)$$

or

$$n_i = N_{i.i} \cdot (1 - \Sigma\, D_i) \qquad (11)$$

The total service-life of the overlay is equal to

$$n_{TOT} = n_1 + n_2 + n_3 \ldots \qquad (12)$$

It must be pointed out that at each calculation level the deterioration of the material also affects the value of the elastic modulus and equation (4) is to be applied to each finite element at each state of the crack propagation.

2.3 Combination of traffic and thermal damage

The assumption is made that the traffic and thermal damages may be calculated separately and additionned (principle of the damage superposition).
It must also be considered that thermal stresses are also cyclic loads (one cycle/day) so that the approach developed in section 2.2. hereabove is also applicable.

Considering that the traffic loads are applied with a frequency of f cycles per day, the thermal loads are applied at a frequency of 1 cycle per day :

traffic frequency = f . thermal frequency   (13)

The combination of the traffic and thermal damages may thus be done as follow :
- for each element, the traffic (σ_{tr}) and thermal (σ_{th}) stresses are calculated separately;
- the Wöhler curves for the overlay permit the evaluation of
  . N_{tr} for σ_{tr} and
  . N_{th} for σ_{th}

- the number of cycles necessary to obtain the failure of the most loaded element of the configuration is calculated by

$$D_i = D_{i-1} + \frac{n}{N_{tr}} + \frac{n/f}{N_{th}} \qquad (14)$$

where
D_i = damage of the most loaded element for a given configuration (= 1 at failure)
D_{i-1} = damage of the same element induced by the stresses supported during the previous configurations

$$\frac{n}{N_{tr}} = \text{traffic damage}$$

$$\frac{n}{f \cdot N_{th}} = \text{thermal damage}$$

Equation 13 might also be presented as follow :

$$n = (1 - D_{i-1}) \cdot \frac{N_{tr} \cdot N_{th} \cdot f}{N_{tr} + N_{th} \cdot f} \qquad (15)$$

It must be reminded that equation (4) is applicable separately for traffic and thermal loadings :

$$E_{tr,i} = E_{tr,o} \cdot (1 - D_i)$$
$$E_{th,i} = E_{th,o} \cdot (1 - D_i) \qquad (16)$$

where
$E_{tr,i}$ = elastic modulus for the traffic loads for the configuration i
$E_{tr,o}$ = elastic modulus for the virgin material for the traffic loads
$E_{th,i}$ and $E_{th,o}$ = same but for thermal loads.

3 Practical analysis of the reflection crack growth

3.1 Materials characteristics

The proposal methodology has been applied to a semi-rigid structure. The basic characteristics of the materials were the next :

- overlay

E (Mpa)	Frequency (Hz)	Temperature (°C)
5400	10	20
3100	0,1	10
2100	0,1	12,5
1330	0,1	15
390	0,1	25
210	0,1	30

The frequency of 10 Hz is supposed to be valid for the case of traffic loading while 0,1 Hz (which is not totally correct but no data are available) is supposed to be valid for thermal loading.

- the interface

Characteristics	Traffic	Thermal
G	2 Mpa	1 Mpa
E_V	1300 Mpa	1000 Mpa
E_H	4 MPa	2 Mpa
ν	0	0

where
- G = the interface elastic shear modulus
- E_V = the vertical (perpendicular to the interface plane) elastic modulus
- E_H = the elastic modulus parallel to the interface plane
- ν = POISSON's ratio.

E_V is equal to the bitumen bulck modulus in order to take into account the container effect of the geotextile. The effect also induces $\nu = 0$.

- The concrete
  E = 15.000 Mpa
  ν = 0,3
  ρ = 2300 kg/m^3
  α = 12 10^{-6}
  The crack in this layer is supposed to be empty.

- The granular sub-grade
  E = 200 Mpa
  ν = 0,5
  ρ = 2000 kg/m^3
  α = 12 10^{-6}
  which corresponds to a rigidity modulus K = 100 Mpa/m.

3.2 The loads

The structure is supposed to present one crack each 5 meters.
The traffic loads correspond to the local standards (13 ton/axle) and are modelled by a pressure of 6,62 bars on a 9 cm width band. This pressure and band width give in the 2D-model approximately the same principal stresses in the overlay in the cross section on top of the crack as those calculated by a 3D-model with 13 T applied with a pressure of 7 bars on a surface of 30 x 30 cm. The loads are applied at a frequency of 10 Hz. Delcuve (1990) and Hendrick (1992) have shown that the most critical position of this load correspond to the position showed on figure 2.

The overlay temperature is distributed as follow : 30°C at the top surface and 10°C at the interface. The concrete slab is subjected to a ΔT = 10°C which corresponds to a total movement of a 0,6 mm on 5 m.

It is supposed that 1000 vehicles are passing on this section per day.

3.3 Traffic loading

The load is applied as shown on figure 2.

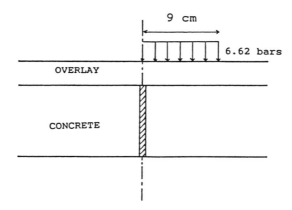

Figure 2 : traffic loads applied to the structure

Only half of the structure is studied and the results are obtained by the superposition of a symetric and a anti-symmetric loading configuration.

On the other hand, for the traffic loading the nodes at the central cross-section of the overlay have no horizontal movement (for symmetry reasons) and the relative vertical movements are blocked.

Figure 3 shows the propagation of the crack in the overlay under traffic loading only.

It must be observed that the load presence influences the crack propagation. This one is deviated transversely and appears to propagate at ± 45°.

3.4 Thermal loads

The movement of the concrete base induces very important stresses in the overlay and the crack propagation is fast and nearly vertical (figure 4).

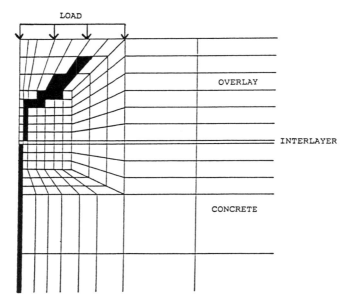

Figure 3 : Crack propagation in the overlay under repeated traffic loads

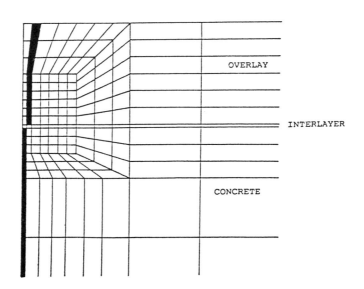

Figure 4 : Crack propagation in the overlay under thermal load

3.5 Combined action

When traffic and thermal effects are combined the crack orientation is intermediate between the two above described situations : the crack propagates more vertically than in the case of traffic loads alone (figure 5).

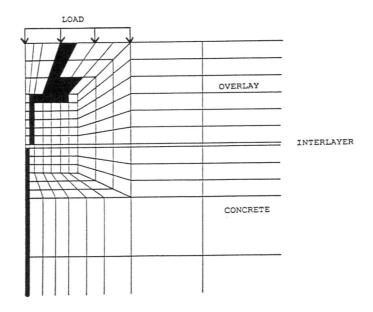

Figure 5 : Crack propagation in the overlay under combined effects (traffic and thermal)

3.6 Crack propagation

For the data presented in section 3.1. it was possible to present in figure 6 the evolution of the crack height as a function of the number of cycles. The results are expressed in days knowing that 1 thermal cycle and 1000 traffic cycles are applied per day.

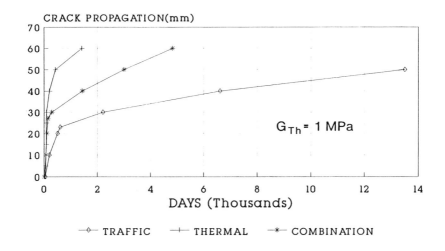

Figure 6 : Crack propagation due to traffic and/or thermal actions as a function of the days of application
(1000 vehicles/1 day, 1 thermal cycle/day)
(see parameters in section 3.1.)

4 Crack propagation as a function of the interface properties

It appeared to be interesting to follow the crack propagation as a function of the G value of the interlayer.
For this purpose the elastic shear modulus was adapted and the next results were obtained.

Figure 7 gives the evolution of the overlay service-life versus the thermal shear modulus of the interlayer. For this exercise, the traffic shear modulus was kept constant and equal to 2 Mpa. All the other data reported in section 3.1. remained constant.

It can be observed that the crack propagation seems to be rapid for the early age of the structure due to an important thermal contribution and then, depending on the $G_{thermal}$ value the crack propagation is more or less slowed down.

The overlay service-life is dramatically decreased when the $G_{thermal}$ value goes up to 1 Mpa.

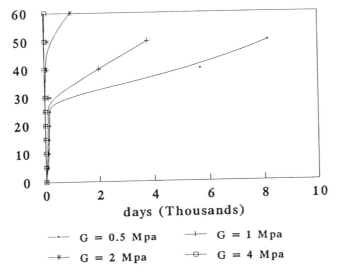

Figure 7 : The overlay crack propagation during its service-life (traffic and thermal effects) versus the thermal shear modulus of the interlayer ($G_{traffic}$ = 2 Mpa, see the data in section 3.1.)

5 Conclusions

This paper is to be considered as a contribution to the understanding of the crack propagation in overlays due to both traffic and thermal effects.
 The presented method might be used in order to evaluate the crack propagation under the combined effects. It is based on a number of assumptions and data. The latter are missing dramatically in certain fields. Emphasis is to be placed on correct data collection in order to decrease the gap between calculation tools and real materials and structures behavior.

6 References

Rashid (1982). ASCE state-of-the-art report on finite element analysis of reinforced concrete. Prepared by a Task Committee chaired by A. NILSON. American Society of Civil Engineers, New York.

Haas and Ponniah (1989). Design oriented evaluation of alternatives for reflection cracking through pavement overlays. Reflective Cracking in Pavements. Assessment and Control, Liege, 1989.

Schmidt (1989). Theoretical considerations and measures for the prevention of reflection cracking in asphalt roads with cement-bound bases. Reflective Cracking in Pavements. Assessment and Control, Liege, 1989.

Clauwaert and Francken (1989). Etude et observation de la fissuration réflective au Centre de Recherches Routières Belge. Reflective Cracking in Pavements. Assessment and Control, Liege, 1989.

Rigo et al (1989). Laboratory testing and design method for reflective cracking interlayers. Reflective Cracking in Pavements. Assessment and Control, Liege, 1989.

Hendrick (1992). Etude de l'emploi des géotextiles pour la réfection des fissures des chaussées. Liege University. Faculty of Civil Engineering, Liege, 1992.

Delcuve (1990). Emploi des géotextiles dans la réparation des bétons routiers endommagés. Liege University. Faculty of Civil Engineering, 1990.

10 DESIGN OF ASHPHALT OVERLAY/FABRIC SYSTEM AGAINST REFLECTIVE CRACKING

B. GRAF
Consulting Engineer, Zurich, Switzerland
G. WERNER
Geosynthetics Consulting, Polyfelt GmbH, Linz, Austria

ABSTRACT

For an asphalt/fabric repaving system, a design procedure for prevention of reflective cracking is developed. Crack propagation due to cyclic traffic - and thermal loading is considered. The design is based on linear fracture mechanics methods and the overlay thickness is determined for some prespecified value of damage accumulation. Semi-analytical solutions are derived for required stress intensity factors.

Consistent with findings from case histories, computational results reveal remarkable reduction of required overlay thickness due to application of a paving fabric. Possible reduction of design overlay thickness is mainly caused by controlled reduction of shear coupling of overlay and old asphalt.

Keywords: Design, Reflective Cracking, Overlay, Paving, Fabric, Fracture Mechanics, Traffic, Thermal, Shear Coupling, Damage Accumulation.

1 Introduction

Asphalt overlays are applied for cost-efficient reconstruction and maintenance work of cracked pavements. One of the major goals of design of overlay thickness is prevention of reflective cracking, i.e. propagation of pavement cracks through the overlay before design life has been reached. Paving fabrics have been applied successfully in this regard besides reducing the required overlay thickness. The main objective of this paper is the proposal of a procedure for design of overlay thickness for a fabric/overlay-reconstruction system against reflective cracking. Traffic - and thermal loading is considered. Modeling of the crack propagation problem is based on linear fracture mechanics methods, whereas semi-analytical solutions will be derived for quantities governing crack propagation. Computational results are demonstrated for a case study.

Reflective Cracking in Pavements. Edited by J.M. Rigo, R. Degeimbre and L. Francken.
© 1993 RILEM. Published by E & FN Spon, 2–6 Boundary Row, London SE1 8HN. ISBN 0 419 18220 9.

2 Formulating the Problem

For the sake of simplicity, the layered cross section shown in Fig. 1 will be considered, whereas h_1, h_2 and h_3 denote thickness of overlay, paving fabric and old asphalt, respectively. Unbound base and subsoil are substituted by continous elastic support, **K** denoting the modulus of foundation support.

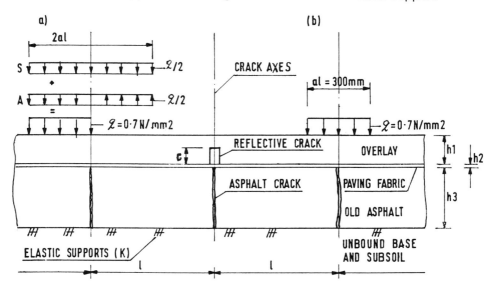

Fig. 1 Cross section and traffic loading configurations: a) before/after crack axes,
b) symmetric to crack axes

The old asphalt is assumed to be fully cracked, "l" denoting the mean crack distance. According to observations it will be assumed, that a reflective crack starts from the base of an overlay and propagates vertically to the surface, **c** denoting current crack length.

Cyclic traffic- and thermal loading will be considered. For a standard axle load configurations "symmetrical" - and "before/after" crack axes may be distinguished. According to Finite Element (FE) calculations, however, only the assymetric portion due to load configuration "before/after" is relevant for crack propagation; cf. also van Gurp and Molenaar (1989). Consequently, reflective cracking due to traffic loading will be studied for the system and loading as shown in Fig. 2.

Arand et. al. (1989) predicted magnitudes (σ_r) and corresponding probabilities (r_σ) of thermal tensile stresses to be expected for the surface; cf. Fig. 2. Prediciton of reflective cracking will be based on these data, available for various climate zones and bitumen characteristics.

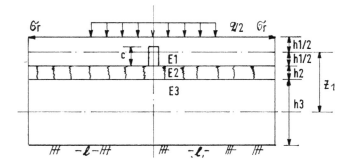

Fig. 2: Idealised system as well as considered mechanical - and thermal loading.

Elastic material behaviour will be assumed for overlay and old asphalt. For the paving fabric PGM, material response due to compression is assumed to be described by modulus E_2, and parameter γ_1 is introduced to describe shear coupling properties; cf. Fig. 2:

$$I_1 = \frac{b\,h_1^3}{12} + \gamma_1(bh_1)\,z_1^2 \qquad (1)$$

I_1: moment of inertia of overlay
b: width of cross section
z_1: distance of center line of overlay from center of gravity of cross section.

According to equ. (1), $\gamma_1 = 0$ and $\gamma_1 = 1$ indicate no- and fully shear coupling of an overlay. Preliminary evaluation of three point bending tests indicate values in the order of magnitude of $\gamma_1 = 0.5$ for a bitumen/geotextile paving fabric system. Also compressible - and incompressible paving fabric systems will be distinguished according to the conditions: $E_2 < E_1$ and $E_2 = E_1$, respectively, E_1 denoting modulus of overlay.

Design of overlay thickness against reflective cracking requires prediction of the number of load cycles needed for a reflective crack to propagate to a certain lenght.

3 Design Procedure

The rate of crack propagation is assumed to be described by the following empirical formula; cf. Paris (1964):

$$\frac{dc}{dN} = A(K_v)^n \qquad (2)$$

whereas N denotes number of load cycles. The calculation of equivalent stress intensity factor K_v due to traffic and thermal loading will be outlined in the next section, and material parameters A, n may be choosen according to Molenaar (1983). For a certain load type (..)$^{(i)}$, the number of load cycles needed for a crack to grow from length c_1 to length $c_2 > c_1$, can be derived by integration of equ. (2):

$$N^{(i)} = \frac{1}{A} \int_{c_1}^{c_2} \frac{dc}{(K_v^{(i)})^n} \qquad (3)$$

A certain value $c_1 > 0$ has to be assumed for the initial crack length. Equ. (3) is insensitive with respect to final crack lengths $c_2 > 0.5\ h_1$. Consequently, the upper integration boundary condition in equ. (3) may be set according to equ. (4):

$$c_2 = 0.5\ h_1 \qquad (4)$$

With $n^{(i)}$ denoting the number of load cycles to be expected during design life, a damage index may be defined:

$$\kappa_s^{(i)} = \frac{n^{(i)}}{N^{(i)}} \qquad (5)$$

According to a proposal stated by Miner (1945), a measure for the accumulated damage due to a load collective: $i = 1, ..., m$ can be derived by summation of damage indices:

$$\kappa_D := \sum_{i=1}^{m} \kappa_s^{(i)} \leq 1 \qquad (6)$$

whereas prevention of reflective cracking requires $\kappa_D < 1$. Consequently, the required overlay thickness h_1 can be derived iteratively for some prescribed value κ_D of damage accumulation. Latter is an easy task, provided an explicit solution for k_v, appearing in equ. (3), is available.

4 Calculation of Stress Intensity Factors

Within the frame of linear fracture mechanics, so called Mode I - and Mode II

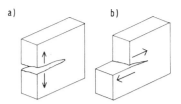

Fig. 3: Crack opening modes. a) Mode I, due to thermal loading, b) Mode II due to assymetric mechanical loading.

crack opening is distinguished; cf. Fig. 3. Crack propagation can be assumed to be governed by corresponding stress intensity factors K_I and K_{II}, respectively.

Due to considered traffic- and thermal loading, Mode I - and Mode II crack opening can be expected; cf. Fig. 2. Assuming simultaneous occurence of both load types, mixed mode crack opening can be assumed to be governed by an equivalent stress intensity factor , cf. Richard (1983):

$$K_v = 0.5 \left(K_I + \sqrt{K_I^2 + (2\alpha_1 K_{II})^2} \right) \tag{7}$$

$$\alpha_1 = \left(\frac{2(1-v)-v^2}{3(1-2v)}\right)^{3/2} \tag{8}$$

v: Poisson's Ratio of overlay material.

Subsequently, calculation of stress intensity factors K_I and K_{II} will be further outlined, whereas calculation of K_I is considered first. Assuming an uncoupled overlay, corresponding to $\gamma_1 = 0$ in equ. (1), crack propagation can be studied for a plate with cracked boundary (cf. Fig. 4a1), and for K_I the following approximation holds; cf. Hahn (1976):

$$K_I\big|_{\gamma_1=0} \cong \sigma \sqrt{\pi c} \; f\left(\frac{c}{h_1}\right) \; ; \; 0 \leq \frac{c}{h_1} \leq 0.5 \tag{9}$$

$$f\left(\frac{c}{h_1}\right) = 0.265 \left(1-\frac{c}{h_1}\right)^4 + \frac{0.857 + 0.265\frac{c}{h_1}}{\left(1-\frac{c}{h_1}\right)^{3/2}} \tag{10}$$

The boundary stress σ, required in equ (9), may be assumed to be related to the known thermal surface stress σ_r by the following relation:

$$\frac{\sigma}{\sigma_r} = F_R\left(\frac{c}{h_1}\right) \leq 1 \tag{11}$$

whereas function F_R has been determined with the help of FE calculations. Consequently, for a given value of σ_r, K_I is fully determined by equations (9) to (11).

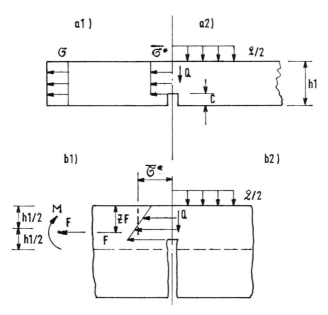

Fig. 4: Modelling of crack propagation: a) uncoupled ovelay, b) coupled ovelay, a1, b1) thermal loading, a2), b2) assymetric portion of standard axle load for configuration "before/after" crack axes

Assuming full shear coupling of an overlay, corresponding to $\gamma_1 = 1$ in equ (1), crack propagation can be studied for a crack near to the boundary of a half space, cf. Fig.4b1. Under this condition the following approximation holds; cf. Hahn (1976):

$$K_I\big|_{\gamma_1=1} \cong 0.952 \frac{F}{\sqrt{\pi(h_1-c)}} + 7.637 \frac{M}{(h_1-c)\sqrt{\pi(h_1-c)}} ; \; c<h_1 \qquad (12)$$

For a known thermal stress σ_r force F and bending moment M, appearing in equ. (12), are determined by equ. (11) and equations (13) to (15); cf. Fig. 4b1:

$$\overline{\sigma}^*/\overline{\sigma} = 1.78 \qquad (13)$$

$$\overline{\sigma} = \sigma(1-\frac{c}{h_1}) \qquad (14)$$

$$z_F/(h_1-c) = 0.68 \; ; \; c < h_1 \qquad (15)$$

Equations (13) and (15) have been derived from the results of FE calcualtions. Equ. (13) is also verified by experiments; cf. Saler (1992). According to equ. (13), the mean thermal stress for the crack axes is remarkably higher for a fully coupled overlay, if compared to the uncoupled case. Full shear coupling of an overlay also causes additional bending; cf. Fig. 4b1 and equ. (15).

Qualitative plots of stress intensity factors, according to equ. (9) and (10), also clearly reveal the difference of system behaviour due to coupling - or uncoupling of an overlay (cf. Fig. 5); i.e. according to equ. (3) one could expect significantly higher values for the number of load cycles for the uncoupled system.

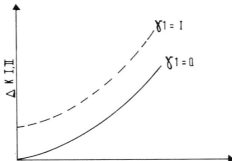

Fig. 5: Qualitative plot of stress intensity factors vs. crack length for coupled ($\gamma_1 = 1$) and uncoupled ($\gamma_1 = 0$) overlay.

Crack propagation due to traffic loading is governed by shear force Q, caused by assymetric loading according to Fig. 2; cf. also Fig. 4a1, 4b1. Based on a modified analytical solution for a beam on elastic foundation, an approximate solution for shear force Q has been derived; modifications are for the influence of cracked hight of beam and compressibility of an inlay. It should be noted, that shear force Q in particular is sensitive towards compressibility of an inlay, whereas increase of compressibility causes decrease of Q.

Considering again uncoupling - and coupling of an overlay, for known shear force Q the following approximations for the stress intensity factors can be derived:

$$K_{II}\Big|_{\gamma_1=0} \cong 0.65 \frac{Q}{h_1-c} \sqrt{\pi c} \; ; \; c < h_1 \tag{16}$$

$$K_{II}\Big|_{\gamma_1=1} \cong 2.31 \frac{Q}{\sqrt{\pi(h_1-c)}} ; \; c < h_1 \tag{17}$$

The patterns of stress intensity factors vs. crack length are qualitatively the same as already discussed for thermal loading; cf. Fig. 5.

For the considered Mode II crack opening due to traffic loading, aggregate interlock for the crack surfaces has to be expected. Consequently, according to a proposal by van Gurp and Molenaar (1989), calculated K_{II} values will be reduced by a factor of 0.5.

So far, full coupling - or uncoupling of an overlay has been assumed. For partially shear coupling of an overlay, i.e $0 < \gamma_1 < 1$, the following interpolation will be assumed for the equivalent stress intensity factor:

$$K_v = K_v\big|_{\gamma_1=0} + \gamma_1 \left(K_v\big|_{\gamma_1=1} - K_v\big|_{\gamma_1=0} \right) \tag{18}$$

5 Case Study

Design of required overlay thickness h_1 of overlay will be demonstrated for a layered old asphalt of total thickness h_3 = 230 mm and mean crack distance of "l" = 1000 mm; cf. Fig. 1. Thickness h_2 = 1 mm is assumed for the paving fabric, and c_1 = 1 mm for the initial crack length.

Temperature dependent material properties of asphalt are calculated according to Francken and Verstraeten (1974), leading to variation of moduli in the range of 1000 N/mm^2 to 30000 N/mm^2, if the surface - and subsoil temperatures are considered according to Mais (1968) and Arand et. al. (1989). Depending on season, foundation moduli in the range of 0.6 N/mm^3 to 2.2 N/mm^3 are assumed. Parameters appearing in equ. (3) are choosen as log A = -9.18 and n = 4.762, respectively. Modulus E_2 for a compressible bitumen/ paving fabric system is set equal to modulus of bitumen. Shear coupling properties of the considered inlay are varied in the range: $0 < \gamma_1 < 1$, whereas γ_1 = 0 is included for the sake of completeness.

Traffic loading due to a standard axle load with frequency f = 10 Hz is considered. Temperature stresses (σ_r) and corresponding probabilities ($r\sigma$) are assumed for climate zone II and softening point T_{AB} = 50 °C of overlay bitumen: 0.005 N/mm^2 ($r_\sigma \sim$ 85 %) $< \sigma_r <$ 0.8N/mm^2 (r_σ = 0.0001 %) ; cf. Arand et. al. (1989).

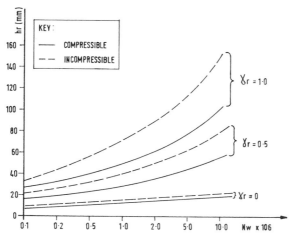

Fig. 6: Design chart for required overlay thickness vs. number of wheel passes (N_w), κ_D = 0.3

Assuming $\kappa_D = 0.3$ for the value of damage accumulation, computational results for the required overlay thickness are plotted in Fig. 6 vs. number of wheel passes. These data clearly reveal the significance of shear coupling properties of a paving fabric on the required overlay thickness, whereas the influence of compressibility for increase of shear coupling parameter γ_1 also becomes more pronounced.

6 Conclusions

According to the predicitons of Fig. 6, application of a compressible paving fabric with controlled reduction of shear coupling of an overlay also results in significant reduction of required overlay thickness, if compared to reconstruction without paving fabric (cf. $\gamma_1 = 1$, "incompressible" in Fig. 6). This prediction is consistent with findings from various case histories; cf. van Wijk and Vicelja (1989), Potschka (1992).

Further development of the considered bitumen/paving fabric system will concentrate on optimization of shear coupling properties, whereas a lower bound for parameter γ_1 is due to required material properties for action of in plane shear forces in the pavement system.

References

Arand, W. Dörschlag, S. und Pohlman, P. (1989): Einfluß der Bitumenhärte auf das Ermüdungsverhalten von Asphaltbefestigungen unterschiedlicher Dicke in Abhängigkeit von der Tragfähigkeit der Unterlage, der Verkehrsbelastung und der Temperatur. Schriftenreihe "Forschung Straßenbau und Straßenverkehrstechnik", Heft 558. Hrsg. vom Bundesminister für Verkehr, Abt. Straßenbau, Bonn - Bad Godesberg.

Francken, L. and Verstraeten, J. (1974): Methods for Predicting Moduli and Fatigue laws of Bituminous Road Mixes Under Repeated Bending. Transp. Res. Record 515, Washington DC, 114 - 123.

Van Gurp, C.A. and Molenaar, A.A.A. (1989): Simplified Method to Predict Reflective Cracking in Asphalt Overlays. Conf. on Reflective Cracking in Pavements, Liege, Belgium, 190 - 198.

Hahn, H.G. (1976): Bruchmechanik. Teubner Studienbücher.

Mais, R. (1968): Ein Beitrag zur Ermittlung der Beanspruchung standardisierter Fahrbahnbefestigungen mit Hilfe der Mehrschichtentheorie. Diss. TH München, Heft 13.

Miner, M.A. (1945): Cummulative Damage in Fatigue. J. of Applied Mechanics, Vol. 12,3.

Molenaar, A.A.A. (1983): Structural performance and design of flexible road constructions and asphalt concrete overlays. Thesis, Delft University of Technology.

Paris, P.C. (1964): The fracture mechanics approach to fatigue. In: Fatigue. An interdisciplinary approach. Proc. 10th Sagamore Army Materials. Research Conference (Ed.: Burke, J.J., Reed, N.L., Weiss, V.) Syracuse, N.Y., 107 - 132.

Potschka, V. (1992): Erfahrungen bei Fahrbahndeckensanierungen mit Asphaltvlies. 2. Kongr. Kunststoffe in der Geotechnik. K-GEO 92, Luzern, Schweiz. 59-62.

Richard, H.A. (1983): Examination of Brittle Fracture Criteria for Overlapping Mode I and Mode II Loading Applied to Cracks. Proc. Int. Conf. on Application of Fracture Mechanics to Materials and Structures, Freiburg i. Brsg, BRD, 309 - 316.

Saler, P. (1992): Tieftemperaturverhalten von mechanisch verfestigten Asphaltvliesen im bituminösen Deckenbau. 2. Kongreß Kunststoffe in der Geotechnik. K-GEO 92, Luzern, Schweiz 85 - 90.

van Wijk, W. and Vicelja, J.L. (1989): Asphalt Overlay Fabrics, a Life Time Extension of New Asphalt Overlays. Conf. on Reflective Cracking in Pavements, Liege, Belgium. 312-319.

11 DESIGN LIMITS OF COMMON REFLECTIVE CRACK REPAIR TECHNIQUES

T.R. JACOB
Owens Corning Fiberglas, Granville, Ohio, USA

Abstract
In this paper, finite element modeling and/or laboratory testing are used to determine the limits of performance of commonly used repair techniques. Results are presented as absolute limits or as stress curves as a function of material modulus and thickness. The results also highlight critical material properties that should be tested for as part of a standard laboratory testing program.
<u>Keywords:</u> Asphalt, Reflection, Cracking, SAMI, Thickness, Stress, Modulus, Reinforcements, Grids, Fiber, Traffic

1 Introduction

Laboratory testing was done on 7.5cm x 7.5cm x 60cm asphaltic concrete beams using the Owens Corning Fiberglas combined traffic and thermal load tester. The finite element modeling was done using the MARC FEM program. Reflective cracking in asphaltic concrete is caused by thermal loading, traffic loading or a combination of the two. The testing of 7.5cm x 7.5cm x 60cm asphaltic concrete test beams in direct tensile tests and simulated traffic loading as a function of temperature have been previously reported.[1] Typical results of loads as a function of temperature and tensile strength as a function of temperature are shown in figure 1. Catastrophic failure occurs when the overlay stress exceeds the overlay tensile strength. The mechanism for traffic load failure is shown in figure 2. Deflection causes part of the asphalt to be in compression and part to be in tension. The neutral point is a function of construction method. The traffic load testing referenced above [1] has shown that asphalt concrete designs and construction types have a critical deflection level. If the amount of deflection from traffic loading exceeds this level, rapid fatigue crack growth occurs. If the deflection levels are kept below the critical level, then no damage occurs to the overlay during the deflection.
   Knowing these characteristics of the overlay (thermal load and tensile strength as a function of temperature and

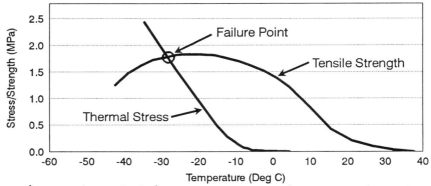

Fig.1. Thermal failure mode for Ohio 404 design mix.

critical deflection level), along with material physical properties such as modulus and thermal conductivity, it is possible to evaluate how different overlay repair techniques will perform or at least to identify the limits of potential performance.

Functionally all repair techniques are intended to affect the overlay point of failure by changing one or more of the following basic characteristic of the overlay:

1. Increasing the overlay tensile strength.
2. Decreasing the overlay stress.
3. Increasing the critical deflection level.
4. Decreasing the amount of deflection.

The intent of this paper is to examine the potential of typical repair techniques to affect these four characteristics. A combination of laboratory test data and finite element modeling (MARC FEM program) is used to conduct the analysis.

2 Overview of thermal stress transfer

An important factor in thermally caused reflective cracking is the transfer of stress from the old pavement. By itself, this layer of payment would expand and contract with temperature changes. At the midpoint between cracks, there is essentially no movement (assuming the resistance to movement at the base is essentially uniform). The maximum movement would occur at the crack edges. However when a new overlay has been bonded to the old pavement, this bond restricts the movement of the old pavement. The constrained movement results in a stress that is transferred to the new overlay. The amount of stress transferred is a function of the material modulus, temperature, and the distance from the midpoint between

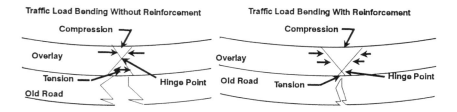

Fig.2. Overlay bending with traffic load

the two old cracks. Since the maximum movement constraint is at the crack edge, this is where the maximum stress is transferred. The stress transfer decreases exponentially to zero at the midpoint between cracks where there is no potential movement to constrain. Figure 3 shows the stress transfer profile at the old road/overlay interface as projected by a finite element model.

3 Specific Evaluations

3.1 Thicker Overlays

The primary affect of a thicker overlay is to decrease the stress per unit area directly above the old crack. A thicker overlay has no effect on the overlay tensile strength per unit area. It has minimal effect on the amount of deflection for a given load and on the critical deflection level. The amount of stress in the new overlay is made up of the uniform stress from its own contraction and the transferred stress from the old pavement. In the extremes, a very thick overlay will experience very little stress transfer because the insulating effects of the thick overlay result in very small short term thermal cycles. At the other extreme, as the thickness of the overlay approaches zero, the stress per unit area above the old crack approaches infinity. The reference stress (without SAMI) in figure 4 gives an indication of the effect of overlay thickness.

3.2 Stress Absorbing Membrane Interlayer (SAMI)

A SAMI is a layer of material normally placed between the old pavement and the new overlay. The purpose of the SAMI is to isolate the thermal effects of the old payment from the new overlay. The SAMI does this by providing a slip plane between the old pavement and new overlay that allows movement of the old pavement. The portion of potential stress from the old pavement that is transferred by the old pavement to the new overlay is determined by the thickness of the interlayer, the modulus of the interlayer, and the temperature of the interlayer. The

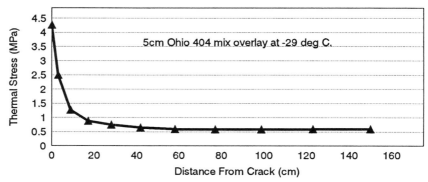

Fig.3. Thermal stress transfer profile at old road/overlay interface.

potential stress that could be transferred by the old pavement is a function of the overlay thickness and temperature as indicated in the section on overlay thickness above.

A SAMI with a very high modulus or a very low thickness will essentially transfer 100% of the potential stress from the old payment to the new overlay and act as if there were no SAMI. The other extreme, 0% transfer of stress, results if there is no
bond between the old payment and the new overlay. This is of course not practical put it does represent the limit of SAMI reduction in transferred stress. In practice, the SAMI has to have a high enough modulus to withstand paver and asphalt truck traffic and to hold the new overlay in place. Curves showing the overlay stress for various overlay thicknesses as a function of SAMI modulus are shown in figure 4. Note that the limits of stress for the various thicknesses of overlay at complete bond (no SAMI, 100% stress transfer) and no bond (0% stress transfer) are given for reference.

A SAMI is not intended to increase the tensile strength of the overlay although it may do this by a small amount. It also does not significantly change the critical deflection level of an overlay or reduce the amount of deflection of the overlay for a given load.

3.3 Underseal
An underseal operation is intended to fill the voids under a pavement to bring the pavement back into contact with the road base. This will reduce the amount of deflection of the pavement for a given load up to the load carrying capability of the base. An underseal operation does nothing to increase the critical deflection level of the payment, the tensile strength of the overlay, or the amount of thermal stress transferred to the overlay by the old pavement.

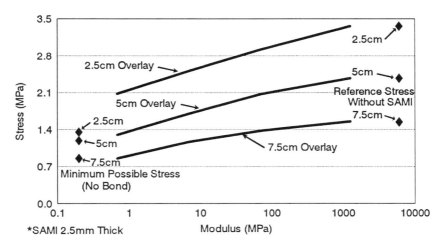

Fig.4. Average overlay stress above crack for various SAMI configurations.

3.4 Reinforcing grid

Reinforcing grids can be placed either at the interface between the old pavement and the new overlay or within the new overlay. The effectiveness of both design strategies are a function of the materials and application.

A reinforcing grid placed within the overlay has two primary effects on the overlay. It reinforces the overlay to increase the tensile strength and it changes the neutral point within the overlay during deflection (see figure 2). Approximately 2 Kn/m of reinforcement tensile strength is needed for each °C of increased thermal performance required. Figure 5 shows how high tensile

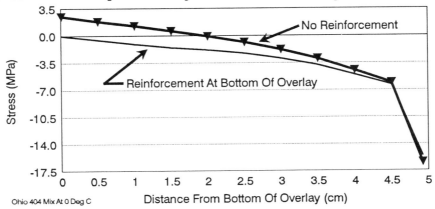

Fig.5. Overlay cross section traffic load stress profile over pre-existing crack.

strength/high modulus reinforcements affect deflection. Because high strength reinforcements (glass, steel, or similar material) have moduli orders of magnitude higher than the asphaltic concrete (at moderate temperatures), the reinforcement becomes the hinge point or neutral point during deflection. The asphaltic concrete above the reinforcement is all in compression. Only the asphaltic concrete below the reinforcement is in tension. Because the asphaltic concrete has excellent compressive properties but very limited tensile properties, the lower the reinforcement is in the overlay the more it improves the critical deflection level and the ability to withstand greater amounts of deflection. It should be noted that although placement of the reinforcement at the bottom is the best location from a deflection standpoint, it may have detrimental effects on the thermal load performance because of its limited ability to increase the tensile strength of the overlay in that location. When placed at the very bottom of the overlay, the reinforcement is dependant on the adhesive strength of bond between the reinforcement and the overlay to transfer all stress. When placed in the overlay, even a few millimeters, then the aggregate can lock into the grid matrix and create a physical as well as a chemical bond.

3.5 Chopped fiber reinforcement
Chopped Fiber asphaltic concrete reinforcement can be mixed into the asphaltic concrete at the asphaltic concrete plant or it can be sight applied. When the chopped fiber is added into the asphaltic concrete at the plant, it is intended to uniformly increase the tensile strength of the entire overlay. The benefit from a thermal load standpoint would be the same as a reinforcing grid. Because of the higher tensile strength, the chopped fiber reinforced asphaltic concrete would also have higher critical deflection level. However this would not change the hinge point as shown in figure 2.

Fiber that is site applied usually is intended to form a reinforcement interlayer between the old road and the new overlay. Therefore the effect would be the same as that described for placing a grid reinforcement at the bottom an overlay.

3.6 Reinforcement at the old crack
A reinforcement at the old crack is intended to hold the cracked pavements together and by preventing movement of the overlays, stopping stress transfer from the old pavement into the new overlay. The degree of success of this strategy is dependant on the properties of the reinforcement and the binder used to adhere the reinforcement to the old pavement and the new overlay.

Obviously the maximum amount of stress that the reinforcement prevents from being transferred cannot

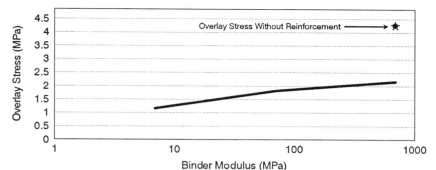

*5cm Overlay At -29 deg C; 1.7mm Binder Thickness Top and Bottom; Ohio 404 Mix; Glass Reinforcement

Fig.6. Peak overlay stress with crack reinforcement as a function of reinforcement binder modulus.

exceed its own tensile strength, otherwise the reinforcement will rupture. The amount of stress in the reinforcement is ultimately transferred to whatever material it is adhered to. Usually it is adhered to both the old pavement and the new overlay. If the same binder is used on the top and bottom of the reinforcement and the thicknesses of the binders are the same, then equal amounts of the stress ultimately get transferred to both the old pavement and the new overlay. Since the purpose of the reinforcement is to prevent stress transfer to the new overlay, special attention needs to be paid to the binder adhering the reinforcement. If the same binder is used on the top and bottom of the reinforcement, then binder thickness can control the ratio of the transfer of stress to the old pavement and the new overlay. If the top binder is twice as thick as the bottom binder, then only about ⅓ of the stress is transferred to the new overlay, the rest is transferred to the old pavement. The same result can be obtained by using the same thickness of binder on the top and bottom but using a binder on the bottom with a modulus twice as high as that on the top. Obviously combinations of moduli and binder thickness can be used to design the ratio of stress that is transferred to the new overlay as compared to the old pavement. Figure 6 shows the thermal stress in the overlay as a function of the binder modulus.

It should be noted that for spot repair systems, the moduli of the top and bottom binders, their thicknesses, and the width of the reinforcement can determine the amount of stress in the reinforcement. If the binders do not transfer sufficient stress across the old crack, the maximum tensile strength of the reinforcement is not being utilized. Again however, if to much stress is transferred, the reinforcement will rupture. Because the modulus of the binders are changing with temperature, this can offer some interesting design challenges.

3.7 Full depth excavation

The purpose of a full depth excavation is to repair damage that has been done to the road base and return it to its design load carrying capability. This repair strategy is intended to reduce the amount of deflection the road endures for a given load to keep it below the critical deflection level. A full depth excavation does nothing to increase the critical deflection level and it does nothing to increase the tensile strength of the new overlay. It also does nothing to reduce the transfer of thermal stress and in fact introduces two new cracks that can lead to thermal load reflective cracking.

4 Summary

Each repair technique will now be assessed on its ability to affect the four basic characteristics of the overlay:

Thicker Overlays - Can reduce the stress per unit area that comes from thermal stress transfer. Does not increase tensile strength or critical deflection level, minimally decreases deflection.
SAMI - Reduces thermal stress transfer. Maximum possible is about 40%. More typical is about 10-15%. Does not increase tensile strength or critical deflection level. Does not decrease amount of deflection.
Underseal and Full Depth Excavation - Both are only intended to reduce the amount of deflection.
Reinforcing Grid - Increases tensile strength of overlay. There is no theoretical limit. Also increases the critical deflection level. Decreases thermal stress only when placed at the road/overlay interface but this negatively affects tensile strength. Minimally reduces deflection.
Chopped Fiber Reinforcement - Increases tensile strength and critical deflection. Does not reduce stress transfer and minimally reduces deflection.
Crack Reinforcement - Reduces thermal stress transfer and increases critical deflection level. Minimally increases tensile strength and minimally reduces deflection.

5 Reference

Jacob, T.R. 1990 **Basic Limits Of Overlay Performance During Simultaneous Thermal And Traffic Loading** Association of Asphalt and Paving Technicians

PART THREE

ASSESSMENT METHODS FOR REFLECTIVE CRACKING IN PAVEMENTS

12 APPARATUS FOR LABORATORY STUDY OF CRACKING RESISTANCE

H. DI BENEDETTO and J. NEJI
Laboratoire Géomatériaux, ENTPE, France
J.P. ANTOINE and M. PASQUIER
Gerland Routes, Corbas, France

Premature cracking in hydraulic binder treated bases and reflective cracking in wearing courses hinder the extension of semi-rigid pavement structures.
 A prototype apparatus allowing laboratory simulation of the reflective cracking phenomenon is proposed.
 Recent modifications have been made to this apparatus through the joint efforts of the company GERLAND and the ENTPE. It now includes a set of force and displacement transducers and an ultrasonic transmission system for monitoring the cracking process on the wearing course over the pre-cracked area of the base course.
 A parametric test series on a conventional wearing course bituminous concrete is then described.
 Two criteria allowing the characterization of damage and/or cracking in the wearing course are proposed. These criteria give consistent results and will be used to define a practical pavement structure sensitivity criterion with regard to reflective cracking.
 A comparison is drawn with test sections set up on French highway RN 20
Keywords : Reflective Cracking, Fissurometer, Apparatus, Damage Characterisation, Ultrasonic transmission, Geotextiles, Anticrack Method.

1 Introduction

They are six different base courses : cement bound granular material, slag bound granular material, slag sand, lime and fly ash bound granular material, pozzolana and lime bound granular material and sand cementitious.
 Because of their technical and economical interests, semi-rigid pavements are numerous in France. According to statistics, in 1987, 44 % of new pavements and reinforced pavements have this kind of structure.
 The main problem of these structures is the premature cracking due to shrinkage. This phenomenon appears systematically in the subgrade.

Reflective Cracking in Pavements. Edited by J.M. Rigo, R. Degeimbre and L. Francken.
© 1993 RILEM. Published by E & FN Spon, 2–6 Boundary Row, London SE1 8HN. ISBN 0 419 18220 9.

The French climate is characterized by daily and/or rather high annual thermal gradients . That favours the cracking in the base course and then the reflection of this latter through the wearing course. This fact is accentuated by the traffic.

In reflective cracking, two elements must be considered : the traffic in mode 1 or 2, depending on the position of the axle and the temperature variations which favours the mode 1 (fig.1).

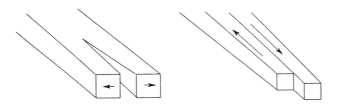

Mode 1 : Bending mode   Mode 2 : Shear mode

Fig.1.

The aim of the research is the study of the reflective cracking from the base course in the wearing course of the semi-rigid pavement.

The fissurometer allows laboratory simulation of the reflective cracking phenomenon due to thermal shrinkages. At present, its elaboration does not allow the introduction mechanical stresses, which cause of cracking in mode 2.

A parametric study aims to explain the reflective cracking properties according to the thickness of the wearing course, to the stress rate, to the cycle amplitude and to the temperature.

The practical objective is to classify the different anti-reflective cracking methods which are proposed.

2 The fissurometer principle

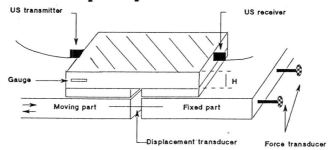

Fig.2. Apparatus description

The fissurometer (fig. 2) put a upper course sample through shear cycles. It includes a moving part and a fixed part on which is stuck the underside of the sample.

The shear is due to the horizontal displacement of the moving part.

The measuring system includes : an ultrasonic transmission system (US), force transducers, strain gauges and temperature probes.

The US system can measure the transmission energy of an ultrasound wave train through the test sample. This wave train goes through the area where the cracking should appear : near a vertical section crossing the pre-cracking.

These two US force and energy measures give a damage characterization (then micro-cracking and cracking) of the material(s) constituting the upper part of the pavement : for the sample, it is the non-initially pre-cracked courses.

One of the objectives is to qualify and to quantify this damage according to the evolution of these measures.

3 Presentation of the parametric test series

A parametric study aims to specify the evolution of the road asphalt degradation according to stress parameters and to a geometric parameter.

The stress parameters are :

the temperature
the opening and closing speed of the moving part : v
the opening amplitude of the moving part : e
the geometric parameter is the wearing course thickness : H

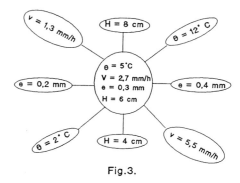

Fig.3.

Considering the number of experiences, the influence of each parameter is studied independently, around an average reference position which has the following characteristics : θ = 5°C, v = 2,7 mm/h, e = 0,3 mm, H = 6 cm.

Three values are considered for each parameter. The figure 3 presents the experimental plan.

4 Results

4.1 Monotonous test

Figure 5 shows the efforts given by the force transducers according to the displacement amplitude relative to the two parts "e".

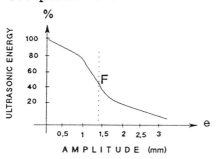

Fig.4. Monotonous breaking　　　Fig.5. Monotonous breaking

The curve has a peak typical of curves observed for weakening materials. In this case, the phenomenon observed does not correspond to rheologic weakening but to structure weakening (the damage moves gradually from the bottom to the height of the stressed section).

Figure 4 presents the percentage of ultrasonic energy transmitted through the sample according to the amplitude "e". The reference value is the one obtained for the material before any stress, but is applied at the test temperature.

At the end of the test, for a value of e = 3,5 mm, the ultrasonic energy transmitted is almost null (under 3 %). So the sample is completely cracked.

This residual value corresponds to transmission through the apparatus. One can notice that it is very low.

The point F (figure 4) corresponds to the point E (figure 5) (same amplitude as the amplitude "e"). At this amplitude, the material changes its behaviour.

As the matter of fact, the "force amplitude" and "US %
amplitude" curve slopes show a radical change at this
moment.

Physically speaking, we can explain this phenomenon by
the beginning of total decohesion of the material. Only a
few aggregates make punctually bridges binding the two
semi-parts of the sample. Then the cracking has totally
gone through the sample section.

One can not detect on the "US % amplitude" curve, the
maximum of the "effort amplitude" curve. So the damage
seems to be characterized more directly by the ultrasonic
process because the decay curve corresponds to the
forseeable damage curve.

4.2 Cyclic test

The sample goes through tension-compression cycles. We
follow the ultrasonic transmission evolution and the
stress evolution.

Figure 6 presents the effort given by the force
transducer according to the number of cycles.

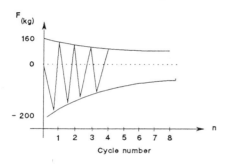

Fig.6. Cyclic breaking

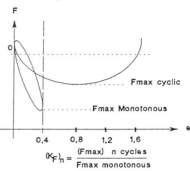

Fig.7. Cyclic breaking

Figure 7 shows the effort supported by the middle
section of the sample according to the amplitude "e"
between the two parts. This figure presents clearly the
"opening-closing" cycles of the moving part. The sample
stressed until the breaking after 80 cycles also appears.

One can observe the cyclic surface deterioration by
the effort diminution during the stress inversion.

Figure 8 presents the ultrasonic transmission rate
according to the number of cycles. For each cycle, the
transmission percentage decreases during the opening
stage and increases during the closing stage. The
inferior and superior envelopes of this curve are
decreasing with the number of cycles. After a total
cycle, the ultrasonic energy transmitted decreases. We
can compare this evolution to the one existing for the
damage.

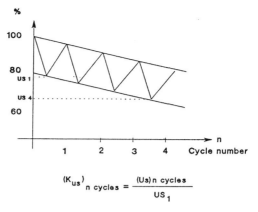

Fig.8.

Two damage criteria have been introduced :

$$(K_F)_n = \frac{(Fmax)n\ cycle}{Fmax\ monotonous} \qquad (1)$$

That gives an idea of the tension modulus weakening after a fatigue of n cycles

$$(K_{us})_n = \frac{(\%\ US)n\ cycle}{(\%\ US)1st\ cycle} \qquad (2)$$

is the ratio between the ultrasonic transmission after n cycles and the ultrasonic transmission after one cycle.

The results of the parametric study are given for $(K_{us})_{80}$: Fig.9 and Fig.10.

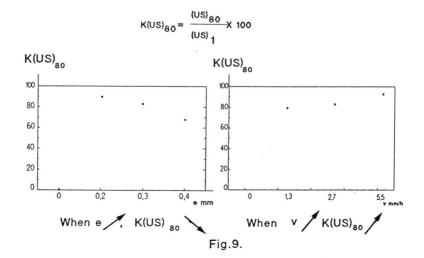

Fig.9.

184

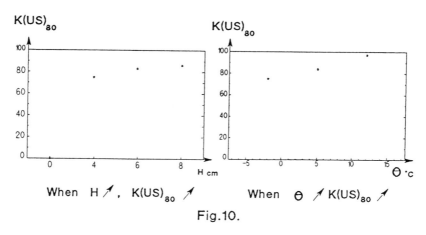

Fig.10.

5 Comparison with the French highway RN 20

5.1 Comparative study

We compared the results of the solutions, implemented on the French highway RN 20 between Angerville and Monnerville with the results given by the fissurometer.
The anti-crack methods which are compared are :

Sand asphalt - 80/100 bitumen
Sand asphalt - Cariphalte binder
Geotextile U24 calendered at 210 g/m2 and tack coat bound with Cariphalte binder
Geotextile U34 and tack coat with 65 % Cariphalte emulsion.
Thick coat bound with Styrelf and chipping 6/10.
Bridging of the cracks with Iraflex and chippings 2/4.
Each process is covered with 6 cm bituminous concrete.

5.2 Classification

1. sand asphalt
2. geotextiles
3. thick coat
4. crack bridging

This classification, made as a result of the measures and observations of the experimental test sections set up on the French highway RN 20, is in the reverse order with the test fissurometer for the two first techniques. Indeed, with this apparatus, the test is based only on thermal shrinkage without considering the effect of the traffic.

6 Conclusion

An apparatus allowing laboratory simulation of the reflective cracking phenomenon in the upper courses - the fissurometer, has been presented. For the moment, concerning the conception of this prototype apparatus, only cracking in mode 1 can be tested.

Its originality lies in the measure of the efforts and displacements in the cracking area and the attenuation of the ultrasonic waves thanks to an ultrasonic transmission system.

From this experimental test series, we can say that :

Some classical results were confirmed (influence of the thickness of the wearing course, of the amplitude of the stress and of the temperature).

The damage appears less important for higher stress speeds. We propose to take into account a "viscous" damage to understand this phenomenon.

The two criteria proposed for the qualification of the damage $(K_F)_n$ and $(K_{US})_n$ give concordant results. However, it seems that the criterion which uses the ultrasounds is the most sensitive to the phenomenon.

A practical qualification criterion of the pavement structure with regard to the sensitivity to reflective cracking can be simply expressed from one criterion or a combination of the two criteria which are proposed.

Nonetheless, the choice of this practical criterion implies a validation from data taken on pavements.

A validation using the test section of the French highway RN 20 is presented.

13 INVESTIGATIONS ON THE EFFECTIVENESS OF SYNTHETIC ASPHALT REINFORCEMENTS

P.A.J.C. KUNST
Netherlands Pavement Consultants, Hoevelaken, Netherlands
R. KIRSCHNER
Huesker Synthetic GmbH, Gescher, Germany

Abstract
Three-point-bending-tests on reinforced and unreinforced asphalt specimens were performed investigating the effectiveness of seven different geogrids and nonwovens to retard reflective cracks. Crack propagation was described by means of fracture mechanics and design curves were plotted for reinforced and unreinforced asphalt overlays.
Keywords: Fracture, Laboratory testing, Overlay, Reinforcement, Three point bending test.

1 Introduction

Cracks in asphalt pavements and their durable elimination present a technical and economic problem, in whose solution the use of polymer asphalt inlays as reinforcements has gained considerably in importance. Efforts are made to approximate the effectiveness of the nonwovens and geogrids generally used under real conditions on test roadways. An objective assessment of the reinforcing effects of the different asphalt inlays on the basis of such test roadway results, however, is very difficult if not impossible, since the individual sections of the test roadways are generally subject to different boundary conditions.

An objective comparison of the reinforcing effect of these asphalt inlays is only possible in laboratory tests. In the summer of 1991, Huesker Synthetic commissioned Netherlands Pavement Consultants (NPC) to carry out such investigations. The effectiveness not only depends on the reinforcing material itself but also on its interaction with the asphalt concrete, the type of loading etc. Although dynamic tests most properly simulate most of the conditions occurring in practice, for this investigation semi-static tests were used that give a first indication of the behaviour of the reinforcement and its interaction with the asphalt concrete.

This paper reports on the results of these investigations.

2 Materials investigated and manufacture of the test specimens

Five geogrids and two nonwovens were chosen for the comparative

Reflective Cracking in Pavements. Edited by J.M. Rigo, R. Degeimbre and L. Francken.
© 1993 RILEM. Published by E & FN Spon, 2–6 Boundary Row, London SE1 8HN. ISBN 0 419 18220 9.

laboratory investigation. The most significant characteristics of these materials are summarised in table 1.

A total of eight two-layer asphalt specimens were manufactured in the laboratory of NPC: Seven specimens with asphalt inlays and one unreinforced specimen as a reference test specimen. An asphalt concrete 0/16, type B for traffic class 3 with bitumen 80/100 and mineral aggregates corresponding to the Dutch Guidelines was used as asphalt mix.

Table 1. Characteristics of the asphalt inlays

Specimen number	Construction	Raw material	Strength long./lateral (kN/m)	Quantity of binder (kg/m^2)
91	Geogrid, woven	PET	50/50	0.4
92	Geogrid, woven	PET	50/50	0.4
93	Geogrid, woven	PET	40/40	0.4
94	Geogrid, woven	Glass	35/56	0.4
95	Geogrid, extruded	PP	14/18	1.3*
96	Nonwoven, mech.	PP filament	8.5	1.6
97	Nonwoven, mech.	PP staple	8.5	1.6
98	No inlay	--	--	0.3

* with 10 kg/m^2 of chippings 8/11.

An asphalt layer of 3 cm thickness was first built up in a special formwork measuring 60 x 60 x 8 cm. A bitumen emulsion U 70 K, type Eshalite, was then applied to this layer as adhesive together with an asphalt inlay. The quantities of binder are also shown in Table 1.

The second asphalt layer with of thickness of 5 cm was applied the following day. Each asphalt layer was compacted to the same degree of compaction using a hand compactor and a hand drawn steel roller. The unreinforced asphalt specimen was produced in the same manner.

After a curing time of one month the specimens were sawn perpendicularly to the direction of compaction into four beam-like test specimens, each measuring 60 x 14 x 8 cm.

3 Performance of the tests and results

The tests were conducted as semi-static three-point bending tests on a computer-controlled multi-purpose testing machine, type FTS Seidner 102/300 HV. The static system of the three-point bending tests is illustrated in Figure 1.

The load was applied with a hydraulic piston at a deformation rate of 0.85 mm/s. Load and deformation were recorded graphically. The test was conducted at a room temperature of 15° C.

Figure 2 shows schematically the results of the tests in the form of load/deformation curves. The reinforced test specimens numbers 91 to 93 exhibit a noticeably different behaviour than the unreinforced test specimen no. 98. The maximum load, and in particular the energy applied until failure of the test specimen which can be determined using the area under the load-deformation curve, is higher for these specimens. Obviously this is due to the "tail" at the load deformation curve. It is noted that this part of the test is performed at deformations that usually not occur in practice.

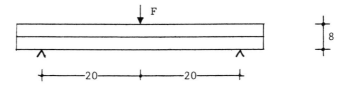

Fig.1. Static system.

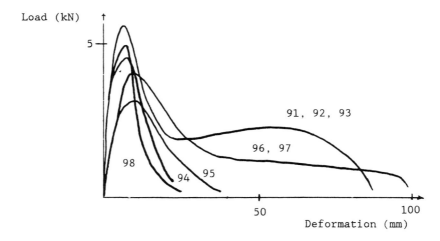

Fig.2. Load/deformation curves.

The test results for the various asphalt inlays and for the unreinforced test specimen are listed in table 2 as the mean values from four individual tests. The deformation shown is the deformation at maximum force; the energy is the area under the load-deformation curve and the maximum stress is calculated at the bottom of the slab from the maximum force and the beam geometry.

All test specimens, with the exception of no. 95, failed under the bending load as a result of fracture in the middle of the beam. For no. 95 the failure occurred in the form of a shearing of the test specimen in the "interlocking layer" of chippings. All the other grids broke under the tensile load. The nonwovens were appreciably elongated at only a low tensile force but remained intact.

Table 2. Results of the three-point bending tests

Specimen number	Max. force [kN] avg.	std.dev.	Deformation [mm] avg.	std.dev.	Energy [Nm] avg.	std.dev.	Max. stress [N/mm^2] avg.	std.dev.
91	5.82	0.44	7.15	0.49	172	10.6	3.49	0.26
92	5.40	0.24	6.19	0.61	183	18.5	3.37	0.03
93	5.77	0.06	6.82	0.46	192	25.4	3.52	0.10
94	4.58	0.23	7.52	0.34	61	5.8	2.97	0.13
95	3.15	0.31	9.95	0.36	61	9.1	1.97	0.08
96	4.34	0.36	10.62	0.88	136	41.1	2.94	0.16
97	4.04	0.10	11.39	0.60	117	25.1	2.76	0.08
98	5.13	0.31	7.35	0.53	51	2.1	3.23	0.22

4 Evaluation

In linear elastic fracture mechanics, the propagation of a crack may be described using the differential equation according to Paris (Molenaar 1983):

$$dc/dN = A * K^n \tag{1}$$

where:
dc/dN = increase in crack length per load cycle
K   = stress intensity factor
A, n = material constants

Parameter A is dependent on the stress at failure σ and the energy W required for the propagation of the crack:

$$A = f \left(\frac{1}{\sigma^2} * \frac{1}{W} \right) \tag{2}$$

Although asphalt concrete is not a linear elastic material and A is calculated from semi-static tests, for a first impression (1) and (2) are used in this evaluation. Using the results from table 2 and (2) it is possible to calculate the ratios for the stresses at failure of the reinforced and unreinforced test specimens, for the energy required untill failure and for the A values. These ratios are listed in table 3.

The ratio of the A values allows conclusions to be drawn as to the reinforcing effect of the asphalt inlay. The smaller the ratio of the A values, the more effective the reinforcement compared with the unreinforced test specimen. On this basis it is possible to make an objective assessment of the different asohalt inlays with regard to their reinforcing effect. The resulting ranking of the asphalt inlays are given in table 3.

Table 3. Ratios for σ, W and A

Specimen number	$\frac{\sigma}{\sigma_{98}}$	$\frac{W}{W_{98}}$	$\frac{A}{A_{98}}$	Ranking
91	1.08	3.41	0.25	2
92	1.01	3.62	0.27	3
93	1.09	3.80	0.22	1
94	0.86	1.21	1.12	7
95	0.61	1.25	2.15	8
96	0.91	2.69	0.45	4
97	0.85	2.32	0.60	5
98	1	1	1	6

5 Design curves for asphalt overlays

The service life of an asphalt overlay or the number of load cycles untill a crack in the existing asphalt layer breaks through the asphalt overlay can be calculated from the Paris equation if the material constants A and n and the stress intensity factor K are known. The factor K was calculated using a finite element program, whereby the state of stress in the immediate vicinity of the crack was simulated by a "crack element".

Using the assumptions listed in figure 3 it is possible to draw up design curves for the various asphalt inlays, and thus to indicate the service life of the asphalt overlay.

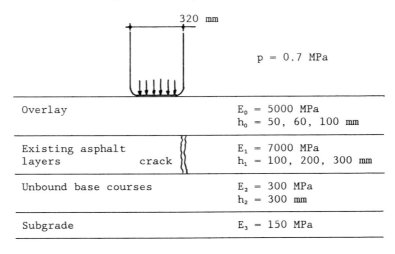

Fig.3. Calculation assumptions

Figure 4 shows an example of design curves for a geogrid (no. 93) and for a nonwoven (no. 96). With an existing asphalt layer of 200 mm thickness and an overlay of 50 mm, the following figures are obtained for the number of load cycles untill a crack breaks through the overlay:

Geogrid no. 93 $10^{7.4}$
Nonwoven no. 96 $10^{7.1}$
Without reinforcement no. 98 $10^{6.8}$

Ratios of approx. 4 : 2 : 1 can thus be calculated for the service life of the asphalt overlay for reinforcements with geogrids, nonwoven and without reinforcement.

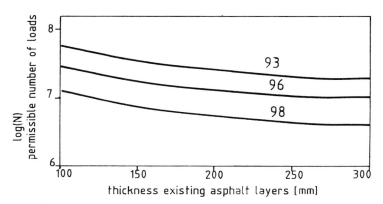

Fig.4. Design curves.

6 Conclusions and recommendations

A laboratory investigation on the reinforcing effect of asphalt inlays produced widely different results. In a semi static three point bending test differences occurred in maximum force, form of the deformation curve and energy to make the specimen fail. Compared to the reference samples polyester grids show an increase in both maximum force and energy to fail, followed by the non wovens. Based on linear elastic fracture mechanics, design curves for overlays were developed. Future investigation should be based on dynamic tests that properly model traffic loading and over a wider temperature range.

7 References

Netherlands Pavement Consultants (1990) **"Results of three-point bending tests on HaTelit-reinforced asphalt concrete beams"**, project No. 91431, Hoevelaken, the Netherlands, Unpublished.
Molenaar, A.A.A. (1983) **"Structural performance and design of flexible pavements and asphalt concrete overlays"**, Dissertation at the Delft University of Technology.

14 A NEW TEST FOR DETERMINATION OF FATIGUE CRACK CHARACTERISTICS FOR BITUMINOUS CONCRETES

S. CAPERAA, E. AHMIEDI and C. PETIT
Civil Engineering Laboratory, University of Limoges, Egletons, France
J-P. MICHAUT
Colas Society, Paris, France

Abstract
This work presents a new testing apparatus, with imposed displacement, for the determination of materials intrinsic strength characteristics in a crack process; the test takes into account rest periods between cycles and mixed-mode loadings. As an example, a first result shows the employed procedure.
Key-words: Crack, fatigue test, rupture, bituminous concrete

1 Introduction

The fatigue strength law for a cracked medium generally relates the propagation speed of the crack (the answer) to one or more mechanical parameters (the action), during a repeated, cyclic loading. For exemple, paris's law relates the mean propagation per cycle da/dN to the variation of the stress intensity factor ($K_{max} - K_{min}$) during the cycle

$$da/dN = A (K_{max} - K_{min})^m \qquad (1)$$

The validity of this law for bituminous mixes has been proved theoretically, Schapery (1975) and experimentally, Molenaar(1983), Germann and Lytton(1979); it is not our purpose to discuss this point.
Surface layers' design regarding the "reflective cracking" phenomenon relies on the integration of this type of law. Computation of road structures and numeric modelisation enable us to predict K's values under climatic and traffic conditions; intrinsic parameters of the material, such as A and m in (1), remain to be determined.
Many apparatus have been designed, Vecoven (1989), Rigo et al.(1989), to test efficiency of various "anti-cracking" processes such as polymer-modified asphalts, soaked geotextiles and grids. In this case, the essentially qualitative results lead to a classification of these processes using the number of rupture cycles.

Reflective Cracking in Pavements. Edited by J.M. Rigo, R. Degeimbre and L. Francken.
© 1993 RILEM. Published by E & FN Spon, 2–6 Boundary Row, London SE1 8HN. ISBN 0 419 18220 9.

The object of this paper is to present a new fatigue test of a material and not a process, with certain characteristics:
- simulation of all the components of the traffic
- possibility of crack's loading in pure mode I or in mixed mode; the mixed mode is rarely explored today, although it is vital for thin layers
- consideration of the "self-repairing" properties of the material.

2 Test Design

2.1 Sample choice
The basic idea is to be able to relate accurately, the value of the measurable load parameter and the value of the mechanical internal parameter (non measurable), such as the stress intensity factor. This entails a mechanically "simple" test and excludes any form of simulation of a multilayered complex road system, for which the interfacial conditions are difficult to treat.

Forms and modes of sample solicitation are shown in figure 1. Initial crack propagation can then be predicted, enabling us to set up the measuring equipment. Optimisation and research of other sample forms remains possible thanks to a detachable and adjustable device.

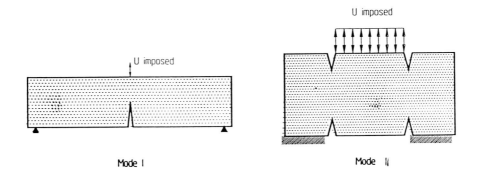

Figure 1. Sample forms and imposed displacements

2.2 Design of the apparatus
The necessity for controlling the crack's propagation has led us to adopt an imposed-displacement test; the main measured parameter is thus the sample's reaction, or its compliance.

In mode I, the stress intensity factor is established in the form:

$$K_I = U_{imposed} \cdot E \cdot f(a) \tag{2}$$

f(a) is obtained by numeric modelisation of the sample, with a modulus E = 1, submitted to an unit displacement, in plane stress conditions; its form is represented in figure 2. Values of K_I are obtained by exploitation of displacements near the crack's tip (cinematic method) and computation of compliance's variation, with differences not exceeding 3%

$$K_I = (E / 2) \cdot F^2 \cdot dC/da \tag{3}$$

C is the global compliance, F is the sample's reaction

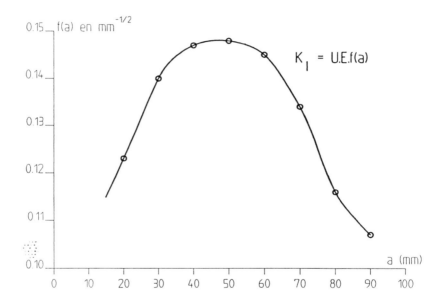

Figure 2. K_I's characteristic function.

One can notice a particular range of variation in the crack's position, between 40 and 60 mm, for which the factor K_I is approximately constant, with constant displacement; this zone will be favored for the speed da/dN measurement. A displacement increase could be necessary in order to obtain total rupture.

Figure 3 shows the mechanical part of the apparatus designed in Egletons Civil Engineering Laboratory, in collaboration with Colas Society.

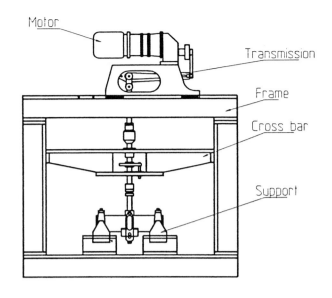

Figure 3. Mechanical apparatus.

The "traffic simulator" - made up of a motor, a clutch-brake unit, a reducer, a connecting rod and a leverage system - enables us to act independently, in a wide range of variations, upon each of the three factors composing the traffic:
- the frequency of loading, related to the vehicle's speed and type of overlay, is adjusted by an electronic modulus acting upon the motor's rotation speed with a constant coupling
- the amplitude, related to the axle load, is simulated by action upon an eccentric rod and leverage system (2/100 mm precision), while a special retaining unit guarantees the return to an identical position at the end of each cycle
- the traffic's "density", i.e the rest period between the passage of two wheels, during which the material's "self-repairing" faculties act, is achieved by programming the motor with work and rest times required by the user.

The test's environment, not represented in figure 3, consists of a controlled atmosphere enclosure, an acquisition and control unit (force, displacement and temperature captors), and a micro-computer.

2.3 Crack propagation measurements

Propagation's measurement is obtained by simultaneously-used methods: direct observation, use of crack detectors bonded on the sample and periodically checked by the control unit, and an indirect method characterising damage to the cracked zone.

This last method consists of measuring, during the test, the sample's global "compliance", defined as the slope of the displacement/force curve; by reference to a theoretical curve, one can deduce an "equivalent position" of the crack. The theoretical curve (representing the product E.C of compliance and modulus, versus the crack's position a), is obtained by numeric modelisation and it characterizes the geometry and the type of the sample (figure 4).

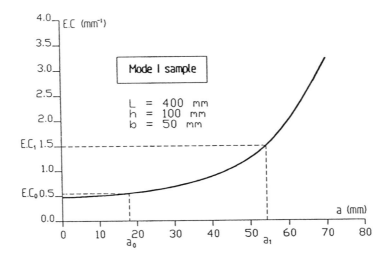

Figure 4. Compliance method.

Modulus E is obtained from the measurement of the initial compliance C_0 corresponding to an undamaged cut a_0. The procedure is only valid if compliance measurements are performed in the same conditions and if the real conditions correspond to the boundary conditions of the model; this last point is facilitated by a patented support system.

3 Results

We herein present our first results to show the procedure followed, in mode I.

The 400*100*50 mm samples are made with a plate compactor and sawed. The material is a reference 0/10 bituminous concrete composed of Meilleraie aggregates and 60/70 bitumen. Mean compaction, obtained by geometric and gamma methods, is about 92%. Presented tests are performed at 0°C and 2 Hz, with no programmed rest periods; imposed displacements are respectively 0.2 mm and 0.3 mm.

Figure 5 represents the evolution of the crack length and number of cycles required; results are obtained from crack detectors and compliance method.

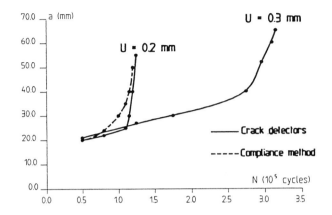

Figure 5. Crack length evolution.

One can notice a most progressive variation of a(N), in the first part of the cracking process, when the compliance method is used; the two methods lead to similar values in the range a = 40 - 60 mm.

Plotting in bi-logarithmic coordinates, of da/dN (deduced from curves in figure 5) versus $K_I' = U.f(a)$ (obtained from figure 2), exhibits a linear variation with a slope m = 6, although the few results provided today do not permit us to propose a linear regression and a significant variance for the fatigue law's form.

4 Conclusion

This apparatus is a research tool, designed to study materials' intrinsic properties regarding crack fatigue, and to analyse the influence of its components (asphalt, aggregates, compaction...). It is also possible to perform comparative studies of anti-crack processes.

A priori, a paris law has been used, but other forms of laws are available, integrating in particular the rest periods.

Finally, as we said before, determination of the initiation time, followed by the propagation of a sheared crack, must enable us to take into account the influence of mode II due to rolling loads.

6 References

Germann, F.P, Lytton R.L.(1979). Methodology for predicting reflective cracking life of asphalt concrete overlays. **Report FHWA/TX**

Molenaar,A.A.A.(1983). Structural Performance and design of flexible road constructions and asphalt concrete overlays. **Doct. Thes.**,Delft

Rigo, J-M et al.(1989). Laboratory testing and design method for reflective cracking interlayers. **Proc. Conf. Reflect.Cracking**,Liège

Schapery,R.A(1981). Non-linear Fracture Analysis of visco. composite materials. **Proc.Jap-USA Conf.on Composite Materials**, Tokyo (Japan)

Vecoven,J.H.(1989). Methodes d'etude de systemes limitant la remontée de fissures. **Proc. Conf. Rilem Reflect. Cracking**,Liège

15 DYNAMIC TESTING OF GLASS FIBRE GRID REINFORCED ASPHALT

M.H.M. COPPENS
Netherlands Pavement Consultants, Hoevelaken, Netherlands
P.A. WIERINGA
Chomarat, Mariac, France

Abstract

On behalf of the French company Les Fils d'Auguste CHOMARAT et Cie, NPC - Netherlands Pavement Consultants - tested the mechanical properties of Roadtex glassfibre grid that is used for reinforcement of asphalt layers. Tests were carried out on beams sawn from large reinforced and unreinforced asphalt slabs. Hydraulic dynamic testing equipment was used to measure stiffness modulus and fatigue behaviour under temperature controled conditions.
Significant differences were found between reinforced and unreinforced beams. The strength and stiffness values can serve to calculate the positive effects of the glassfibre grid.

<u>Keywords:</u> Glassfibre Grid, Asphalt Reinforcement, Initial Stiffness Modulus, Fatigue.

1 Introduction

The materials testing laboratory of NPC developed a dynamic strength testing method to evaluate the performance of Roadtex asphalt reinforcing grid.

This grid is an open glassfibre structure, the mechanical stability of which is secured•by a polymer coating. It has been used on a large scale in The Netherlands the last five years. It is successfully applied on asphalt pavements that show fatigue cracking, which is an important failure mechanism in asphalt roads having unbound base courses or subgrades with relatively low bearing capacity. The grids are applied over the full pavement width and overlaid
with normal asphalt mixes.

An essential feature of the system is that complete bonding exists between the old asphalt layer and the new overlay. So normal quantities of tack coat are used and the aggregate of the asphalt overlay gives good interlocking with the old pavement due to the open structure of the grid.

In the period 1987-1990 semi-static bending tests were set up to study the reinforcing effect of such grids[1]. Furthermore these tests were used to optimize adhesion to the asphalt mortar and embedding in the asphalt mix. This resulted in a specific mesh size and a deliberate choice of warp and weft strength.

Reflective Cracking in Pavements. Edited by J.M. Rigo, R. Degeimbre and L. Francken.
© 1993 RILEM. Published by E & FN Spon, 2–6 Boundary Row, London SE1 8HN. ISBN 0 419 18220 9.

A laboratory method for preparing large asphalt slabs was standardized in which the grid was applied on a compacted asphalt layer first and covered with an asphalt overlay. From these asphalt slabs beams could be sawn very accurately. In the tests always reinforced and unreinforced beams were compared, using the same asphalt mix from a laboratory batch mixer. The three-point bending test turned out to be a useful screening method by which significant effects could be measured.

However, these bending tests could only give information about ultimate strength. The deflections often reached high values, that could not be compared with those measured in real pavement structures. Dynamic testing was considered indispensable for further evaluation of the behaviour of a reinforced road structure.

For the asphalt slabs an asphaltic concrete 0/16 mm mix was used consisting of: 62% crushed stone;
32% crusher sand and natural sand;
6% filler   and 6.4% bitumen pen. 80/100.

2 Choice of dynamic testing method

Several dynamic testing methods were studied in view of their suitability for the specific samples under study. It is essential that a sufficiently large sample can be tested because of the mesh size of the grid.

Testing cylindrical or prismatic samples was rejected although these would be easier to prepare. Relatively big beams were chosen in the end.

To have a good simulation of the phenomena that can be observed in pavements, it is advantageous to test samples in the same way as the asphalt pavement is loaded in practice. Therefore it was decided to use a dynamic (four-point) bending testmode. The beam dimensions were approximately 8 x 8 x 60 cm^3 in which the grid is positioned at 2 cm height to have it well belowe the neutral line. It turned out necessary to prepare thicker slabs (6 + 6 cm) first and then remove 4 cm of the lower asphalt layer by accurate sawing.

The dynamic testing machine used is equipped with double hydraulic plungers so two samples can be tested simultaneously. The clamps to fix the beams are designed in such a way that they allow for deflections of 20 mm, which is important when testing asphalt at temperatures above 10°C. Temperatures can be controlled with an accuracy of plus minus 0.5 °C. In figure 1 the layout of the bending device is given.

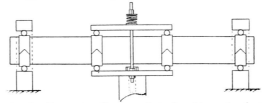

Fig.1. Dynamic four-point bending device

3 Dynamic stiffness measurements

The measurements were carried out at 15°C using a force level of 400 N. For calculation of the initial stiffness modulus some 20 values are sampled during the first 20 seconds of loading. These data are represented graphically and the average value of the stiffness modulus in this time interval is calculated.
   The values for the reinforced and unreinforced beams are given in table 1.

Table 1. Initial Stiffness Modulus

	reinforced beams	unreinforced beams
nr. 1	6700 MPa	6400 MPa
2	6200	6100
3	6700	4900
4	7000	5500
average	6700	5600

As can be observed the modulus of the reinforced beams is significantly higher.

4 Fatigue measurements

The beams were loaded in the four-point bending mode. Force was applied at only one side of the sample at a frequency of 30 Hz. Temperature was kept at 15°C. Tests were done under stress controlled conditions, so the initially chosen force level was maintained during the complete test duration. This varied from several hours to more than one day.
The beams will of course show creep under repetitive loading and the deflection will increase with time.
   In figure 2 and 3 the graphs are given that result when the deflection of the beam is plotted versus the number of load repetitions.
At the start of the test the beam was in a 20 mm off-set position. When loaded the beam shows increasing deflection and the vertical position of the centre of the beam decreased rapidly.
At the 5 mm level the unreinforced beam was loaded approximately $0.1 \cdot 10^5$ times while at that same level the reinforced beam withstood $1.8 \cdot 10^6$ load repetitions.
The first tests were carried out at a stress level of 0.40 MPa.
At 15 mm deflection following numbers of load repetitions were found:

unreinforced beams	reinforced beams	
0.12	1.8	10^6
0.11	1.3	10^6

Further tests at a stress level of 0.26 MPa gave following results at 10 mm deflections:

unreinforced beams	reinforced beams	
0.10	1.6	10^6
0.12	1.2	10^6

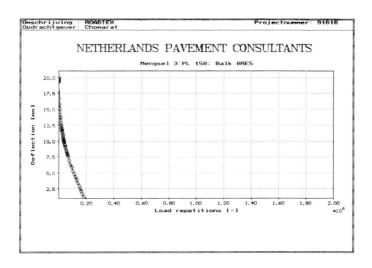

Fig.2. Dynamic bending test on unreinforced beam

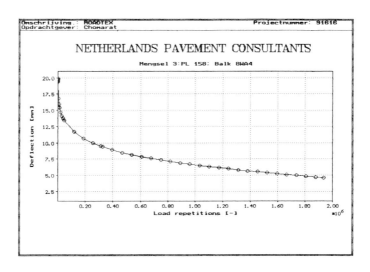

Fig.3. Dynamic bending test on reinforced beam

5 Crack propagation tests

The next step was to study the effect of the glassfibre grid in an asphalt structure in which the old asphalt layer had already failed. For that purpose beams were prepared in which the lower layer was sawn from the bottom up to the level of the grid.
To avoid too high deflections tests were carried out at 5°C.
An example of the resulting graphs is given in figure 4.

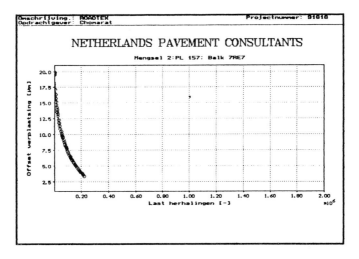

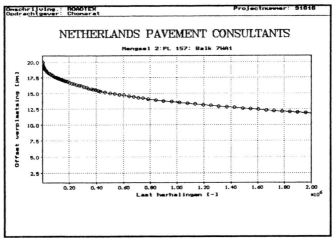

Fig.4. Unreinforced and reinforced beam in crack propagation test

In the unreinforced beam deflection increased rapidly due to crack growth. The beam failed at approximately $0.25 \cdot 10^6$ load repetitions. In the reinforced beam the grid prevented the crack to propagate vertically. After more than 2.10^6 load repetitions slight delamination of the upper and lower asphalt layers occurred but still the beam did not crack.

6 Conclusions

Reinforcement of asphalt overlays with glassfibre grids gives a significant increase in dynamic stiffness due to the high elasticity modulus of the glassfibre.
Fatigue behaviour of a 2 layer grid reinforced asphalt construction is far better than that of an unreinforced one.

References

1. M.H.M. Coppens and C. Renaud. Glassfibre reinforced asphalt. Proc. of the Conf. on Reflective Cracking in Pavements. Liège 1989.

16 ON THE THERMORHEOLOGICAL PROPERTIES OF INTERFACE SYSTEMS

L. FRANCKEN and A. VANELSTRAETE
Belgian Road Research Centre, Brussels, Belgium

Abstract
Thermal effects in overlay systems over rigid slabs containing discontinuities are considered in this paper. Criteria and evaluation models are discussed, results of laboratory simulation tests are described and interpreted. It is shown that the stress-strain relationship derived from experiments allows an accurate description of the rheological behaviour of overlay systems.
Key words : Reflective cracking, Thermal stresses, Testing, Interface systems.

Introduction

The distribution of stresses in a road structure is the result of a complex combination of environmental factors, loading conditions, geometry of the structure and properties of the component materials.
 This paper deals with the behaviour of a pavement structure containing a pre-existing discontinuity in the base layer under the influence of external temperature variations with emphasis on the case of low mean temperature conditions.
 It will be shown that the resistance to thermal effects under winter conditions is a primary requisite of a road structure and that a preliminary analysis must be made to evaluate the effective strength which remains available for the structure to resist the additional influence of traffic loads.
 Traffic loads will not be considered in detail in the present study.

Description of the structure

The road structures considered in this paper consist of four layers (see figure 1):

1. The overlay layer is a continuous layer of bituminous material.

Reflective Cracking in Pavements. Edited by J.M. Rigo, R. Degeimbre and L. Francken.
© 1993 RILEM. Published by E & FN Spon, 2–6 Boundary Row, London SE1 8HN. ISBN 0 419 18220 9.

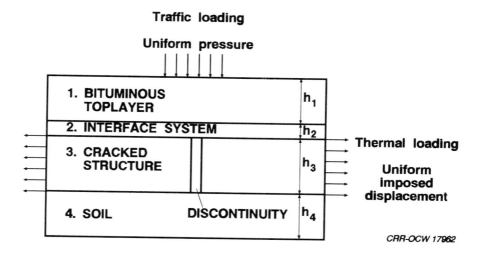

Figure 1 : Cross section of the type of road structure under study

2. A functional interlayer. Similar structures without interlayer will also be considered as references.
3. A stiff base layer containing a discontinuity (joint or crack) which may be either a cracked cement stabilized base material or an old pavement made of jointed concrete slabs.
   This layer is assumed to move frictionless over the subgrade.
4. The subgrade or soil is a semi infinite space. Its role in the present study is not very important as long as traffic loads are not considered.

Layers 1 and 2 may be considered as an overlay system placed over a rehabilitated or strengthened road.

On the origin and initiation of cracks

The onset of a crack in the upper asphalt layer of such a pavement structure takes place at any location where the breaking strength of this material is exceeded by tensile stresses whatever may be their generating mechanism.
Several mechanisms can act independently with additive results.
For thermal stresses one must take into account :
. the stresses due to the shrinkage of the upper asphalt layer

- stresses created by the opening of the discontinuity as a result of the shrinkage of the concrete slabs.
- the concentration of the stresses around the discontinuity.

Criteria for crack initiation

One of the main criteria which determine the initiation of a crack is the tensile strength of the asphalt mixes used in the overlay. Many results on this performance characteristics can be found in the literature (ref 1,2). An extensive study recently carried out by Arand (ref 3) on the basis of tensile tests and cooling tests has pointed out effects to be attributed to the nature of the bitumen and to the composition of the aggregates. Stress resistance values presented in this study were derived from tensile tests carried out on 4 bitumen types (see table 1) at 4 temperatures at a rate of loading of 100 µstrain/second). For example figure 2 gives the critical tensile stresses for the case of a 0/11 dense mix, similar to the compositions used in experimental tests reported at the end of this paper.

Table 1 : Characteristics of bitumens studied by Arand

Bitumen	B45	B65	B80	B200
Penetration	36	50	89	189
Ring & Ball T	55.5	51	45.5	39
A	.04416	.04631	.04768	.04476
C	.4524	.5412	.7573	1.575
Strain Rate	Critical temperatures for Xr=-2.5 in 0/11 asphalt concrete			
10^{-7}	-15.3	-16.6	-20.6	-31
10^{-6}	-9.2	-10.7	-14.9	-24
10^{-5}	-2.7	-4.5	-8.9	-18
10^{-4}	4.1	2	-2.6	-11.7
10^{-3}	11	8.8	4	-4.6

The different stress resistance values obtained for the 4 bitumens have been transposed for a 45pen reference. Tr is a calculated temperature at which the reference bitumen has the same viscosity as the considered bitumen. Therefore use was made of the well known relationship between penetration temperature and thermal susceptibility (ref 4): log pen = A T + C
Tr becomes thus for a given bitumen x:

$$T_R = \frac{(A_x \cdot T_x + C_x - C_R)}{A_R} \qquad (1)$$

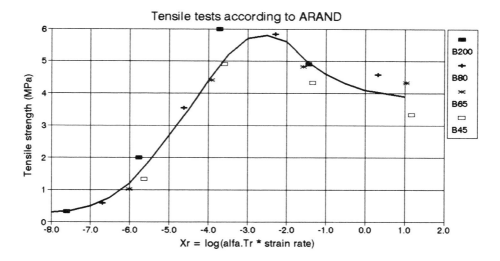

Figure 2 : Tensile strength values of 0/11 asphalt mixes containing different bitumen types

The presentation adopted in figure 1 is based on the temperature-loading time equivalency principle.
Xr is a reduced variable taking account simultaneously of temperature and strain rate Sr through the use of formula 2.

$$Xr = 11000 \cdot (\frac{1}{(Tr+273)} - \frac{1}{288}) + \log(Sr) \qquad (2)$$

These results are similar to those published earlier by Heukelom (ref 2). The strength curve displays a maximum value for a critical Xr value which corresponds to the limit between the viscoelastic behaviour of the mix (left hand side of the curve) and the purely elastic behaviour. The corresponding value of the Xr parameter is in this case:
X = -2.5 which for a rate of strain of 100 µstrain/s corresponds with critical temperatures given in table 1.

The advantage of this representation is that the tensile strength can be derived for any combination of temperatures and strain rates and for any bitumen type provided it is a straight-run bitumen in the penetration range covered by the four cases mentioned in table 1.

A similar procedure could also be followed for the determination of the tensile strain at break.

Coefficient of thermal expansion ß

This parameter is very important for the calculation of the thermal stresses in the case of thermal cracking. It can be estimated with good accuracy from the thermal expansion coefficients of the base components (binder $\alpha b \approx 2.10^{-4}$; aggregates $\alpha a = 6.10^{-6}$) and the mix composition by means of the averaging formula :

$$\beta_{MIX} = \frac{(\alpha_b \cdot V_b + \alpha_a \cdot V_a)}{100} \qquad (3)$$

where V_a: volume % of the aggregates
where V_b: volume % of binder

Thermal stresses induced in the asphalt layers

The evaluation of thermal stresses σ_{Th} taking place after a time t in an asphalt layer under the influence of a temperature variation of the form T(t) is given by the general formula :

$$\sigma_{Th} = -\beta \int_{\Theta=0}^{\Theta=t} E(\frac{t1}{\alpha_T(t1)} - \frac{\Theta}{\alpha_T(\Theta)}) \frac{\partial T(\Theta)}{\partial \Theta} d\Theta \qquad (4)$$

The crack will be initiated when this stress exceeds theresistance in tension of the asphalt or when the induced strain becomes higher than the yield strain.
Equation 4 allows the evaluation of the curve giving the building up of stresses as a function of the temperature and hence of the reduced parameter Xr.If the cooling rate is constant then the corresponding strain rate can be evaluated from the expansion coefficient : Sr= ß .δT/δt
    Under high cooling rates this curve can cross the strength curve given in figure 2.
In this case a crack will be initiated.If the cross point lies on the left hand side of the curve, the crack will propagate according to the fracture mechanics models.
On the other hand if the cross point is located beyond the maximum value (i.e in the low temperature/high strain rate region) a brittle fracture will occur and the asphalt layer will crack at once over its full thickness.
    This last situation is well known in northern regions where mean winter temperatures below -10C or lower can be observed over long periods.

Movements of the base layer in the vicinity of a crack or joint

The magnitude of the displacements at the edges of the crack opening depends on the value of the expansion coefficient of the concrete slab (usually taken to be 10^{-5}), on their length and on the amplitude of the thermal variations.
A short calculation shows that for slabs of some meters long this variations are of the order of 1 or 2mm. The maximum rate of variation is between a few hours and one day (seasonal variations are not taken into account here).

Stress concentrations above the pre-existing discontinuity

Stress and strain distribution analysis have been carried out by 2D modelling of this type of structure (ref 5).
The results obtained allow the derivation of a stress concentration factor CF which is equal to the ratio between the maximal stress computed close to the discontinuity and the average stress over the cross section of the base layer.(Not to be confounded with a stress intensity factor K). Table 2 gives some values of this factor for the case of a structure submitted to the opening of the discontinuity under the effect of horizontal tensile stresses applied at both sides of the base layer (with modulus E2=20000MPa).
The magnitude of this concentration factor is such that even for low average stresses in the base layer the critical stress levels can be exceeded. In any case they will reduce the available reserve of stress resistance by an important fraction.

Table 2: Stress concentration factors CF for an asphalt layer placed on a base layer with a f mm wide discontinuity without (A) or with (B) soft inter layer (E=20MPa)

Overlay thickness (mm):		40		70		120	
E1(MPa)	f(mm)	A	B	A	B	A	B
1000	2	18.3	2.3	16.5	2.4	14.6	2.2
1000	4	13.4	5.9	11.5	4.9	10.4	4.2
1000	8	8.9	6.1	8.0	5.0	7.0	3.8
10000	2	10.8	2.6	8.5	1.9	6.9	1.7
10000	4	8.1	3.4	6.5	2.1	5.4	1.9
10000	8	5.9	3.1	4.9	2.0	3.9	1.9

Experimental simulation of thermal cracking

Testing device

An experimental device already described in the former conference proceedings (ref 6) allows a very accurate control and monitoring of the width of a crack created in the base layer of a test specimen. Periodical variations of the crack opening can be achieved at very slow deformation rates (of the order of some tenth of mm/hour) on specimens maintained at a constant temperature (usually -10°C).

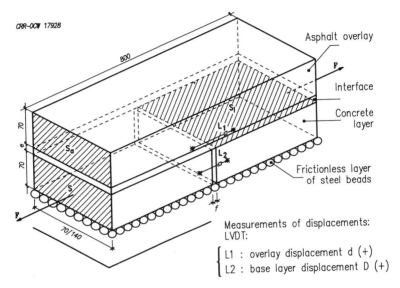

Figure 3 : Schematic view of the thermal test device and experimental setup

The specimen which is conform with the structure type defined hereabove (see figure 3) rests on top of a layer of steel balls allowing a frictionless displacement at the bottom of the concrete base.

During the test, transducers fitted on the specimen allow the measurement of the following parameters during the time t: displacements between the edges of the discontinuity : D(t); displacement at the base of the overlay : d(t); resulting force exerted on the cement concrete base : F(t). The observation of the onset and development of the crack is made automatically by means of a video recorder.

Experimental conditions

Tests were carried out under the following conditions :
    Displacement Wave form D(t) : triangular

Amplitude : 1mm under standard conditions.
Period   : from 6000 to 8000 seconds (7200 typical).
Temperature : -10 and -5°C .
Initial opening of the discontinuity : f=4mm.

Meaning of the measurements and typical cases

The strain rate in the asphalt layer depends on the place where the deformation actually takes place. It is given by the rate of variation of the overlay displacement divided by the length of the affected zone.

The test is able to give straightforward information regarding the behaviour in some typical cases described hereafter :

If there is a perfect bond between the base layer and the overlay then the displacements $D(t)$ and $d(t)$ must be equal and the total deformation will be restricted to the opening f of the existing crack. The strain rate will then be : $r=36$ μstrain/s.

If some **differential movements** take place at the level of the interface then the displacement measured in the overlay $d(t)$ will be less than that of the base plate $D(t)$ and the strain rate will be reduced accordingly. The difference $D(t)-d(t)$ can then be considered as the part of the displacement which is taken over by the interface.

If a **vertical crack** is initiated in the overlay then the displacement $d(t)$ will in fact give the crack opening displacement (COD). The development of the crack propagation can then be accurately observed and described by the $F(t)$, $D(t)$ and $d(t)$ curves completed with the crack length $c(t)$ measured on the video-recorded images of the test.

If a **horizontal crack** develops in the interface due to a lack of resistance to shear stresses then debonding will occur. The displacement $d(t)$ will be negligible in this case and only very low forces (dry friction plus elastic forces exerted by the remaining interface) will be measured. The test can then proceed forever without any cracking of the overlay ! But the resulting structure will be the worst possible for what concerns bearing capacity, resistance to traffic loads and waterproofing.

Experimental results

During the experimentation on a wide variety of interface systems, different types of behaviours have been clearly distinguished on the basis of the results of this type of test, and some examples are given here.
Figures 4 to 6 give the force deformation relationships of a reference specimen without interface and two different interlayer products.

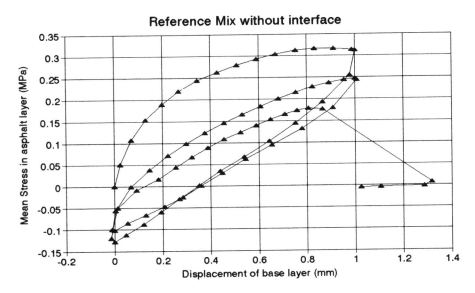

Figure 4 : Reference mix

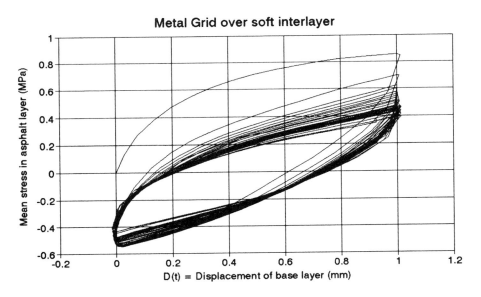

Figure 5 : Interface 1

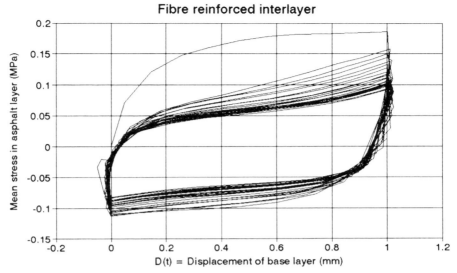

Figure 6: Interface 2

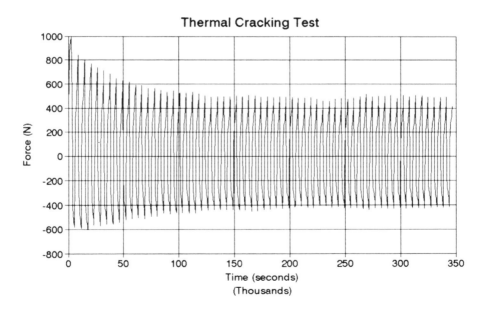

Figure 7 : Force vs-time relationship during a termal cracking test

These figures are very different from one interlayer to another and hence they constitute a real finger print from which many information regarding the rheology and the mechanical characteristics can be derived. It is clear that interlayers can increase the life of the structure at least for what concerns the type of loading simulated in this test. In many cases no cracking was observed in the limits of some 30 or more loading cycles whereas the reference structure cracked after only 3 cycles. However it can be seen from the F(t) curves that the stresses are decaying. The force can become stationary in some cases but the decrease can also go on, which is the sign of a weakening of the overlay system in the long term.

Interpretation of the results

The net force F(t) measured in function of time can be represented by the following general function:

$$F(t) = K_1 \exp(-K_2 . N) \int_{\theta=0}^{\theta=t} \phi(\frac{(\theta-t)}{\tau}) \frac{\partial D(\theta)}{\partial \theta} d\theta \quad (5)$$

with adjusting parameters :

K_1 = Initial stiffness
K_2 = Decay constant
ϕ = Relaxation function
τ = Relaxation time

test characteristics :

N = Number of cycles
θ = Integration variable
D (t) = Displacement of crack edges

It is possible for example to show on the basis of the adjustment of such a function with experimental results that the mechanical behaviour of the system presented on figure 4 can be modelized on the basis of a maxwell model (exponential relaxation function).
The total deformation recorded at the level of the crack discontinuity can be divided into three components :

1) Di(t)=D(t)-d(t); the deformation of the interface; resulting from horizontal shear stresses in the interlayer.

2) d(t); the deformation of the overlay resulting from the strain of the asphalt layer or the opening of a crack.

The total deformation amplitude remains constant during the test but the relative proportions taken by these components are changing together with the reaction force when a crack propagates in the upper layer.

This separation makes it possible to analyze the behaviour of the different elements of the system independently.

Shear in the interface layer

If the effective thickness e of the interface is known, one can derive the value of average shear stress and the shear modulus from the measured force, displacement and contact surface Si :

$$\tau(t) = \frac{F(t)}{S_i} \qquad G(t) = \frac{F(t) \cdot e}{S_i \cdot D_i(t)} \tag{6}$$

Average tensile stress and stiffness modulus of the overlay

In a similar way, it is possible to derive the mean values of tensile stresses and stiffness modulus of the asphaltic overlay from formulas :

$$\sigma(t) = \frac{F(t)}{S_a} \qquad S(t) = \frac{F(t) \cdot d_o}{S_a \cdot d(t)} \tag{7}$$

Compliance with the criteria

Under very low temperatures, stresses created into the asphaltic layers by shrinking have such a long relaxation time that they remain almost constant. They must be added to the stresses induced at the same time by the shrinkage of the base layer by taking into account the stress concentration factors CF defined in table 2. This situation can give rise to the onset of unstable cracks which can propagate instantaneously through the full thickness of the structure.

In the case where these stresses are lower than the resistance of the asphalt overlay they will reduce the effective resistance by an equivalent amount. This will leave what Arand defines as the reserve capacity of tensile stress. The knowledge of this reserve capacity is absolutely necessary for the assessment of the effective resistance of a structure against traffic loads.

Conclusions

The cracking procedure of an overlay layer is the result of the combination of three factors :
- the thermal stresses induced by the restrained expansion of the overlay
- the stress concentration in the vicinity of the discontinuities.
- the thermal movements of the concrete base layers.

This study leads to the following statements :

1) The primary requisite in the design of an overlay system is that the upper asphalt layer has a sufficient resistance against tensile stresses.
   The risk of unstable shrinkage cracks depends on the climatic conditions and on the performance of the asphalt mix irrespective of the underlying layers.
   This risk is high in cold winter conditions and for stiff mixes displaying a limited relaxation capacity.
2) The ability of an overlay material to resist additional stresses is given by the difference between the tensile strength of the upper asphalt layer and the thermally induced stresses.
3) In many cases the presence of a crack in the stiff base layer can be the source of local stress concentrations which exceed by far the stress reserve capacity.
4) An interlayer product is able to favourably influence the magnitude of stresses around discontinuities and may improve the ability of the layers to deform separately from each others while keeping an acceptable bond.
5) The resistance of the structure with respect to traffic loads is essentially determined by the reserve capacity of tensile strength in the vicinity of discontinuities and cracks.
6) The Experimental test procedures used at the BRRC together with the interpretation proposed allow the description of the behaviour of the two components of the overlay system : the inter-layer and the asphaltic overlay.
   The research in progress will be oriented towards the development of models taking into account the combined effect of traffic and thermal stresses.

References

1. **C.L. Monismith, C.A. Secor and K.E. Secor** : Temperature induced stresses and deformations in asphalt Concrete. Proceedings of the Annual Meeting of the Association

of the Asphalt Technologists. Vol 34 1965.
2. **W.Heukelom :** Observations on the rheology and fracture of bitumens and asphalt mixes. Proc A.A.P.T february 1966.
3. **W.Arand :** Behaviour of Asphalt Aggregate Mixes at Low Temperatures. Proceedings of the Fourth International RILEM.Symposium on Mechanical Tests for Bituminous Mixes. Budapest October 1990.
4. **W.Heukelom :** Observations on the rheology and fracture of bitumens and asphalt mixes. Proc A.A.P.T february 1966.
5. **L.Francken and A Vanelstraete :** Interface systems to prevent reflective cracking. Proceedings of the 7th International Conference on asphalt Pavements. Nottingham 1992 Vol 1 pp 45-60.
6. **L.Francken :** Fissuration thermique de structures semi-rigides; essais de simulation. Proceedings of the Fourth International RILEM Symposium on Mechanical Tests for Bituminous Mixes. Budapest October 1990.

17 INFLUENCE OF MODULUS RATIO ON CRACK PROPAGATION IN MULTILAYERED PAVEMENTS

C. PETIT and S. CAPERAA
Civil Engineering Laboratory, University of Limoges, Egletons, France
J-P. MICHAUT
Colas Society, Paris, France

Abstract

The aim of this work consists in studying the influence of a low modulus interlayer upon the decreasing of a crack propagation velocity. Efficiency of Virtual Crack Extension Method in modelling a thin layer is used for the determination of energy release rate G near the interfaces. An example shows the practical interest of the method, by comparing the results of a Paris law integration for two different structures.

Key-words: Fracture mechanics, Multilayered materials, Fatigue Cracking in Pavement Structures, Finite Element Numerical Model, Traffic load.

1 Introduction

This paper results from research between Colas Society and Civil Engineering Laboratory, University of Limoges (Egletons). The aim of this work is to investigate how a crack grows up to the pavement surface through the overlay. A lot of people try to limit the crack propagation velocity with differents solutions. So, it has been already proved that with the interposition of an interlayer such as a thin film between the base layer and the upper layer, C.Petit (1990), the crack grows more slowly.

Here, we are interested in the effect of an interlayer with a thickness more than 1 cm. Cases of very thick interlayer, C. Petit et al. (1991), can be numerically treated with joint finite elements. The pavement structure which is represented by finite element model is associated to a new numerical method to determine fracture mechanic parameters. Secondly, we will use a fatigue law to get the failure's cycle number. Numerically speaking, the difficulties are to represent a thin layer with finite elements and to calculate the stress field with accuracy in the crack tip area. The method proposed enables us to obtain a good representation of the phenomena.

2 Presentation of the problem

2.1 The pavement structure hypothesis

The structure is represented on figure 1; we can see a soil layer under a cement traited layer. We consider the existing cracks in this layer to be due to shrinkage behaviour. Then, one or two intact upper layers are considered. First, we will give results of a 6 cm-thick overlay made with a reference bituminous concrete. Secondly, we will take into account a 2 cm thick special sand asphalt layer with above, a classical 4 cm thick layer.

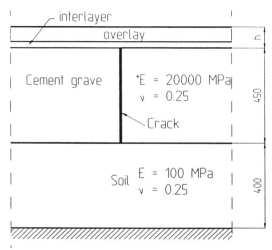

Figure 1 : Characteristics of the pavement structure

In both cases, the thickness is the same but the crack behaviour may be different. The mechanical behaviour is assumed as linear and elastic. Characteristics of soil layer have been defined from experimental measurements of pavement deflexion. In this paper, we will take into account only the crack growth at 15 °C and under traffic load with a frequency of 10 Hz. Traffic load has been defined as an uniform pressure of 0.67 Mpa on a 25 cm width area, with plane strain hypothesis, Caperaa S. and Petit C.(1990). Jayawickrama P.W and Lytton R.L. (1988), Vergne A.et al. (1989), consider that the maximum stress field around the crack tip is obtained when the load acts just before or after the crack direction. In this case the crack propagates in shear mode.

2.2 The finite element mesh

We have chosen herein to represent only an half mesh (fig. 2), to limit the number of finite elements. The superposition principle shown in figure 3 has been used to get the solution.

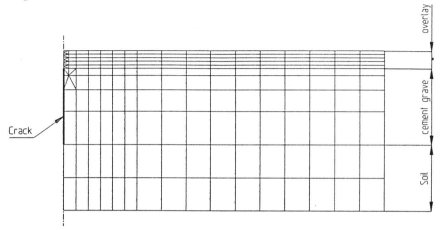

Figure 2 : Half finite element mesh

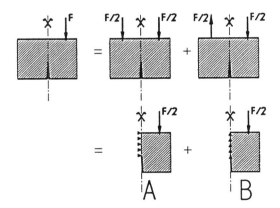

Figure 3 : Superposition principle.

2.3 The crack propagation method
For both structures we propose to propagate the crack tip using a Paris (1961) fatigue law:

$$\frac{da}{dN} = A \cdot (\Delta K_I^*)^m \qquad (1)$$

In this formula, a is the crack length, N the number of cyles. The parameters A and m are determined with fatigue tests, Molenaar (1983), Caperaa and al.(1993). K_I^* is the stress intensity factor in the crack growth direction. Numerical values are obtained from a finite element model, using the method proposed in 3.

3 Fracture mechanics aspects

In this part of the paper we review some theoretical and numerical elements of fracture mechanics which are necessary in order to go on.

3.1 Stress intensity factors
Stress intensity factors are defined in singular stress field near the crack tip. We assume herein plane conditions such as opening mode K_I and shear mode K_{II}. The problem of stress intensity factor determination is the poor accuracy and the great number of finite elements particularly when multilayered materials are studied.

$$K_I = \lim_{r \to 0} \frac{\mu}{k+1} \cdot \sqrt{\frac{2\pi}{r}} \cdot [v] \qquad (2)$$

$$K_{II} = \lim_{r \to 0} \frac{\mu}{k+1} \cdot \sqrt{\frac{2\pi}{r}} \cdot [u] \qquad (3)$$

$$k = 3 - 4\nu \qquad \text{plane stress}$$

$$k = \frac{(3-\nu)}{(1+\nu)} \qquad \text{plane strain}$$

[u] is the relative displacement of both crack surfaces in the crack direction.
[v] is the relative displacement of both crack surfaces perpendicular to the crack.
    In general, we observe mixed mode (K_I and K_{II}) which leads us to bifurcation and the crack tip does not propagate in straight direction. It has been proved that the crack goes on in a new direction Ox^* where $K_{II}^* = 0$, Amestoy et al. (1979). In our particular case, mode I does not exist, then $K_I^* = 1,2.K_{II}$.

3.2 The Virtual Crack Extension Method (V.C.E.M)
The V.C.E.M consists of a numerical procedure to obtain the energy release rate G, Griffith.

$$G = -\frac{\partial \Pi}{\partial a} \qquad (4)$$

G is the variation of entire potential energy against the crack tip propagation.

It has been proved that in linear and elastic cracked solids, G is a function of the stress intensity factor components :

$$G = \frac{K_I^2 + K_{II}^2}{E'} \qquad (5)$$

$E' = E$ \qquad plane stress

$$E' = \frac{E}{1 - \nu^2}$$ \qquad plane strain

Here, we would rather calculate G, in order to get the stress intensity factor. Effectively, K_I can be obtained from several numerical methods such as the cinematic method (2)(3), but this necessitates a high density of finite elements near the crack tip. Then, V.C.E.M enables the user not to have too many finite elements. G Finite Element formula (by finite differences method) is :

$$G = -\frac{1}{2} \cdot \sum_{i=1}^{n} \{U_i\}^T \cdot \left[\frac{\Delta K_i}{\Delta a}\right] \cdot \{U_i\} + \left\{\frac{\Delta F}{\Delta a}\right\}^T \cdot \{U\} \qquad (6)$$

{Ui} is the displacement vector of finite element i
[Ki] is the stiffness matrix of finite element i
{F} is the external force vector.
n is the finite element number of (S2) area.

The computation consists in moving crack area inside (S2). So, only S2 elements have been distorted.
The calculation does not depend on the distorted area chosen (fig 4). In our case, the distorted area is only made with the few elements at the crack tip, because only one material must be taken into account, particularly in thin layers. Results are accurate enough, Petit (1990), if quarter node elements are required, Roshdy and Barsoum (1976).

☐ distorted area (S2)

☐ undistorted area (S1, S3)

○ moved nodes

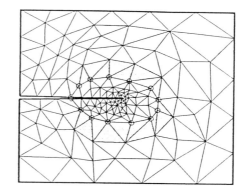

Figure 4 : Virtual distorted area

4 Results

Case A in figure 3 leads to mode I, but the crack does not open. However, case B leads to mode II, so we only have to calculate this model.

$$K_{II} = \sqrt{\frac{E \cdot G}{1 - \nu^2}} \qquad (7)$$

$$K_I^* = 1.2 \cdot \sqrt{\frac{E \cdot G}{1 - \nu^2}} \qquad (8)$$

4.1 Energy release rate evolution versus crack length

Studies have already shown that the energy release rate G is a parameter quite easy to calculate when we are in composite multilayered materials. The half structure is convenient to use V.C.E.M. There are only two finite elements associated with the crack tip, according to 3.2. The computed results are obtained only from these two elements.

In figure 5, one can observe the evolution of G versus a the crack length. The crack tip starting is supposed such as on figure 1 (a = 0). The classical technique - 6 cm thick bituminous concrete alone - shows us that G increases from a = 0 to a = h. In the other case, we replace the two first centimeters by sand asphalt. A lower Young modulus than the bituminous concrete one is the reason why G decreases in this interlayer. Then, both KI^* and crack length velocity da/dN decrease. Now the relationship between the crack length and the number of cycles, can be determined.

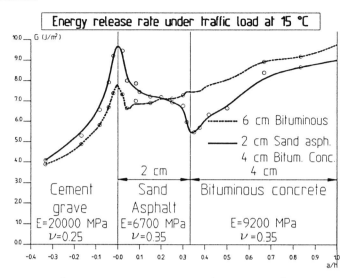

Figure 5 : G curves in upper layers

4.2 Influence on fatigue crack propagation

The formula (8) associated to G results given in figure 5 enables the user to obtain the velocity da/dN (1). So the crack length versus the number of cycles can be computed and results are represented in figure 6.

Here, A and m numerical values are not taken into account because they are no experimentally defined in this case.

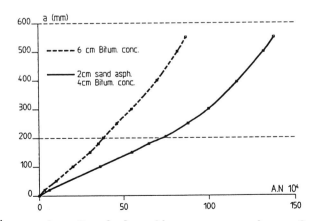

Figure 6 : Crack length versus number of cycles

Figure 6 only gives the relative evolution of crack because we have supposed that A and m are the same in both layers. m has been fixed equal to 4. So in this case, the crack growth of the bi-layered overlay takes 1.5 time more than the reference structure. Moreover, if we suppose A parameter 10 times less than the reference structure, a more realistic hypothesis then the coeficient becomes more than 2.

5 Conclusion

Sand asphalt interlayer is a very interesting solution because, under traffic load and in the conditions given before, the life time is at least twice as much as for a classical pavement structure.

The main result of this paper is about multilayered overlay of pavement structures. The effect of the interlayer Young's modulus, which is lower than the surface layer's one is a very important parameter to increase the life time of structure.

6 References

Amestoy, M. Bui, H.D. Dangvan, K. (1979), Déviation infinitésimale d'une fissure dans une direction arbitraire, **Comptes rendus de l'académie des sciences**, B, 289, p.99.

Caperaa, S. and Petit C. (1990). Fissuration des chaussées en fatigue, **Research report for Colas Society**

Caperaa S., Ahmiedi E., Petit C. (1993). A new test for determination of fatigue crack characteristics for bituminous concretes, **Second Conf. on Reflective Cracking in Pavement**, RILEM.

Jayawickrama, P.W. and Lytton, R.L.(1988). Methodology for predecting asphalt concrete overlay life against reflection cracking, **University of Texas.**

Molenaar, A.A.A.,(1983). Structural Performance and design of flexible road constructions and asphalt concrete overlays. Doct. **Thesis, Delft, The Netherlands.**

Paris, P.C. and al, (1961). A rational analytical theory of fatigue, **University of Washington**, Wa, Vol. 13, n°1.

Petit, C., (1990). Modélisation de milieux composites multicouches fissurés par la mécanique de la rupture, **Thesis of University of Clermont Fd, France.**

Roshdy and Barsoum, S (1976), On the use of isoparametric finite elements in linear fracture mechanics, **Int. J. of Num. Method in Engineering**, Vol. 10, 25-37.

Vergne, A. and al., (1989), Simulation numérique de la remontée d'une fissure dans une structure routière. **First Conf. on Reflective Cracking in Pavement**, (Ed. Rigo and R. Degeimbre), RILEM.

18 REFLECTIVE CRACKING IN ASPHALT PAVEMENTS ON CEMENT BOUND ROAD BASES UNDER SWEDISH CONDITIONS

J. SILFWERBRAND
Swedish Cement and Concrete Research Institute, Stockholm, Sweden

Abstract

Thin asphalt surfacings on cement bound roadbases would be a technically and economically good solution, if the risk of reflective cracking could be proved to be insignificant. Test sections with surfacing thicknesses of 35 and 100 mm work well after seven years. Crack mechanisms such as tensile fatigue and tensile yield are discussed. Trials are made to estimate fatigue life and risk of tensile yield. Recently observed cracks at the test site can be explained by an increasing obvious risk for tensile yield.

<u>Keywords:</u> Reflective cracking, Thin surfacings, Swedish test sections, Fatigue life, Tensile yield, Beam model.

1 Introduction

The Swedish use of asphalt surfacings on cement bound roadbases has been more frequent during the latest decade. The use of cement bound roadbases would, however, increase rapidly if the problem with reflective cracking could be solved definitively. Today, the Swedish pavement code (1984) punishes cement bound roadbases in relation to asphalt bound roadbases through prescribing a thicker asphalt surfacing in the cement case. The reason for this is said to be the presence of cracks in the cement bound roadbase and the assumed reflective cracking. According to the usual opinion on reflective cracking, the reflective crack will reach the upper surface later in a thick asphalt surfacing than in a thin one. There are, however, both empirical and theoretical evidences that even a thin asphalt surfacing can be used successfully on a cement bound roadbase.

In 1988, Foulkes published an extensive study on reflective cracking. He identified three crack mechanisms:

   i) **Tensile fatigue**. Horizontal movements in the cement bound roadbase due to daily temperature cycles cause stresses in the asphalt pavement. The stresses are concentrated to the zone close to the roadbase crack. The reflective crack starts at the bottom surface of the surfacing and propagates to the upper surface.

   ii) **Tensile yield**. A rapid temperature drop during winter conditions leads to shortening and upward curling of the roadbase slabs (roadbase between cracks). The great crack width that has its maximum value on the

Reflective Cracking in Pavements. Edited by J.M. Rigo, R. Degeimbre and L. Francken.
© 1993 RILEM. Published by E & FN Spon, 2–6 Boundary Row, London SE1 8HN. ISBN 0 419 18220 9.

upper surface of the roadbase, induces tensile strain in the asphalt surfacing. Long term shrinkage may increase the induced strain. The asphalt strain reaches its annual maximum while the asphalt has a minimum ductility. If the strain exceeds the tensile yield capacity, a crack is formed. According to Foulkes, this crack propagates rapidly from the upper surface down to the crack in the roadbase.

iii) **Shear fatigue**. As traffic crosses a roadbase crack, vertical differential movements may appear between adjoining slabs. Shear stresses are induced in the asphalt surfacing. This mechanism acts accelerating during the end of the crack propagation. It is of minor interest for the life cycle determination and thus it will not be treated in this paper.

In the following, I will first briefly describe two Swedish test sections and then evaluate their behaviour with respect to Foulkes' crack mechanisms.

2 Test sections

On a road (Lambohovsleden) outside Linköping in south of Sweden, some test sections were built in August 1985. The objective was to study thin asphalt surfacings on cement bound roadbases. The road was designed for 500 to 1500 heavy vehicles every day. The subgrade varies between silt and clay. Five different cross sections were studied, see Petersson, Karlsson and Ydrevik (1987).

I have limited my study to the two extreme cases. The asphalt surfacing is 35 (Test Section 4) or 100 mm (Test Section 1), respectively. The thin surfacing consists of a wear course (HAB). The thick surfacing consists of 35 mm wear course on top of 65 mm base course (AG). In both cases, the surfacing is placed on a 150 mm cement bound roadbase (CG) (Fig. 1).

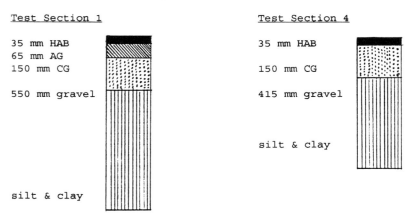

Fig. 1. Test sections.

Before the asphalt surfacings were placed, transverse cracks were observed in the cement bound roadbase. After 18 months, the pavement still behaved satisfactorily. No reflective cracks were observed

(Petersson et al. 1987). After seven winters, the pavements still work well. Some tiny reflective cracks can be seen on both Test Sections 1 and 4. On Test Section 4 with a thin asphalt surfacing, the cracking is more extensive (Carlsson 1992).

3 Estimation of tensile fatigue life

3.1 Calculation procedure
Foulkes has developed a computer programme for calculation of fatigue life. I have used his hand calculation model, described in Foulkes (1988), to calculate the life length for the test sections. Following data is needed:

1) Layer thicknesses
2) Material data of the asphalt materials
3) Distance between transverse cracks in the roadbase (=slab length)
4) Monthly or seasonal values of average and daily variation of temperature
5) Monthly or seasonal values of daily crack opening width cycles in the roadbase
6) Relationship between temperature, loading time and binder stiffness
7) Relationship between binder stiffness and asphalt stiffness
8) Stress intensity factors
9) Material constants A and n i Paris' law
10) Knowledge of Miner's law

Based on the material description in Swedish pavement code (1984), I have assumed that the wear layer has a binder content of 6,5 %, a binder penetration of 85 (mm/10) and a void content of 3 %. Corresponding values for the base layer are 4,5 %, 85 and 6 %, respectively. I have assumed a slab length of 15 m.

Comparing Foulkes' British temperature data with Swedish temperature values according to Taesler (1972), I have drawn following conclusions:

i) The average temperature in Linköping is 5°C below the average temperature in Great Britain.
ii) The range of the daily temperature variation in Linköping is 2/3 of the British range.

The first conclusion means that a certain asphalt material is stiffer in Linköping than in Britain. The second conclusion leads to minor daily crack opening widths.

Foulkes uses van der Poel's nomograph (1955) for determination of binder stiffness as a function of material properties, temperature and loading time. I have used the same nomograph, determining stiffness values for January, April, July and October, every month representing one season. I have then used Foulkes' relationship between binder stiffness and asphalt stiffness to calculate stiffness values for the two surfacing materials in current study. This relationship can be used because these materials are similar to the materials in Foulkes' examples.

Using the Finite Element Method (FEM), Foulkes has determined stress intensity factors k_0 for a standard pavement. k_0 is dependent on the crack length c. Foulkes' surfacing might consist of two layers (40+60 mm). k_0 is dependent on the ratio between the stiffness values of the two layer materials. For other pavements, following expression gives the modified stress intensity factor k_r:

$$k_r = \frac{k_0 \cdot u \cdot S_m(w)}{10^8} \cdot \sqrt{\frac{h}{100}} \qquad (1)$$

where k_r and k_0 are stress intensity factors in $N/mm^{3/2}$, u is crack opening width in mm, $S_m(w)$ is stiffness of wearing course in N/mm^2 and h is the total surfacing thickness.

The relationship between rate of crack growth and stress intensity factor is given by Paris' law:

$$\frac{dc}{dN} = Ak_r^n \qquad (2)$$

where N is the number of cycles (=number of days) and dc/dN is the crack growth per cycle. A and n are two material constants. Foulkes have tested different asphalt mixes and developed graphical relationships between binder stiffness and the constants. I have used these curves.

The life length for the mechanism of tensile fatigue can be obtained by integrating Paris' law:

$$N = \int_0^h \frac{dc}{A\{k_r(c)\}^n} \approx \sum \frac{\Delta c}{A\{k_r(c_i)\}^n} \qquad (3)$$

where c_i is the mean value of the crack length during step No. i. Similar to Foulkes, I have divided the surfacing thickness into 10 steps, i.e., $\Delta c = h/10$.

The rate of the crack growth varies during the year. Miner's law has been used to regard this fact:

$$\frac{N_1}{N_{1,u}} + \frac{N_2}{N_{2,u}} + \frac{N_3}{N_{3,u}} + \ldots \leq 1 \qquad (4)$$

where N_1 is the number of cycles at damage level 1 and $N_{1,u}$ is the ultimate number of cycles at damage level 1, etc. Here, the number of available cycles (=number of days for the crack to propagate through the surfacing) is dependent on the season. The months January, April, July and October represent the four seasons. From Eq. (4) follows:

$$N = \frac{1}{\frac{0.25}{N_{Jan,u}} + \frac{0.25}{N_{April,u}} + \frac{0.25}{N_{July,u}} + \frac{0.25}{N_{Oct,u}}} \qquad (5)$$

3.2 Calculation results

For Test Section 1 (35 mm wear layer + 65 mm base layer), the calculated life length was 9 years. For Test Section 4 (35 mm wear layer, no base layer), it was 190 years. At first, the results seem very surprising. It is true that the total crack path is longer in the thick pavement. A thick pavement also insulates the roadbase reducing the daily crack opening width. There are, however, theoretical explanations.

As shown in Eq. (1), the stress intensity factor is dependent on both crack opening width and surfacing thickness. Here, the thickness effect predominates, because the calculated insulating effect only reduces the crack width with 25 to 30 %. The thin surfacing has the lowest stress intensity factor. The prolonged crack path of the thick surfacing cannot compensate the stress intensity factor, because the exponent n here has such a high value as n=6.6.

According to Foulkes' FEM results, the stress intensity factor increases as the stiffness ratio between the two materials in a two layer surfacing increases. That punishes the thick two layer surfacing in this study.

According to the calculation results, reflective cracks are likely to reach the upper surface of the thick surfacing (Test Section 1) in a few years.

4 Risk determination for the mechanism tensile yield

4.1 Calculation model

In order to compare the risk of tensile yield in different surfacing, I have developed a beam model. The study by Foulkes (1988) cannot be used because his number of examples is very small. Consider a long continuous asphalt surfacing on a cracked cement bound roadbase. The distance between the transverse cracks is $2L=2(L_1+L_2)$. Assume that the pavement can be substituted by a continuous beam with varying stiffness. The stiffness has a high value along the length $2L_1$ and a low value along the crack width $2L_2$ (Fig. 2). Following assumptions are made:

1) The pavement extension in the transverse direction is neglected.
2) The dead load of the pavement and the subgrade reaction are assumed to neutralize each other.
3) The friction between roadbase and subgrade is neglected.
4) The pavement edges in the longitudinal direction are fixed.
5) The stiffness of the cement bound roadbase is much higher than the surfacing stiffness.

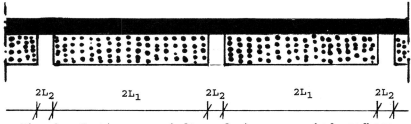

Fig. 2. Continuous asphalt surfacing on cracked roadbase.

The weak point of the beam model is that the ratio between beam depth and beam length is too high along $2L_2$, i.e., along the crack width. The real value is perhaps 10, whereas a ratio of 0.1 is desired for beam theory. This fact makes a calculation of absolute strains impossible. On the other hand, relative comparisons between different surfacings can be done easily. Factors like surfacing thickness, stiffness values and coefficient of thermal expansion can be varied.

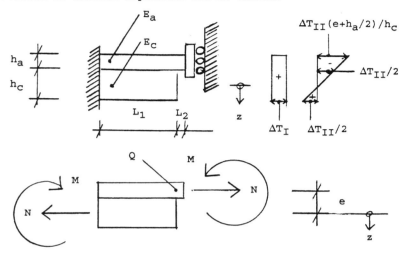

Fig. 3. Beam model, temperature distribution, forces and moments.

Using symmetry, a two-sided clamped beam is considered (Fig. 3). The beam length is $L=L_1+L_2$. The beam has the tensile stiffness of $EA_1=E_ch_c$ and the flexural rigidity of $EI_1=E_ch_c^3/12$ along L_1. Corresponding stiffnesses along L_2 are $EA_2=E_ah_a$ and $EI_2=E_ah_a^3/12$, respectively.

Assume a linear temperature distribution $\Delta T=\Delta T_I+\Delta T_{II}$ according to Fig. 3. The coefficients of thermal expansion of the cement bound roadbase and the asphalt surfacing are α_c and α_a, respectively.

Along L_1 at the level z=0, the strain ε_1 equals:

$$\varepsilon_1 = \frac{N}{EA_1} + \alpha_c\Delta T_I \qquad (6)$$

The curvature κ_1 is:

$$\kappa_1 = \frac{M}{EI_1} + \frac{\alpha_c\Delta T_{II}}{h_c} \qquad (7)$$

Along L_2 through the gravity centre of the surfacing (z=-e), the strain ε_2 equals:

$$\varepsilon_2 = \frac{N}{EA_2} + \alpha_a \Delta T_I - \alpha_a \Delta T_{II} \frac{e}{h_c} \tag{8}$$

The curvature κ_2 is:

$$\kappa_2 = \frac{M+Ne}{EI_2} + \frac{\alpha_a \Delta T_{II}}{h_c} \tag{9}$$

The moment M and the normal force N are temporary unknown. They can, however, be determined by using the continuity demand along the common border-line of the beam parts. Following relationship is valid for a point Q situated at the level z=-e on the border-line:

$$\varepsilon_1 L_1 - \kappa_1 L_1 e = -\varepsilon_2 L_2 \tag{10}$$

For the slope of the border-line, following relationship is valid:

$$\kappa_1 L_1 = -\kappa_2 L_2 \tag{11}$$

That gives us six equations and six variables (ε_1, κ_1, ε_2, κ_2, M and N). M and N can now be determined. To save space, the long expressions of M and N are omitted in this paper. Stresses and strains in the uppermost and lowermost fibres of the surfacing above a roadbase crack are given by following expressions:

$$\sigma_{au} = \frac{N}{h_a} - \frac{6(M+Ne)}{h_a^2} \tag{12}$$

$$\sigma_{al} = \frac{N}{h_a} + \frac{6(M+Ne)}{h_a^2} \tag{13}$$

$$\varepsilon_{au} = \frac{\sigma_{au}}{E_a} + \alpha_a \Delta T_I - \alpha_a \Delta T_{II} \frac{e+h_a/2}{h_c} \tag{14}$$

$$\varepsilon_{al} = \frac{\sigma_{al}}{E_a} + \alpha_a \Delta T_I - \alpha_a \Delta T_{II}/2 \tag{15}$$

4.2 Calculation results

Calculations have been carried out for three main cases (Table 1). Two cases correspond to the test sections described in Section 2 above. The third one corresponds to a case in Foulkes (1988). According to Foulkes, the maximum tensile strain occurs at the top of the asphalt surfacing. With the beam model, I get the same result if I neglect the temperature drop ΔT_I that is constant in all the pavement. A disregard of ΔT_I can be justified because a constant temperature drop develops first after a long time giving time for stress relief due to creep. Consequently, I have discerned between two cases, A and B, excluding and including ΔT_I.

The temperature is assumed to drop 22°C at the upper surface of the pavement and with 7°C at a depth of 300 mm. It means that ΔT_I and ΔT_{II} depend on the level of the gravity centre of the roadbase and roadbase thickness, respectively.

Table 1. Calculation results

Case	1 A	2 A	3 A	1 B	2 B	3 B	4 A	5 A
h_c (m)	0.15	0.15	0.175	0.15	0.15	0.175	0.15	0.15
E_c (GPa)	17	17	17	17	17	17	**8.5**	17
h_a (m)	0.035	0.1	0.175	0.035	0.1	0.175	0.1	0.1
E_a (GPa)	0.5	0.5	0.5	0.5	0.5	0.5	0.5	1
ΔT_I (°C)	0	0	0	-16	-13	-9	0	0
ΔT_{II} (°C)	8	8	9	8	8	9	8	8
$\alpha_c/10^{-6}$	10	10	10	10	10	10	10	10
$\alpha_a/10^{-6}$	2.5	2.5	2.5	2.5	2.5	2.5	2.5	2.5
L_2/L	0.001	0.001	0.001	0.001	0.001	0.001	0.001	0.001
$\varepsilon_{au}/10^{-3}$	2.90	1.67	1.11	-14.4	-18.3	-9.2	1.04	1.04
$\varepsilon_{au}/10^{-3}$	-1.34	-0.58	-1.01	27.1	25.5	13.5	-0.95	-0.95

According to the calculation results (Table 1), the tensile strain in the thin surfacing (Test Section 4) is higher than the strain in the thick surfacing (Test Section 1). The strains in both Swedish test sections are higher than the strain in Foulkes' British example.

As mentioned previously, the absolute value of the strain cannot be determined with the beam model. Calculations also show that the strain values are sensitive to the L_2/L ratio. The relationship between different pavements changes, however, only slightly.

Temperature statistics close to the test site (SMHI 1992) show that the maximum temperature drop since construction has been 15°C. (Drop is measured from average temperature one day to minimum temperature the following day). This occurred during the two first winters. Thereafter, the temperature drop has not exceeded 12°C. On the other hand, asphalt materials are exposed to oxidization. Their strain capacity diminishes.

Foulkes calculated maximum strain values around 0.3 % for a 15°C temperature drop. Assume that Table 1 can be used for comparison. That means strain values between 0.5 and 1 % in the Swedish test sections. The highest value occurs in the thin surfacing. According to Foulkes' test results, the yield strain for British asphalt materials varies between 0.1 and 2 % at low temperatures. No similar tests have been done on Swedish materials. According to Isacsson (1992) the discrepancy is not that great. Approximately, the British data can be used on Swedish materials. Consequently, there is a risk of the tensile yield mechanism in the two Swedish test sections. The risk is increasing due to ageing (oxidization). The cracks are most likely in the thin pavement.

Cases 4A and 5A in Table 1 show that a decreased ratio between the roadbase stiffness and the surfacing stiffness reduces the maximum tensile strain in the surfacing.

5 Conclusions

Asphalt surfacings with different thicknesses have been placed on cement bound roadbases on a road in south of Sweden. After seven years, the pavements still behaved well but a certain number of transverse cracks have been visible during the latest years. The cracking is more extensive in the (35 mm) thin surfacing than in the (100 mm) thick one.

A trial has been made to estimate life length for the crack mechanism tensile fatigue. The calculated fatigue life is longer in the thin surfacing. The theoretical explanation is minor stress intensity factors caused by less thickness and the beneficial solution with only one homogeneous material.

A beam model has been developed to compare the risk of the crack mechanism tensile yield for different surfacings. The risk is reduced in thick surfacing with high stiffness values compared to the roadbase stiffness. Estimated tensile strain values in the same range as the tensile yield strain at low temperatures can be compared with recent observed cracking.

Reflective cracks may be visible in a 35 to 100 mm thick asphalt surfacing on a cement bound roadbase after 5 to 10 years. Theoretically, a homogeneous surfacing has the superior behaviour.

6 References

Carlsson, B. (1992) Unpublished correspondence between B. Carlsson, research engineer at the Swedish Road and Traffic Research Institute, Linköping, and the author. January 1992.

Foulkes, M.D. (1988) Assessment of Asphalt Materials to Relieve Reflection Cracking of Highway Surfacings. **Ph.D. thesis**, Plymouth Polytechnic, U.K.

Isacsson, U. (1992) Unpublished correspondence between U. Isacsson, Professor in Highway Engineering, Royal Institute of Technology, Stockholm, Sweden, and the author. May 1992.

Petersson, Ö., Karlsson, B., and Ydrevik, K. (1987) Test Road Lambohovsleden. **Bulletin** No. 524, Swedish Road and Traffic Research Institute, Linköping. (In Swedish).

Swedish National Road Administration (1984). Swedish Pavement Code. **Bulletin** No. TU 154. (In Swedish).

Swedish Meteorological and Hydrological Institute (SMHI) (1992) Temperature Statistics for Malmslätt, Sweden, 1985-1992. **Reprint** from data base.

Taesler, R. (1972) **Climate Data for Sweden**. National Swedish Institute for Building Research and Swedish Meteorological and Hydrological Institute. (In Swedish).

Van der Poel, C. (1955) Time and Temperature Effects of Asphaltic-Bitumen and Bitumen-Mineral Mixtures. **SPE Journal**, September 1955, pp. 47-53.

19 STUDY ON REFLECTION CRACKS IN PAVEMENTS

K. SASAKI
Civil Engineering Research Institute, Hokkaido Development Bureau, Japan
H. KUBO
Faculty of Engineering, Hokkaigakuen University, Japan
K. KAWAMURA
Civil Engineering Research Institute, Hokkaido Development Bureau, Japan

Abstract
Hokkaido has an area of about one fifth of Japan, and road traffic is the main means of land traffic. Hokkaido locates the northernmost part of Japan, and days with temperatures below -20 ℃ in winter are common; air temperatures often rise above 30 ℃ in summer. These factors give the severe climatic conditions with an annual temperature range of more than 50 ℃.
These climatic conditions are largely responsible for the transverse low temperature crackings observed on the roads in Hokkaido. Reflection cracks of asphalt-overlayed pavements bring about the main problems of road maintenance costs. Various control methods are employed to prevent reflective cracks in many areas, using repair materials such as sheets, resinous nets and high penetration of asphalt.
However, none of methods may be established as generally accepted permanent countermeasures for the transverse reflective crackings.
This paper describes the performance of test repair works for preventing low temperature cracks of pavements in Hokkaido.
Keywords: Low temperature crackings, Field test performance, Repair method

1. Introduction

Hokkaido has the area of about one fifth of Japan, and road traffic is the main means of land traffic. Hokkaido locates the northernmost part of Japan, and days with air temperatures below -20 ℃ in winter are common; in summer temperatures often rise above 30 ℃. These factors give the severe climatic conditions with an annual temperature range of more than 50 ℃.
These climatic characteristics are largely responsible for the transverse low temperature crackings observed on the roads in Hokkaido. Countermeasures to transverse crackings have been attempted for the charactristics of cracks with the field test pavements on national roads. The countermeasures are consist of preventions for these cracks on newly established pavements, crack sealing and reflection crack preventions[1].
The object of these investigations is to determine treatment of cracks and preventive methods for reflective crackings through field tests. It reports the findings of test sections on national roads and the investigations of prevention and repair of the cracks in asphaltic pavements.

Reflective Cracking in Pavements. Edited by J.M. Rigo, R. Degeimbre and L. Francken.
© 1993 RILEM. Published by E & FN Spon, 2–6 Boundary Row, London SE1 8HN. ISBN 0 419 18220 9.

2. Field investigations of crackings

2.1 Field investigations

The first field investigations of cracks in Hokkaido was conducted by Civil Engineering Research Institute on national and prefectural roads in 1978[2]. Figure 1 (a) shows four classified distributions of transverse crackings in asphalt road pavements, and it is possible to recognize the distributions of low temperature crackings at the intervals of 100 m and below in Hokkaido from this figure[3]. Further field investigations were conducted on national roads located in the eastern regions of Hokkaido, where transverse cracks were very common in 1990. Figure 1 (b) is a composite diagram overlapping the distribution chart made in 1978 with that prepared from the findings of this investigations. The areas with numerous cracks in 1978 were reportedly repaired by the overlay method, however, the figure shows that the cracks reappear in the same places, seemingly more numerous and in more areas.

Figure 1
Crack Distribution

(a)

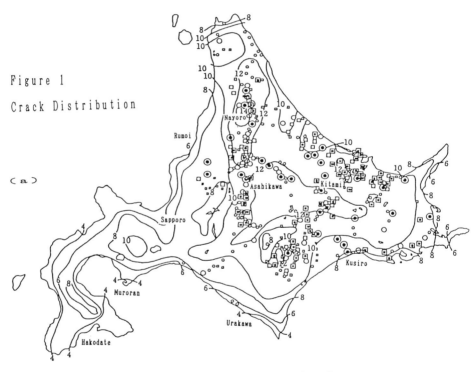

National Highways	Prefectural Highways	Description
●	■	Crack Interval below 10m
○	☐	Crack Interval 10-30m
∘	▫	Crack Interval 30-100m
——10——	:	Freezing Index (×100℃/days) FY 1967-1976 maximum

Legends

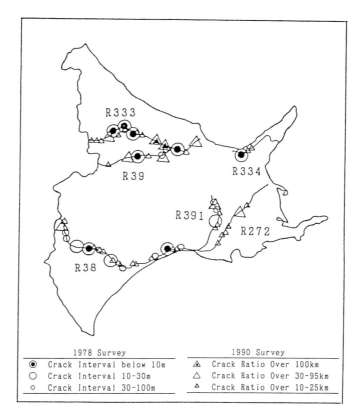

Figure 1   Crack Distribution   (b)

2.2 Existing conditions of crackings
Figure 2 (a) represents the relationship between the pavement cross section and the number of cracks. This figure shows the number of invastigated areas where traffic volume surveys were conducted. There are very few traffic blocks on national roads under the jurisdiction of Hokkaido Development Bureau, and the cross sections of the investigated pavements are shown in figure 2 (b). This figure indicates that cracks are restrained by thicker pavements.

2.3 Factors causing cracks
Various factors contribute to the appearance of cracks. The investigation and the analyses were conducted with four factors: 1) freezing index, 2) period of freezing days, 3) traffic volume and 4) pavement structure. Table 1 and figure 3 show that the freezing index provides the strongest indicator of crack formation with multiple regression analysis. The correlation coefficient determined from the freezing index and cracking ratio is $R^2 = 0.63$ in logarithm regression.

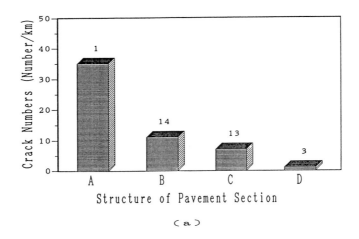

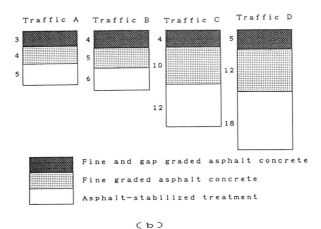

Figure 2　Crack Numbers by Pavement Structure

Table 1 Multiple Regression Analysis Results

Factor	F value of regression coefficient	Multiple correlation coefficient (R^2)
Freezing index	2.686	
Days frozen	0.002	⟩　0.38
Traffic volume	0.399	
Pavement Structure	0.924	

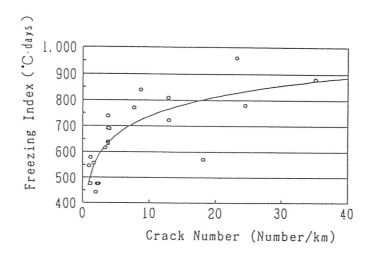

Figure 3 Crack Number NS Freezing Index

3. Performance of crackings

3.1 Investigated locations
Reflection cracks were invstigated on Route 275 in Shintotsukawa and Route 237 in Sohubetsu, and the longitudinal deformation survey was on Route 242 in Rikubetu.

3.2 Investigation method and interval
The measurements of transverse crack width were made in places from one shoulder to other. The field investigations were made with the buried point guages and 1/20mm slide calipers in Shintotsukawa for 7 months from October to the following March, at approximately one month intervals. The point gauges were buried at 3 points at the center and both sidelines of roads.
The longitudinal deformation was measured at 10cm intervals, determining cracks within 50cm, using one meter long steel ruler which was placed at a right angle and every 10 cm from front to back in 0.1 mm units. These measurements were conducted in June, August and November[4].

3.3 Investigation results
The changes in crack width are a maximum of 0.16 cm in Shintotsukawa and 0.7 cm in Soshubetsu, as shown in figure 4 and figure 5, respectively. There are different amounts of expansion and contraction along cracks, varying as much as 2 times in a season. Figure 6 shows the longitudinal deformation. Deformation appears at the center of cracks and the defference of deformation at each side of crack depends on the traffic direction. It is considered that the humps are created by the strong impact traffic loads as shown in figure 6. The performance of cracks depends the conditions of the location, showing no regular features, except possibly that the width increases as air temperature is lower in winter.

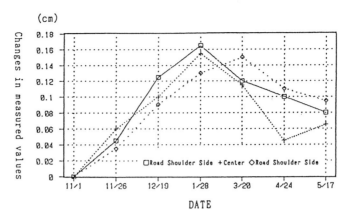

Figure 4 Changes in Crack Widths (Shintotsukawa)

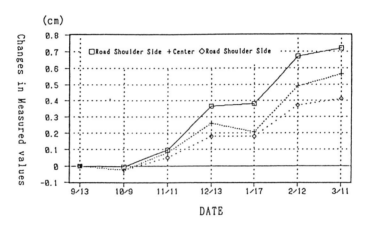

Figure 5 Changes in Crack Widths (Soshubetsu)

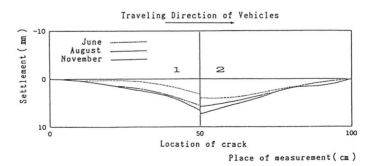

Figure 6 Settlement of Cracks

4. Test of repair work

4.1 High penetration asphalt method
In cold regions, high penetration asphalt is very effective to prevent reflective cracks by buffering low temperature stress.

1) Work location
Test repair work were made at Route 242 in Rikubetsu and at Route 391 in Koshimizu in 1989.

2) Assignment of work sections
Test work sections were divided into 4 sectors: I) straight asphalt of penetration grade 140, with the reconstructed subbase course, II) straight asphalt of penetration grade 140, with the overlay method, III) rubberized asphalt of penetration grade 160, with the overlay method, IV) straight asphalt of penetration grade 80, with the overlay method in control section.

3) Performance after 3 years
No effects of the overlay method using straight asphalt appeared with 20 cracks /km, as shown in table 2. However, the sections using high penetration asphalt appear to have restrained reflection cracking. There were no reflection cracks on Route 391 in Koshimizu after 3 years.

Table 2 Rikubetsu Town, Rikubetsu

Item	Block No. 1	2	3	4
Surface layer	Fine and gap-granded asphalt concrete			
Types of Asphalt	Straight	Straight	Improved	Straight
Length (m)	200	300	300	300
Penetration angles of asphalt	140	140	160	80~100
Repair conditions	Reconstruction	Overlay	Overlay	Overlay
Freezing index (°C/day)	Numbers of cracks (Number/Km)			
863 (1990)	0.0	10.3	0.0	20.0
808 (1991)	0.0	10.3	0.0	20.0
(1992)	0.0	10.3	0.0	20.0

4.2 Non-fiber sheet method
The conventional non-fiber sheet methods were conducted to investigate the effects for the comparison of treated and untreated sections with non-fiber sheets.

1) Work location
Sheet test works were conducted at Route 38 in Makubetu in 1985.

2) Work section
The test sections were divided into 4 sectors: I) control section of no crack treatment, II) cracks were filled with straight asphalt and placed with 33cm wide sheets on the road surface, III) the same pretreatment as in II), with the entire width of the sheet, IV) only 30 cm wide surrounding the cracks was agitated by hand-raking and recompacted down to 3 cm thick. After completing the treatment in II and IV, the overlay method was conducted with straight asphalt as shown in table 3 and figure 7.

Table 3 Makubetsu Town, Makubetsu

Year	Section 1	2	3	4
1986	△	△	△	○
1987	×	△	×	△
1988	×	△	×	△
1989	×	△	×	△
1990	×	△	×	△
1991	×	△	×	△

○ Some cracks
△ Traces of cracks
× No cracks

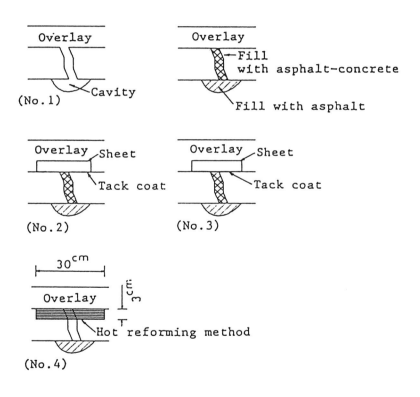

Figure 7 Details of Test Sections

5. Conclusions

The following is a summary of the investigation results and conclusions:
1) The appearance of reflection cracks corresponds to pavement thickness.
2) Low temperature crackings are largely dependent on the freezing index, and the performance is different by the regional surrounding conditions.
3) High penetration asphalt is effective in restraining reflection cracks.

6. References

1) H. Kubo and S. Kumagai: Site Experiments on Reflective Crack Retarding Measures in Asphalt Pavements, Reflective Cracking in Pavements, Assessment and Control, March 1989
2) H. Kubo, S. Kumagai and M. Oguri: On Temperature Stress Cracks in Asphalt, Civil Engineering Research Institute, Monthly Bulletin No. 328, July 1980
3) A. Moriyoshi, A. Hirama, K. Kawamura·and H. Sugawara: Study of on Surface Characteristics of Transverse Cracks in Asphalt Pavements, Proceedings of Hokkaido Branch Convention of Civil Engineering Institute, Feb. 1991.
4) T. Ishida, K. Kawamura and S. Nakagawa: On Tests of Compacted Concrete, The 35th Technology Research Presentation of Hokkaido Development Bureau, Feb. 1991

20 PROCESSES REDUCING REFLECTIVE CRACKING; SYNTHESIS OF LABORATORY TESTS

Ph. DUMAS and J. VECOVEN
Public Road Laboratory of Autun, France

Abstract

This paper is a synthesis of 210 skrinkage-bending tests performed on various kinds of processes. The processes tested can be devided into 4 sets :
- Paving fabric interlayer,
- Layer of high content filler and binder sand (named rich sand),
- Gritted thick binder membrane,
- Some specific processes (geogrids, slurry, etc.),
  - all sets covered with an AC. We notice that, for all processes, the AC overlays must be taken into account in the anticrack properties. Overall, for similar overlays, paving fabric interlayers act by increasing the starting time of the crack while rich sand interlayers act by reducing the speed of the crack. The question to be solved for these processes is the balance between the global behaviour on site and their maximum anticrack efficiency.

<u>Keywords</u> : Cracks, Fabric, Interlayer, Laboratory, Tests.

1 Introduction

A shrinkage bending test was settled in 1988 to access the efficiency of anticracks systems. This test makes a comparison between two reference processes, the behaviour of which is known on site, and the process to be tested. It simulates at once the pavement thermal contraction, and the heavy traffic sollicitations at a constant temperature (+5°C).

The sollicitation simulation is made in the following way (Fig. 1) :
- The thermal skrinkage of precrack semi-rigid pavement is simulated by the opening of the movable plate (average speed 0.6 mm/h).
- The action of a track axle is simulated by a cyclic loading at a frequency of 1 Hz monitored by the deflexion set to 0.2 mm.

Reflective Cracking in Pavements. Edited by J.M. Rigo, R. Degeimbre and L. Francken.
© 1993 RILEM. Published by E & FN Spon, 2–6 Boundary Row, London SE1 8HN. ISBN 0 419 18220 9.

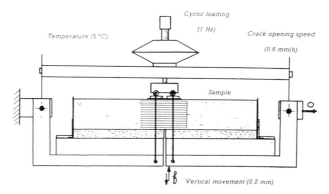

Fig. 1

The measure during the test consists in following the crack (initiation, propagation, breaking). The results of the test is then compared with two references :
* a 6 cm thick classical AC, considered as the less efficient.
* a 2 cm thick rich sand (using pure bitumen) overlayed with a 6cm thick classical AC, considered as the most efficient.

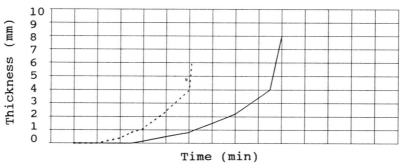

Fig. 2   --- AC 0/10 6 cm
        ___ Rich sand with 80/100 + AC

Figure 2 shows an example of a test giving the initiation time of the crack, its propagation speed in the first centimeter and the breaking time considered as the efficiency of the process tested, compared with the reference.

2 Appraisal after 4 years of testing

2.1 Studied processes
This laboratory test has permitted to compare 210 anticrack systems. Among them, there are :
- Paving fabric interlayer, 8 various kinds were tested with different binders (15) and rates of binder overlayed with various AC (formulation, thickness).
- High filler and binder content sand interlayers (named rich sand) covered with various AC overlays. 26 types of rich sand were tested linked with different overlays.
- Thick binder membrane, 10 cases, were tested with various natures and rates of binder protected with chipping or slurry and covered with various AC overlays.
- Geogrids, 2 cases were tested.
- Textil fibers thrown down on a binder layer and covered with chipping and AC overlay.
- Combination of different basic processes like rich sand and paving fabric.

2.2 Trends for the main anticrack systems
It emerges from these studies that the AC overlays characteristics (nature and content of binder, thickness, mortar content) are of a noticeable effect on the anticrack properties of the process, whatever the process is.

Tables I and II set a comparison between the three most used processes : Table I is a compilation of all the processes, whatever the AC overlays are, and table II compares only the processes having received the same AC overlay (6 cm thick of AC 0/10).

Table I shows the large scale of the process characteristics, assessed through the shrinkage bending test.

This large scale is explained by the number of parameters which varied between each test. Nevertheless, when we compare rich sand interlayers with paving fabric interlayers, we notice that both systems act differently:

- Paving fabrics are inclined to increase the crack initiation time.
- Rich sand tends to decrease the crack propagation speed in the first centimeter.

As for the binder membrane interlayers, their behaviour, assessed through this laboratory test, is halfway between rich sand and paving fabric but with a high breaking time. Nevertheless, the process tested in laboratory had a wide range of binder rate (1,2 to 3 kg/m^2).

Table II which compares strictly the processes (same AC overlay), confirms the tendancy shown in Table I, that

Table I. Comparison of each kind of processes whatever AC overlays are.

Average / mini-maxi	Initiation time (min)	Crack speed in the 1st cm (µm/min)	Breaking time (min)	Effectiveness
Paving fabric	180/85-350	90/35-180	525/360-700	1,15/0,8-1,5
Rich binder	160/25-385	75/45-155	495/385-660	1,1/0,85-1,45
Binder membr.	150/75-290	85/45-175	565/255-700	1,2/0,55-1,5

Table II. Comparison of each kind of anticrack processes overlayed with the same AC 0/10 (6 cm)

Average / mini-maxi	Initiation time (min)	Crack speed in the 1st cm (µm/min)	Breaking time (min)	Effectiveness
Paving fabric	170/85-280	90/40-180	520/360-700	1,10/0,8-1,5
Rich binder	95/25-175	65/45-85	535/385-660	1,15/0,85-1,45
Binder membr.	155/75-290	85/45-175	580/255-700	1,25/0,55-1,5

is to say paving fabrics delay the crack initiation and rich sand slows down the crack propagation. As binder membranes are concerned, their behaviour is more similar with paving fabrics. But, if we consider only binder rate of about 2,5kg/m², the results are different. Table III shows that, for such binder membrane, the behaviour is both that of paving fabric (high initiation time) and that of rich sand (low crack propagation speed).

Table III.

Average / mini-maxi	Initiation time (min)	Crack speed in the 1st cm (µm/min)	Breaking time (min)	Effectiveness
Binder membr. 2 to 3 kg/m²	155/75-290	70/45-105	630/560-700	1,35/1,2-1,5

2.3 Composition factor effects

2.3.1 Rich sand interlayer

The study of the effect of the mortar content (in a constant ratio of 60 % filler + 40 % binder) layed in 2

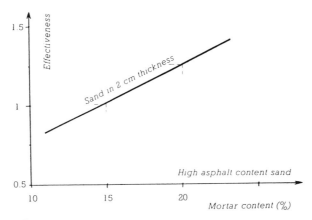

Fig. 3 - Effect of mortar content in a rich sand interlayer process

cm thick and covered with a constant AC overlay, shows (figure 3) that a 5 % drop of the mortar content leads to a 20 % comparative loss of effectiveness.

Furthermore the effect of the thickness of the rich sand is assessed and shows that (for that study with one type of filler, one type of binder, and the same grading curve) one centimeter of a rich sand interlayer is equivalent to 5 % of mortar, for example a 2 cm thick rich sand containing 22 % of mortar is equivalent to a 3 cm thick rich sand containing 17 % of mortar.

2.3.2 Paving fabric interlayer
Each fabric has its own minimum impregnated binder content under which the AC overlay fails to complete the bond. This critical content is estimated in laboratory under constant conditions. Nevertheless this minimum binder content has to be adapted to the site taking into account the texture of the support. All the shrinkage-bending tests performed on paving fabric processes show the predominance of the binder effect. The fabric must be considered as a tank which should be comoletely filled (saturated). The anticrack characteris- tic is given by the binder (nature and quantity. The conti- nuity of the interlayer and its preservation in time is realised by the fabric. Table IV illustrates the behaviour of the fabric + binder system, measured in the same conditions (constant AC overlay) in laboratory :

Table IV - Effect of nature of binder and characteristic of the fabric + binder system (r1 < r2 < r3) on the anticrack effect.

		Binder	
		Pure bitumen	Modified bitumen
F A B R I C	F1 with r1%binder	+ 30 %	+ 14 %
	F2 with r2%binder	↓	↓
	F3 with r3%binder	↓	↓

We note that a fabric taking up higher binder contents leads to a higher effectiveness. However, for these kinds of fabrics the impregnation must be total and not entail a compressible interlayer. The paving fabric interlayers are never broken after the complete cracking of the AC overlay.

Thus they make up a protection against the ingress of water in the pavement. This protection will depend on the type of fabric and its correct impregnation by the binder. Permeability measures performed on cores sampled on site, show that the flow rate of water through the system paving fabric + binder and AC can be reduced by 2 to 10. Other tests, realised on site at the place of crack, show that the system is watertight (under the condition of the test).

2.3.3 Binder membrane interlayers
The effectiveness of these processes is due to the nature and quantity of binder as for the paving fabric. The limitation of their efficiency depends on the system used to protect the binder membrane before laying the AC overlay. The gritting with chipping of the membrane tends to decrease the anticrack effectiveness of the process by punching the membrane. Other systems like slurry on the membrane allow the laying of the AC without disturbing the membrane. Such process has a high anticrack effectiveness. Laboratory tests also show that binder membrane is very sensitive to a decrease of binder rate (see table V).

2.4 The compromiser
Mainly for rich sand interlayers and paving fabric interlayers, a compromise has to be made. The principle of these interlayers is to diffuse the stress concentrated at the head of the crack. This can be solved by using soft bitumen at a high content. But such

Table V - Effect of binder rate on the characteristic of a binder membrane interlayer.

		Modified binder membrane covered with 6/10 chipping		
		Initiation time (min)	crack speed in the 1st cm µm/min	Effectiveness
Binder rate	2kg/m²	110	60	1,2
	1,2kg/m²	100	175	0,55

interlayers will induce other problems :
-For rich sand interlayer a balance should be made between the rutting characteristics and the anticrack effectiveness. The use of modified binder (high content of polymer and soft bitumen base) allow to solve the problem. The way consisting in using a too rigid AC overlay is very harmful.
-For paving fabric interlayer a balance should be made between the ability to be layed on the site and the anticrack effectiveness. The rate of binder applied on the site should allow the laying of the AC overlay. The difficulties met on the site can be due to an excessive rate of binder taking into account the texture of the support or to the use of a too soft binder because of a laying in hot period. To choose the optimum composition of the paving fabric interlayer, one must take into account :
- The period of word, the climatic conditions (in particular the extreme temperature along the year) to choose the type of binder (pure bitumen, soft or hard, modified bitumen).
- The texture of the support and the period of work to go at the highest binder rate compatible with the site.

2.5 Site findings

The correlation between the shrinkage-bending test and the actual behaviour on site is hard to assess since the process really used is often different from the one tested in laboratory and most of the time because of the lack of reference site. Nevertheless some lessons can be drawn from few sites :
- A process using the same kind of anticrack interlayer but with various AC overlays has been used on motorways. The classification obtained in laboratory is confirmed by the site and has led the contractor to avoid the use of too much thin layer of AC.
- The effect of vertical novement at the edge of crack is clearly shown on some sites (cement concrete

pavement) where differential vertical movements higher than 0.4 mm lead quickly to a crack more or less taking into account the AC thickness. Such findings show the limitation of this laboratory test which does nos include mode 2 sollicitation of the crack.

3 Conclusions

The shrinkage-bending test allows to classify the processes, but the relative position of each process inside the classification is different from the reality : differences of effectiveness between each system measured in laboratory are less videspread than on site. Furthermore, so as to criticize objectively the test we should completely control the site conditions. The following step now is to realize a testing machine capable of :

- Sollicitation of the crack in mode 1 ou 2.
- In mode 2, simulating a coupling (partial or total) of the beam near the crack.
- Simulation of the trafic at frequency from 0.5 to 20 Hz.
- Crack opening speed from 0.1 to 25 mm/h.
- Maintaining a temperature from -20°C to 50°C.

4 References

J.H. VECOVEN (1989), Méthode d'étude de systèmes limitant la remontée de fissures dans les chaussées. Conference on reflective crackings in pavements (Liège).
G. COLOMBIER, G. MOREL, J. VECOVEN (1990), Lutte antifissure, R.G.R.A. n°679 (France).
J.H. VECOVEN (1990), Crack reflection treatment testing machine for hydraulic treated pavement, R.G.R.A. n°680 (France).

21 PAVING FABRIC SPECIFICATION FOR ASPHALT OVERLAYS AND SPRAYED RESEAL APPLICATIONS

W. ALEXANDER
Geofabrics Australasia Pty Ltd, Melbourne, Australia
R. McKENNA
Geosynthetic Testing Services, Albury, Australia

Abstract

It is commonly perceived that paving fabrics perform three main functions in asphalt overlay and bituminous sprayed reseal applications ie reinforcement, stress absorption and waterproofing. This investigation examines a range of material specifications currently in use, and analyses their relevance to a paving fabrics functions.

A paving fabric specification, to its relevant required functions, is proposed incorporating the data gained from the investigation for A/C overlays and sprayed reseal applications.

Keywords: Reflective cracking, Reinforcement, Stress absorbing, Waterproofing, Paving fabric, Ply adhesion strength, Asphalt Concrete overlay (A/C), Bituminous sprayed reseal.

Introduction

The use of geotextiles as paving fabrics is widespread in Australia and has been increasing rapidly over a number of years.

There are a number of paving fabrics utilised in Australia for A/C overlay and sprayed reseal applications. Four of the most commonly used fabrics were chosen for this investigation:

Fabric A: nonwoven polyester needlepunched @ 140 g/m²
Fabric B: nonwoven polyester needlepunched @ 160 g/m²
Fabric C: nonwoven needlepunched/thermally bonded one side polypropylene @ 155 g/m²
Fabric D: nonwoven needlepunched polypropylene @ 140 g/m²

Three stages of testing were undertaken:-

Stage one: consisted of establishing the properties of each of the four fabrics, both before and after the application of a bitumen tack coat (class 170 straight bitumen) to establish the effects of bitumen on the fabrics.

Stage two: involved examining the waterproofing effectiveness of paving fabrics, utilising a falling hydraulic head permeability test. The adequacy of the bond between the wearing course and overlay was established by means of a ply adhesion test.

Stage three: involved examining the longterm performance in the field of the paving fabric which best met the specification based on the functions for which it was intended.

Reflective Cracking in Pavements. Edited by J.M. Rigo, R. Degeimbre and L. Francken.
© 1993 RILEM. Published by E & FN Spon, 2–6 Boundary Row, London SE1 8HN. ISBN 0 419 18220 9.

Stage one Testing
The testing initially was aimed at developing an understanding of the fabric's properties and behaviour before and after the applications of bitumen and identifying those properties that were relevant to a paving fabrics performance. Refer to Figure 1 & Table 1. Summary of Testing Results.

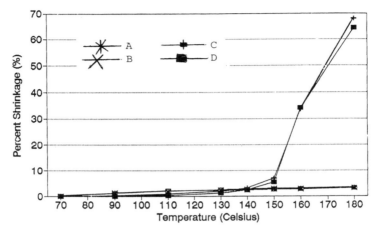

Figure 1. Paving fabric shrinkage test

The percent area change was determined to establish whether the application of bitumen at elevated temperatures (typically 160° - 180°C) leads to an adverse reduction in area of the fabric, thus affecting its ability to perform a waterproofing function. As the degree of bitumen absorption and retention effects the tack coat application rate, the bitumen absorption/retention of the fabric is critical to the performance of the bitumen/fabric interlayer.

The trapezoidal tear test was used to verify that for a paving fabric to act as a waterproofing membrane it must remain flexible and resist tearing, thus allowing cracks underneath the paving fabric to expand and contract without propagating through.

The Geometric Mean Strength Rating is used in Australia to determine the overall robustness of a geotextile and therefore its ability to withstand damage during construction and installation. The G-rating is calculated from the CBR Burst test and the Drop Cone Puncture Resistance test results.

Discussion
It can be concluded that generally the bitumen has a positive affect on a paving fabrics strength, as this was the case for the majority of the tests performed where the bitumen enhanced the properties of the fabrics. There were however cases where the properties of the fabric did change after the application of bitumen. Specifically samples C & D showed evidence of creep at low elongation similar to which it would be subjected to in a road pavement. Also at temperatures greater than 150°C samples C & D undergo considerable

changes in area and shrink excessively, therefore for sprayed reseal applications where a spray coat is applied between 160 - 180°C these fabrics would not meet the functions required. The thickness of a paving fabric has a direct influence on the amount of bitumen it needs to absorb to become a waterproofing membrane and therefore also has an influence on the fabrics ability to remain flexible. This flexibility is essential in assisting the fabric to resist crack propagation and absorb the stresses in the pavement caused by cracking.

Table 1: SUMMARY OF TESTING - MEAN RESULTS

TEST METHOD		A		B		C		D	
		Pre	Aft	Pre	Aft	Pre	Aft	Pre	Aft
MASS (g/m²)		140.4	-	166.9	-	143.7	-	164.2	-
THICKNESS (mm) 2kPa		1.40	-	1.82	-	1.17	-	1.54	-
A.C. (%)~		-	1.63	-	2.83	-	9.36	-	9.45
BIT. RET # (l/m²)		-	1.86	-	2.38	-	2.10	-	1.91
G.T. (N)	-MD	650	-	725	-	717	-	553	-
	-CMD	554	-	626	-	511	-	592	-
ELONG (%)	-MD	67	-	63	-	46.8	-	72.9	-
	-CM	72	-	74	-	82.8	-	70.3	-
WST (kN/m)	-MD	8.7	11.7	10.9	13.3	9.7	11.0	9.1	11.7
	-CMD	9.0	13.3	10.7	13.2	9.0	12.2	8.1	10.5
ELON (%)	-MD	59.1	76	47	88	38.1	65	76.6	125
	-CMD	48.1	73	54	72	45.0	76	43.1	87
T.T (N)	-MD	323	427*	333	514*	359	441*	313	391*
	-CMD	274	403	297	483	261	396	254	359
CBR (N)		1854	1898*	2012	2225*	1598	2285*	1370	1546*
STRAIN (%)		33	47	31	50	37	52	41	62
D.C.	d500	31	-	26	-	25	-	31	-
	h50	990	-	1330	-	1410	-	1010	-
G RATING		1354	-	1636	-	1501	-	1176	-

Key:

Pre/Aft   Before or After bitumen application to fabric
*   Tack coat application rate 0.81 l/m²
A.C.   Area Change Test - Task Force 25, Method 8 [1]

	Area Change % = after application of Class 170 bitumen
BIT.RET	Bitumen Retention Test - Task Force 25, Method 8 [1]
G.T.	Grab Tensile Test - ASTM D4632
WST	Wide Strip Tensile Test - AS 3706.2
ELON	Elongation (Extension) of the sample as a percentage AS 3706.2
T.T.	Trapezoidal Tear Test - AS 3706.3
CBR	California Bearing Ratio Test - AS 3706.4
D.C.	Drop Cone Test - AS 3706.5
MD	Machine Direction (Fabric tested in this direction)
CMD	Cross Machine Direction (Fabric tested in this direction)
G RATING	Geometric Mean Strength Rating - Qld D.O.T. [2]

Stage Two Testing

For a paving fabric to function as intended two key performance factors are important:

1   the homogeneity of the bitumen/fabric interlayer determines the waterproofing effectiveness of the pavement
2   the homogeneity of the bond between pavement layers determines the bearing capacity and stability of the pavement

To evaluate the performance of the trial fabrics, testing was conducted to examine the:

1   waterproofing ability of the paving fabric interlayer in 40mm A/C overlay and sprayed reseal applications
2   bond strength of the paving fabric to A/C layers

Note: the bond strength of a paving fabric does not historically cause debonding problems in reseals due to the application of a spray seal prior to stone laying, therefore the bond strength was not tested.

Testing Methodology

A test pavement was constructed incorporating each of the four trial paving fabrics and a control section without fabric. Existing was a 40mm A/C pavement containing fine cracks and random larger longitudinal cracks of 2 - 3mm. The sub-grade was a sand clay which was poorly compacted, with poor drainage. A 40mm A/C overlay and a 10mm single coat spray reseal were constructed side by side from which 100mm core samples were taken for laboratory testing.
   The trial conditions were:

   Bitumen : class 170 (straight / no cutters)
   Bitumen Temperature : 160°C (@ spray tank)
   Tack coat application rate : 0.8 l/m^2
   Spray coat applicaiton rate : 1.6 l/m^2
   Crushed rock aggregate : 10mm diameter
   Asphalt concrete : standard
   Asphalt aggregate : 10mm nominal fraction

Testing initially involved the evaluation of core samples for waterproofing by subjecting the cores to a 100mm water head for a period of 24 hours and measuring head loss against time. Where head loss occurred the paving fabric layer was later investigated to analyse the structural integrity of the fabric.

After this core permeability test, an evaluation of the bond strength between the fabric and the two adjacent layers in the A/C overlay was performed. This test involved laying and supporting the core sample on its side and shearing at the fabric/pavement interface by means of a wedge shaped steel head fitted to a constant rate-of-extension testing machine, ie ply ahesion strength test.

Discussion
Table 2: SUMMARY OF STAGE TWO TESTING

SAMPLE	WATERHEAD @ 24hr mm RTA NSW T168	FAILURE STRENGTH NEWTONS PLY ADHESION TEST	DISPLACEMENT OF FAILURE mm PLY ADHESION TEST
A	100	1595.0	6.57
B	90	735.6	3.75
C	80	1686.0	3.94
D	60	1581.0	6.29
CONTROL	28	1646.0	6.97

From the results it can be seen samples A, C, & D were similar in strength to the core sample of the A/C on A/C pavement structure. This shows that the introduction of a paving fabric in A/C overlays does not significantly reduce the bond strength between layers where sufficient bitumen saturation enables effective bonding through all layers. However sample B showed a marked decrease in bond strength which can be attributed to the incomplete saturation of bitumen into the paving fabric. Further from the test results, samples B & C were determined to be a brittle failure as once failure had occurred, the sample delaminated completely. Samples A & D behaved similarly to the control sample in that even after failure occurred, some strength was retained and that separation was only achieved through the continuence of force.

The waterproofing effectiveness of each trial sample was greater than the control sample without fabric, however only sample A achieved total waterproofing after 24 hours of testing at an initial head height of 100mm. After examination of each sample the following causes of failure of the paving fabric to provide an effective waterproofing membrane were established:-

      Sample B - incomplete bitumen saturation
      Sample C - rigid fabric structure
      Sample D - homogeneity of fabric structure altered

Paving fabrics specification based on laboratory experiments

Based on the results of the Stage 1 and Stage 2 laboratory testing Sample A was chosen as the paving fabric which best performed the necessary funtional requirements.

The dense fibre bonding of Sample A improved bitumen retention through decreasing the voids volume, whilst maintaining sufficient thickness to reduce compressibility which has the effect of inducing flexural movement within the pavement. Tearing strength was included in the specification, as mentioned previously, this property reflects a paving fabrics ability to resist crack propagation. A minimum G rating was determined to enable the paving fabric to resist damage during installation. Minimum elongation properties were evaluated to enable the paving fabric to remain flexible within the pavement thus inhibiting crack reflection.

Paving fabric Specification Table

Nonwoven, needlepunched, polyester, continuous filament geotextile.

Thickness : 1.4 / 1.5mm
Bitumen Retention : >1.80 l/m²
Melting Point : >250°C
Trapezoidal Tear Strength : >250N for both directions
W/Strip tensile strength : >8.0 kN/m for both directions
W/Strip elongation : >45% for both directions
CBR Burst Strength : >1700N
CBR Strain % : >30%
Drop Cone Puncture Resistance : d500 <31mm
　　　　　　　　　　　　　　　　　 h50 >1009
G Rating : >1309

Stage three – Field Trials

Trial No 1

Carried out on Wyndham Street, Shepparton, Vic. Australia – Resealed 24 Feb 1990

The condition of the pavement surface pre-resealing was badly cracked. The central 4.0m of the pavement was resurfaced using Sample A beneath a 12mm size chip. Table 3 below gives a summary of the relevant technical data. The remainder of the pavement width was resurfaced using a rubberised bitumen spray seal treatment, without paving fabric.

Table 3

Pavement surface condition	Existing asphalt. Pavement surface significantly cracked, block or 'crocodile' cracking
Date of reseal	24th February 1990
Weather condition	Low Temp. 20°C, High Temp 34°C. Overcast, fine conditions
Paving Fabric	Sample A 140g/m²
Tack Coat (L/m²)	0.7 L/m² Class 170 Bitumen 2% cutters
Seal Spray (L/m²)	1.28 L/m² Class 170 Bitumen 2% cutters
Stone Chip (mm)	12mm, 100m²/m³

Fig. 2 Transverse cracking eliminated in areas with Paving Fabric

Fig. 3 Block cracking deterred at interface with PF area

Observations

The area incorporating Sample A was much less cracked than the remainder of the pavement. Reflective cracks that have propagated through the non-fabric area are clearly terminated at their point of interception with the boundary of the areas with Sample A. Refer figure 2 & 3.

The surface of the pavement is in good condition with no stripping of the pavement evident, even though the road is a main highway subject to heavy truck traffic. In the most trafficked areas the stone chip surface has been pushed into the surface of the pavement to give a smooth, flush surface.

Road Coring

Road coring was carried out at a number of different sites, both with and without paving fabric. Tests were carried out to determine waterproofing effects. Refer to Table 4 for results.

Table 4

SAMPLE	DESCRIPTION	PERMEABILITY TEST RTA NSW METHOD T168
1	Nil PF, cracked	Flow 123ml/20 sec
2	PF, cracked surface	No leak
3	PF, longitudinal crack	No leak
4	PF, nil crack	No leak

Trial No 2

To identify if paving fabrics do provide reinforcement two stages of longterm testing and field trials are proposed. Stage one involved determining the 5% secant modulus of the fabrics tested and, as proposed by R.M. Koerner, relating this value to pavement based design on the fabric effectiveness factor (FEF[3]).

Table 5. 5% Secant Modulus Values : Wide-Strip Tensile Test (kN/m) AS 3706.2

SAMPLE	PRIOR TO BITUMEN APPLICATION	AFTER BITUMEN APPLICATION
A	6.0	38.8
B	14.5	32.8
C	17.4	28.6
D	13.0	21.3

Long term flexural fatigue testing, waterproofing testing and ply adhesion tests are being conducted on A/C overlay trial pavements to validate the paving fabric specification as proposed. The relationship between a paving fabrics 5% secant modulus and its cycles to failure in a flexural fatigue test will be statistically validated prior to monitoring the longterm in-situ performance of the paving fabric.

Preliminary waterproofing and bond strength tests indicate that for the paving fabric to function as intended in A/C overlay applications, the following criteria must be satisfied:-

1 cracks greater than 5mm are filled
2 existing surface condition is considered when determining tack coat rate
3 effective fabric rolling occurs to ensure complete bitumen saturation of the fabric
4 effective overlay rolling occurs to ensure an adequate bond to the paving fabric interlayer, ie > 10 passes with multi-tyred roller

Acknowledgements

The authors wish to gratefully acknowledge the contributions of:

D. Tice, Swinburne Ltd - School of Civil Engineering
J. Spears, Swinburne Ltd - School of Civil Engineering
R. Kang, Geosynthetic Testing Services
B. Grant, Geosynthetic Testing Services
R Cheetham, Vic Roads

Bibliography

1. AASHTO - AGC - ARTBA Joint Committee, (1990), Task Force 25 Guide Specifications - Test Procedures for Geotextiles Washington D.C., U.S.A., American Association of State Highway Transportation Officials (AASHTO), August
2. Austroads, (1990), Guide to Geotextiles - Technical Report, Sydney, Austroads, January
3. Koerner, R.M., (1990) Designing with Geosynthetics, New Jersey Prentice Hall, Second Edition

22 NEW TESTING METHOD TO CHARACTERIZE MODE I FRACTURING OF ASPHALT AGGREGATE MIXTURES

E.K. TSCHEGG
Technical University of Vienna, Austria
S.E. STANZL-TSCHEGG
University of Agriculture, Vienna, Austria
J. LITZKA
Technical University of Vienna, Austria

Abstract
The wedge splitting method for fracture tests on asphalt aggregate mixtures and other bituminous materials is outlined. Cubic as well as cylindrical specimens may be tested with this method. The load-displacement curve is measured during stable crack growth and provides all information to completely characterize the fracture behavior of the material. Asphalt aggregate mixtures with crushed gravel and with natural gravel were tested with this method and the results are discussed.
Keywords: Mode I Fracture Behavior, Fracture Energy Concept, Splitting Test, Low Temperature Fracture Behavior.

1 Introduction

Knowledge of the physical processes and concepts to characterize the fracture processes of bituminous materials, like for example asphalt aggregate mixtures, are important for civil engineers. Such knowledge facilitates decisions on constructive and material-technological steps to prevent the formation of cracks (like for example reflection cracks) in road pavements. It is also of interest for failure analysis purposes.
Various fracture mechanical concepts and testing procedures have been developed to characterize different materials, like metals, ceramics, polymers, concrete and other composite materials in the past. It is difficult however, to apply most of the so far developed fracture mechanics concepts and testing methods to asphalt aggregate mixtures, as these materials have an anisotropic and non homogeneous structure and in addition change their mechanical properties drastically with temperature changes: They are viscoelastic at room temperature and brittle at low temperatures. The fracture energy concept, which has been used for concrete (for example by Hillerborg, 1983, and Bazant, 1985, for FE modelling) and other materials, seems to be very useful also for asphalt aggregate mixtures. It completely characterizes the system and does not depend on details of the stress and strain distribution close to the crack front. Therefore, the concept is appropriate for FE-modelling of asphalt aggregate mixtures, in particular for crack propagation studies (Haas and Ponniah, 1989) and for an easier experimental determination of fracture properties.
It is the aim of this paper to present a new, simple and useful method to characterize the fracture properties of asphalt aggregate mixtures in a temperature range which is of interest for the engineer, and to discuss first results, which were obtained with the help of the fracture energy concept.

Reflective Cracking in Pavements. Edited by J.M. Rigo, R. Degeimbre and L. Francken.
© 1993 RILEM. Published by E & FN Spon, 2–6 Boundary Row, London SE1 8HN. ISBN 0 419 18220 9.

2 Principle of the testing method

The principle of the testing method is shown schematically in Fig. 1a. It is the measuring procedure according to Tschegg (1986, 1991) which is applied to asphalt aggregate mixture materials in this work. Cubic (or prismatic) specimens are placed on a narrow linear support in a compression testing machine. The specimen has a rectangular groove with a starter notch at

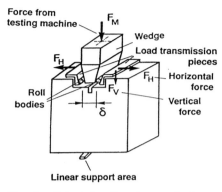

Fig. 1a. Principle of testing method according to Tschegg 1986.

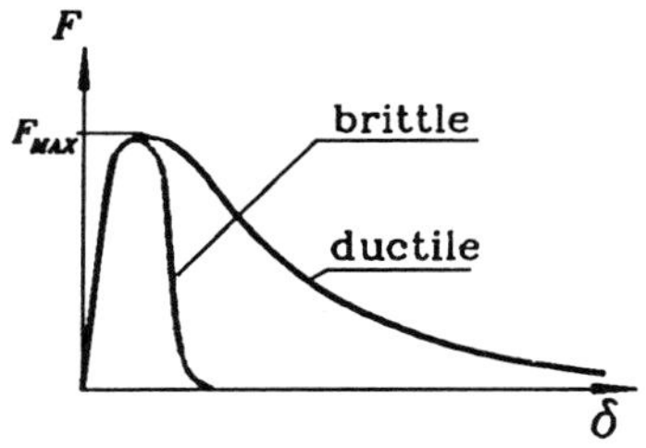

Fig. 1b. Schematic load-deformation curve.

the bottom of the groove. Two load transmission pieces are placed in the groove. A wedge split is formed by inserting a wedge. Wedge, starter notch and linear support area are in the same vertical plane. The wedge transmits a force F_M from the testing machine to the specimen. The slender wedge exerts a large horizontal force component F_H and a small vertical force component F_V on the specimen. The horizontal component splits the specimen similar as in a bending test. The force F_M is determined with a load cell in the testing machine. Knowing the wedge angle (approximately 5 - 10°), one can calculate the force F_H in horizontal direction which causes splitting of the specimen. The vertical force component F_V is small and does not influence the result if the wedge angle is small enough. This has been verified by measurements on concrete (Tschegg and Stanzl-Tschegg 1992). Therefore it is assumed that the same is true for specimens of asphalt aggregate mixture.

Splitting must take place during stable crack propagation until complete separation of the specimen. This is possible with a stiff loading system only (wedge-loading system - testing machine). The wedge loading part is extremely stiff, i.e. it stores little deformation energy. Therefore, such experiments can be performed with usual mechanical spindle driven machines or hydraulic machines under strain or stroke control together with the described loading device.

In order to reduce friction betweeen wedge and the load transmission pieces, roll bodies (e.g. roller bearings) have been introduced in between (Tschegg, 1991). Measurements show (Tschegg and Stanzl-Tschegg, 1992) that the influence of friction is small enough to be neglected.

Induction displacement gauges are mounted on both ends of the starter notch in order to measure the displacement δ on the line of force application (crack opening displacement). For more experimental details see reference (Tschegg, 1991).

Force F_M and displacement δ are registered with an X-Y1, Y2 recorder or an electronic data logger; from this, the load-displacement curve (F_H-δ -curve) is obtained. This load-displacement curve characterizes the fracture behavior of the tested material. In Fig. 1b, such a curve is shown shematically for a "brittle" and a "ductile" asphalt aggregate mixture. The area under the F_H-δ curve corresponds to the energy which is necessary to separate the specimen. Dividing this energy by the fracture area (plain projection of the area was used in this work), yields the specific fracture energy G_F. This value characterizes the crack propagation resistance of the tested material. The G_F value is a material characteristic and does not depend on specimen shape and size. Care must be taken, however that the specimen is not too small, in order to avoid an influence by the "size-effect" (Brameshuber and Hillsdorf, 1990, Bazant and Pfeiffer, 1987) The specimen size as used in this work is larger than such a minimum specimen size.

Notch-tensile strength or notch-bending tensile strength may be calculated from the maximum load K_{max} in the load-displacement curve (Fig.1b).

3 Specimen shape

Different specimen shapes may be used with the above described method, as shown schematically in Fig. 2. All specimens have a rectangular groove with a starter notch. Usually cubic or rectangular shaped specimens (Fig. 2a) are used as generally produced in the laboratory. The rectangular groove and the starter notch may be produced by inserting an appropriate mold into the cubic form during casting or by cutting with a stone saw.

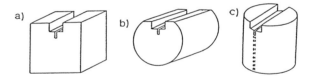

Fig. 2. Specimen shapes according to Tschegg (1986).

Cores from a pavement are appropriate as well, as shown in Figs. 2b,c. Groove and starter notch need not be worked too accurately. Special care must be taken, however that the vertical support areas of the force transmission pieces are parallel to the wedge. If specimens are used as shown in Fig. 2c it is advantageous to continue the starter notch on the cylinder surface on both sides (drawn as broken line), to avoid failure by side-tracked cracks.

Cubic and cylindrical specimens under discussion are handy and insensitive against damage during transport or installation into the testing machine; no lifting arrangements are necessary. In comparison with three point bending specimens, cubic or cylindrical specimens show considerable advantage as to the ratio of specimen volume (material quantity) to fracture surface area. For more details on specimen shape and size see Tschegg (1991).

4 Material and testing procedure

The fracture properties of bituminous base material (i) with crushed gravel (CG)and with (ii) natural gravel content (NG) has been determined with the above described method. The mixture quality of the specimens which were produced with the hot-mixing method, are listed in Table 1. (This asphalt aggregate mixture satisfies the conditions of the specification for bitumen road basis RVS 8.05.14, 1991). Specimen shape and size are shown in Fig. 3. The specimens were cut from 150 mm thick plates (which were compacted by a vibration roller) with a stone saw. The grooves were produced with inserts in the formwork and the starter notches were cut in with a 3 mm thick saw.

Testing was performed with a mechanical compression testing machine with a load capacity of 20 tons. Unstable crack growth was not observed in any of the tests. The cross-head velocity was 1.3 mm/ min in all tests and is approximately equal to the loading velocity as recommended for concrete fracture tests (RILEM Tech. Comm.50-FMC, 1986). Before testing, the specimens were stored approximately 14 hours in a cooling chamber with a control accuracy of +/- 1° C. Testing temperatures were 8, 3, -0.5, -10 and -21° C. As the whole tesing procedure (mounting plus testing) did not last more than 5 to 10 minutes, the specimens did not warm up during the tests, though no cooling equipment was provided. Three identical specimens of each material and for each testing temperature were tested in order to have enough data for statistical evaluation.

Table 1: Mix design of the test material
　　　　　CG ...crushed gravel, NG ...natural gravel

	CG	NG
aggregate gradation	%	%
< 0.09 mm (filler)	10.6	7.7
0.09 - 2 mm (sand)	23.7	32.0
2 -11.2 mm (sheps resp. gravel)	65.7	60.3
> 11 mm (coarse particel)	15.6	21.0
binder content		
100 pen-bitumen	5.0	4.7
properties		
Marshall stability (kN)	7.2	9.0
flow -value (mm)	5.2	3.0
voids content (vol.%)	3.0	5.6

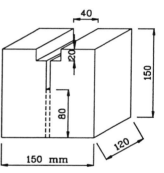

Fig. 3. Shape and dimensions of the specimens

5 Results and discussion

Characteristic load-deformation diagrams from splitting tests on different types of materials and at characteristic temperatures are shown in Fig. 4 (Note that all diagrams are drawn in the same scale). The six diagrams show clearly that the deformation and fracture behavior of the two materials is changing with temperature. The slope of the curves is determined by the elastic and plastic behavior; steep curve and linear increase point to a high modulus of elasticity and few plastic deformation. Pronounced deviation from the linear behavior and deflection of the curves at low F_H values respectively, point to low strength and high deformation of the material.

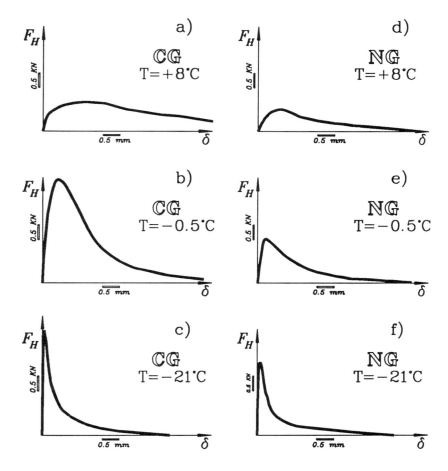

Fig. 4. Typical load-displacement curves (splitting force F_H versus crack opening δ). CG - crushed gravel, NG - natural gravel. (Note that all diagrams are drawn in the same scale).

The post peak behavior characterizes the fracture process. If the curve decreases quickly, as shown in Figs. 4c and f for T = -21°C, few energy is consumed for crack propagation, and "brittle" fracture is observed. Contrary to this, shallow curves are found at + 8°C which point to a "ductile" fracture. The force F_H approaches zero only at high δ values (crack openings). This behavior is determined mainly by three phenomena, which are: i) formation of a process zone ("plastic zone"), ii) bridging effects of the aggregates, which transfer the forces from one to the other fracture surface, though a crack is present, and iii) the deformation properties of the matrix and the tendency of intergranular separation. Fig. 4 shows in addition qualitatively that the curve type of the CG material is similar to that of the NG material, though the F_{max} and the G_F values are higher.

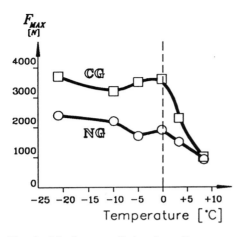
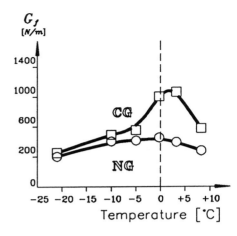

Fig. 5. Maximum splitting force F_{max} versus test temperature.
CG - crushed gravel
NG - natural gravel

Fig. 6. Specific fracture energy G_F versus test temperature
CG - crushed gravel
NG - natural gravel

It has been tried in other works to obtain the "strain-softening" behavior during a tensile test with the aid of an FE-program. Such a material characterization has been successfully performed for concrete already (Roelfstra, 1986). The strain-softening curve of asphalt aggregate mixture will be determined in a similar way with a bilinear approach by FE modelling, in order to make the numerical calculation easier. The bilinear curve characterizes the material and the fracture behavior of road constructions can be determined for all crack geometries during mode 1 crack propagation. In order to characterize the material fracture properties alone (without characterization of construction parts), it is enough to know the load-displacement curve (Fig.4). Results are reported elsewhere.

The F_{max} values are plotted in Fig. 5 and the G_F values in Fig. 6 versus the test temperature. These values were obtained from the load - deformation curves. The mean standard deviation (resulting from three measuring results) is approximately 12% for the F_{max} values and 7% for the G_F values. F_{max} does not change in the same way as the fracture energy, as shown in Fig.5. At temperatures between -21°C and +8°C, CG has higher F_{max} values. The difference between CG and NG decreases with increasing temperature and becomes almost zero for +8°C. As the F_{max} value may be considered as being proportioal to the tensile strength, this result is consistent with the prevailing observations.

The G_F values are between 200 and 1100N/m. The dependence of the specific fracture energy G_F on the temperature as plotted in Fig. 6 shows that CG is superior through the whole tested temperature range. At -21°C, CG and NG are not too different; at +8°C, the GF value of CG is almost twice of that of NG. Between -5°C and +5°C, the G_F values of CG exhibit a pronounced maximum, which is 2.5 times higher than that of NG. The fractographic observations may serve as an explanation for this result. In the CG material, the crack propagates around the edged aggregates in 70 % of the cases and through the aggregates in 30 %. The fracture surface looks more "rough" than at +8°C and -21°C. In the NG material, 100 % of the cracks propagate within the matrix and do not pass the spherical grains. The fracture surface is less rough than the fracture surfaces of the CG material at the same temperature. From this it may be concluded that the good mechanical properties of the binder together with the aggregate shape lead to especially properties, so that much energy is needed for the fracturing process.

The results of this work show that the fracture properties of asphalt aggregate mixture cannot be described adequately by the tensile strength or another similar value alone. This may be well demonstrated with Fig. 1b. In this figure, the load-displacement curve is shown for same F_{max} values and different fracture behavior (brittle and ductile). If the F_{max} values were used to characterize the fracture behavior, it could be concluded that the two materials are similar. This is only true for the crack initiation process, which takes place at identical F_{max} for both materials. F_{max} however, does not give any information about crack propagation and about the whole fracturing process of the material under discussion.

Complete characterization of the fracture behavior of asphalt aggregate mixtures is given by the strain-softening diagram (respectively load-deformation curve). On the other hand, the fracture behavior may as well be characterized by the use of <u>several</u> values, like specific fracture energy, tensile strength and modulus of elasticity. More experimental and theoretical studies are necessary to characterize the fracture properties by a single material characteristic value; this is the aim of a future study.

6 Conclusion

The present work has shown the following:

1. The described method is appropriate to characterize the fracture behavior of asphalt aggregate mixture and other bituminous materials. Specimens are notched cubes (prisms) or cylinders (cores), which may be easily and inexpensively produced. The testing procedure is simple and can be performed in a straight forward and inexpensive way.

2. The testing sequence is: First, determination of the load-deformation curve. Then, from this the maximum load (and from this the notch tensile strength) and the fracture energy (G_F-value) are obtained. The strain-softening curve as determined from the load-deformation curve may be considered as being characteristic for fracture mechanical behavior of the material. Thus it serves as an important base for numerical calculations.

3. The standard deviation of the measured values is approximately 7 % for the specific fracture energy and about 12 % for the maximum load.

4. The fracture behavior of asphalt aggregate mixture with crushed gravel (CG) and of a mixture with natural gravel (NG) has been determined. The measured load-deformation curves show large differences in shape and size for CG and NG in the tested temperature range of -21°C to +8°C.

5. The specific fracture energy and the F_{max} values of CG are higher than for NG in the whole tested temperature range. CG displays a pronouced maximum of the specific fracture energy, which is approximately two times higher than for NG in the temperature range between -5°C and +5°C.

6. The fracture energy concept is appropriate to characterize the fracture behavior of asphalt aggregate mixture and other bitumous materials. The recorded load-displacement curves contain all information needed for a complete material characterization. The tensile strength alone is not appropriate to describe the fracture behavior and must be

supplemented by the specific fracture energy which characterizes the material resistance against crack propagation.

7 References

Bazant, P.Z. and Pfeiffer, A. (1987), Determination of fracture energy from size effect and brittleness number. ACI Mat. Journal, Nov.-Dec., 463-480.

Bazant, P.Z. (1985), Mechanics of fracture and progressive cracking in concrete structures, in: Fracture mechanics of concrete structural application and numerical calculation (eds. C. Shi and A.D. Tonmaso), Martinus Nijhoff Publ. Dordrecht/ Boston/ Lancaster

Brameshuber, W. and Hilsdorf, H.K (1990) Influence of ligament length and stress state on fracture energy of concrete. Eng. Fract. Mech., vol. 10, 148-156.

Haas, R. and Ponniah, E.J. (1989), Design oriented evaluation of alternatives for r eflection cracking through pavement overlays. In: Reflective cracking in pavements, (eds. J.M. Rigo and R. Degeimbre) RILEM Conf. Liege, Belgium, 23-46

Hillerborg, A. (1983), Analysis of one single crack. In Proc. "Fracture Mechanics of Concrete, Developments in Civil Engineering" (ed. F.Wittmann), Elsevier Amsterdam, Vol. 7, 223-249

RILEM DRAFT RECOMMANDATION (50-FMC) (1985), Determination of the fracture energy of mortar and concrete by means of three-point bending tests on notched beams. Materials and Structures, 18, 287-290

RVS 8.05.14, 1991, Oberbauarbeiten, bituminöse Tragschichten im Heißmischverfahren, Richtlinien und Vorschriften für den Straßenbau, Forschungsgesellschaft für das Verkehrs- und Straßenwesen, Wien.

Roelfstra, P.E. and Wittmann, F.H. (1986), Numerical method to link strain softening with failure of concrete, in: Fracture toughness and fracture energy in concrete (ed. F.H. Wittmann), Elsevier Appl.Sci.Pub., Amsterdam, 163-175

Tschegg, E.K. (1986), Prüfeinrichtung zur Ermittlung von bruchmechanischen Kennwerten sowie hiefür geeignete Prüfkörper, patent AT-390328.

Tschegg, E.K. (1990), Lasteinleitungsvorrichtung, Patent application, No 48

Tschegg, E.K. (1991), New equipments for fracture tests on concrete, Materials testing, 33, 338-342.

Tschegg, E.K. and Stanzl-Tschegg, S.E. (1992) Development and experience with the wedge splitting test on concrete, subm. to Fatigue and Fracture of Eng. Materials and Structures.

Acknowledgement

The authors thank Ministerialrat Dipl.Ing. G. Fichtl and OR.Dipl.Ing.Dr. H. Tiefenbacher, Federal Ministry for Economic Affairs, Vienna, Austria, for stimulating discussions. For experimental assistance we thank Mr. P. Hüttner and Mr. G. Mikulosch., T.U. Vienna. Ing. N. Nievelt (Nievelt laboratory, Stockerau, Austria) is thanked for providing the specimens. This work is the initial part of a road research project financially supported by the Federal Ministry for Economic Affairs.

PART FOUR
RETARDING MEASURES FOR REFLECTIVE CRACKING IN PAVEMENTS

23 THE PRECRACKING OF PAVEMENT UNDERLAYS INCORPORATING HYDRAULIC BINDERS

G. COLOMBIER
Laboratoire Régional des Ponts et Chaussées, Autun, France
J.P. MARCHAND
Cochery Bourdin Chausse, Nanterre, France

Abstract

Pre-cracking consists of creating shrinkage cracks in predetermined positions in the pavement underlay at the time it is laid. This reduces the variation in the width of cracks which would have occurred in the ordinary course of events, and reduces the tensile stresses at the base of the bituminous mix wearing course.
With the CRAFT process (the French acronym for Automatic Creation of Transverse Cracks) which has been developed in France since 1988, a transverse trough is created every three metres in the pavement underlay before compaction. The trough is 30 cms deep and runs right across the course. A bituminous emulsion is sprayed into the open trough, which is subsequently closed during compaction. A purpose-designed machine has been developed which automatically performs 2 pre-cracking operations per minute over a distance of 3.50 to 5.00 metres and works at the rate of 3,000 to 5,000 t per day.
More than 250 km of pre-cracking operations have already been performed (corresponding to over 100 km of pre-cracked pavements). Monitoring over a period of three years reveals that this is a technically proven process at the present time for the prevention of reflective cracking in pavements. It effectively replaces, at moderate cost, most crack prevention systems. It opens up the prospect of a new approach to the structural design of pavements and the construction of thick single-layer structures (40 cms), as was the case in 1991.
Keywords : Pavements, Crack, Pre-cracking, Equipment

1 Introduction

During the construction of pavement underlays incorporating hydraulic binders, several techniques are available to prevent, or at least considerably to delay, the rise to the surface of transverse shrinkage cracks. These techniques may be employed either separately or in combination, and they are based on the following principles:
* The insertion between the underlay and the wearing course of a product or a process involving a high proportion of bituminous mix which acts as a crack-resistant membrane.

Reflective Cracking in Pavements. Edited by J.M. Rigo, R. Degeimbre and L. Francken.
© 1993 RILEM. Published by E & FN Spon, 2–6 Boundary Row, London SE1 8HN. ISBN 0 419 18220 9.

* The use in the wearing course of suitably adapted bituminous mixes which resist the tensile stresses accompanying the movement of cracks.
* The reduction of the width and movements of the edges of cracks by pre-cracking.

2 The pre-cracking of pavement courses

Pre-cracking consists of creating shrinkage cracks in predetermined positions in the pavement underlay at the time it is laid. This speeds up the initiation of cracking (which usually occurs during the winter following completion of the pavement) and makes it possible to control the number of cracks.

To this end, the space between the artificially produced cracks is chosen so as to be very short (2 to 3 m) by comparison with that observed between cracks which occur normally, namely 15 to 30 m after the first winter and 5 to 10 m in the final stage of cracking.
The following advantages may be expected of pre-cracking:
The large number of cracks produced reduces their variation of width and better preserves the meshing between the edges of the crack.

Control of the variation of crack size makes it possible to limit the tensile stresses it creates at the base of the bituminous mix wearing course. In some cases this obviates the need for specific bituminous mixes to resist the propagation of cracks to the surface of the wearing course.

3 The CRAFT process

3.1 Description
The pre-cracking process known as CRAFT (the French acronym for Automatic Creation of Transverse Cracks) consists of (see fig. 1):
* Creating, before compaction, across the entire width of the course being treated, and at selected regular intervals (every 2.5 or 3 metres) a transverse trough whose depth is equal to the thickness of the course.
* Automatically injecting a bituminous product into this temporarily open trough.
* Closing the trough at the time of compaction.
Currently, the bituminous product used is a cationic emulsion of bitumen of the rapid-breakdown type. It serves a two-fold purpose:
Its low pH aqueous phase creates a weaker zone in the course, helping to localize shrinkage cracks.

Its bituminous phase creates a discontinuity and enables the crack to be accurately pre-localized, while at the same time reducing the material's sensivity to water and abrasion.

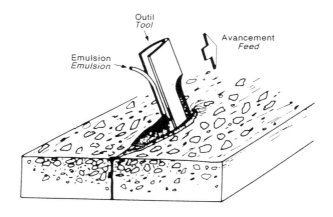

Fig 1 : Operating principle

3.2 Technical advantages

The bituminous emulsion maintains discontinuity between the edges of the crack.

The pre-cracking operation is performed on the material when it has been simply spread and levelled, and throughout the entire thickness and width of the course.

Opening the trough requires only a slight mechanical effort and no heavy mechanical equipment is necessary.

Compacting and levelling operations are not interfered with, and the process has no effect on riding quality.

3.3 Validation trials and patents

Back in 1986 laboratory tests were carried out on cylindrical samples to verify the sound basis of the principle, and on-site trials were conducted to substantiate the effectiveness of the process.

Following these preliminary stages, the method was patented in 1987, and in 1988 it was developed on the industrial scale with the development of patented purpose-built equipment.

4 Pre-cracking equipment

The purpose-built equipment (fig. 2 and fig. 3) is mounted on a conventional vehicle. It is in three parts, each performing specific functions.

The tool (1) opens the trough and injects the binder.
The articulated arm (2) inserts the tool into the layer of material and drives it through the material transversely, across the pavement.

The tank (3), which is lagged, heated and regulated, maintains the reserve of binder at a constant temperature, while the pump feeds the tool with the treatment product.

The hydraulic unit (4) supplies the necessary power and controls the operating cycle determined by a programmable automat (5).

Fig 2 : Machine CRAFT (general view)

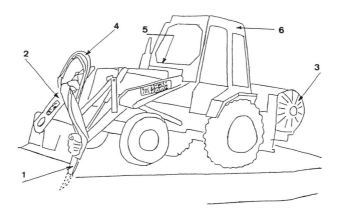

Fig 3 : Machine CRAFT Sketch

The vehicle (6) drives the equipment around the working site, and its engine is also used as a primary energy source.

The principal innovative aspects are:

The vertical, tapering blade which opens the trough is fitted with the spray bar which injects the treatment product.

The tool is fitted with a vibrator to facilitate its penetration in the material. It is not rigidly locked to a given depth, but the penetrating force is kept constant.

The translatory speed is constant.

The articulated arm is not a permanent obstacle on the pavement during treatment. Its rate of working enables the pre-cracking machine to fit in with other on-site operations without interfering with normal progress.

A crack is produced, and the vehicle moves on to the next operation, in less than 30 seconds.

The machine and all its components and accessories are mounted on the same vehicle, constituting a completely self-contained unit.

Up to the present, two machines have been built. Their principal characteristics are given in table 1.

Table 1 : Characteristics of CRAFT machines

	Prototype	New machine
Working width	3.50 m	5.00 m
Emulsion capacity	500 l	1.000 l
Working rate	2 pre-cracks/mn	
	500 t/h	800 t/h

5 Results of four years of CRAFT pre-cracking

5.1 Sequence of operations

After the earliest applications, the following sequence of operations was adopted:
* Spreading of the material.
* Levelling.
* Pre-cracking (before compaction).
* Compaction.
* Finishing.

Fears that the machine would slow down the rate of working on the site were soon allayed, for it immediately follows the grader or finisher. Even if pre-cracking involves several passes, at no time does the machine slow down progress.

As expected, compaction after pre-cracking (Fig. 4) eliminated the ridge associated with the creation of the trough without prejudice to the riding quality.

Fig 4 :   Organisation of a jobsite with compacting after pre-cracking

The interval between artificially produced cracks was set at 3 m. This is a compromise between minimizing the risk of shrinkage cracks due to temperature changes and enabling on-site working to proceed at a normal rate.

A monitoring programme was systematically applied to all sites. It made it possible to establish the results of CRAFT pre-cracking in early 1992 after four years of application.

5.2 The mechanical behaviour of a pre-cracked pavement   underlay

Apart from the visual monitoring which will be referred to below, several mechanical tests have been carried out in line with artificially produced cracks in order to assess the degree of load transfer on either side of them.

The tests measured the differential deflexion between the upstream slab and the downstream slab, either with an inclinometer or with a falling weight deflectometer. The average difference of deflection was found to be 5. 10^{-2} mm.

Furthermore, core samples in line with the cracks showed that the film of bitumen does indeed ensure a discontinuity, but sufficiently fine to preserve a load transfer.

5.3 Sites treated

Altogether, 34 sites and more than 40 sections have been pre-cracked in France (along with 8 reference sections which were not pre-cracked), plus one site in Spain.

a) Total length pre-cracked
All the sites treated correspond to a total length of 250 km of
artificially produced cracks.

1988	1989	1990	1991	1992	(up to June)
1	22	38	105	84	

Linear kilometres of artificially produced cracks since 1988.
This is equivalent to nearly 100 km of pre-cracked pavements.

1988	1989	1990	1991	1992	(up to June)
0.5	13	15	38	32	

Linear kilometres of pavements pre-cracked since 1988.

b) Breakdown of work performed
Routes Nationales (major highways) accounted for 69% of the surface
treated, and Routes D.partementales (secondary roads) for 28%. The
percentages are much the same for new pavements (60%) and overlays
(34%).

c) Nature of the underlays
In the majority of cases, the underlays were continuously graded
aggregated aggregates treated with cement or binder, and a not
insignificant proportion of compacted concrete. The machine was tested
on a wide variety of underlays in order to assess the effectiveness of
the process.

d) Nature of the wearing course
Between 1988 and 1990 several types of wearing course were associated
with pre-cracking, so as to discover whether pre-cracking was a
substitute for, or a complement to, a crack-resistant wearing course.
    Since 1991 the two-layer concept, combining a bituminous mastic
rich in bituminous concrete (thin or very thin) has no longer been
current. The same is true of geotextiles impregnated with polymer
bitumen, except in specific cases where the wearing course is in
porous bituminous concrete.
    Pre-cracking has gradually replaced crack-resistant wearing
courses.

5.4 The effectiveness of CRAFT
It was possible to assess the effectiveness of pre-cracking after four
years by monitoring the number of cracks which rose to the surface of
the pavement.
    In the report <1> mention is made of a study of the rate at which
transverse cracks appear in function of the age of the underlay and
its bituminous overlay (bituminous concrete or bituminous continuously
graded aggregates).

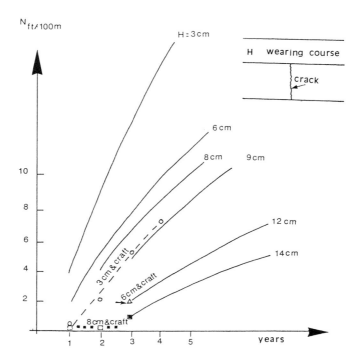

Figure 5: Comparison of the evolution, with and without pre-cracking, of the number of transverse cracks (per 100-metre section) in function of the thickness of the bituminous overlay and the age of the underlay.
-------- with CRAFT   ─────── without pre-cracking

Transferring on to this curve (fig. 5) the visual observations made on CRAFT sites, we see that quite apart from any structural aspect, CRAFT pre-cracking is just as effective as 5 or 6 cm of bituminous mix in delaying the rise of cracks.

This is further accentuated if account is taken of the seriousness of the crack. With pre-cracking, the cracks are fine (< 0.5 mm), whereas natural cracks are wider than 1.0 mm.

This highlights the value of pre-cracking in maintaining the imperviousness of the wearing course.

The more so since there has been no sealing of cracks on any pre-cracked section, nor is any such sealing programmed in the coming years.

5.5 Modelling of pre-cracking

A numerical simulation <2> was performed on a pre-cracked pavement whose temperature was lowered by 10°C or 20°C.

The parameter calculated was the width of the crack in function of the spacing between cracks and the temperature drop. A spacing of 12 m corresponds to natural cracking.

The width of the crack diminishes with the spacing between cracks. A spacing of 3 m corresponds to a technical and economic optimum. Without pre-cracking, no wearing course can withstand a stretch of 4 mm.

6 Conclusions and future prospects

After four years experience of pre-cracking, corresponding to 250 km of artificially produced cracks and 100 km of pre-cracked pavements, it can be asserted that:

The effectiveness of pre-cracking with the interposing of a film of bitumen has been confirmed in terms of the prevention of the rise of cracks to the surface, since it is equivalent (leaving the structural approach out of account) to 5 or 6 cm of bituminous mix.

The machines are capable, working at the rate of one operation every 30 seconds, of pre-cracking all types of pavement underlays to a depth of 25 cm across their entire width.

Thanks to CRAFT, pre-cracking is an operation which, in France, is included in the recommendations for the construction of underlays incorporating hydraulic binders, in accordance with the SETRA information note <3>.

The monitoring of sites and the record of their behaviour leads to a conception of pre-cracked pavements in which the base course and the sub-base constitute a single entity, and in which the wearing course is structurally designed not in the light of the rise of cracks, but uniquely in the light of safety and riding quality.

7 References

<1> Fissuration de retrait des chaussées à assises traitées aux liants hydrauliques. Bulletin de Liaison des Laboratoires des Ponts et Chaussées, n° 156 and 157, 1988.
<2> Vers une modèlisation des systèmes antifissures. RGRA, n° 685, May 1991.
<3> Techniques pour limiter la remontée des fissures. Note d'Information du SETRA, n° 57, March 1990.

24 TREATMENT OF CRACKS IN SEMI-RIGID PAVEMENT: COLD MICROSURFACING WITH MODIFIED BITUMEN EMULSION AND FIBERS: SPANISH EXPERIENCE

R. ALBEROLA
Ministry of Public Works and Transport, Madrid, Spain
J. GORDILLO
E.S.M. Research Center, Madrid, Spain

Abstract
Within the range of procedures for coping with the problem of reflective cracking in hydraulic bases, one of the procedures being used in Spain is a special type of slurry seal, with modified bitumen emulsion and reinforced fibers. The combination of fibers and a highly modified residual binder makes for an asphalt mix which is very high in binder content, flexible and with high tensile strength.
   When applied in a double layer to cracked roads, the system seals the cracks thus avoiding the harmful effects of water in the underlaying layers and provides the tensile strength to resist the stresses between the cracked edges, delaying the reflective cracking of the wearing course.
   This paper describes the system procedure, the characteristics of both the binder and the microsurfacing, the effects of the elastomeric product and the fibers in the mix flexibility. A large project made on a freeway outside Madrid, with a Daily Average Traffic over 55.000 vehicles (110.000 vehicles on the nearest area to Madrid), 18% of them heavy trucks, is also mentioned.
Keywords: Reflective Cracking, Cold Microsurfacing, Modified Bitumen Emulsions, Fibers, Flexibility Test, SAM Membrane.

1   Introduction

At present, one of the main preoccupations of the technical experts of highways is the development of cracks, in wearing courses, especially those cracks generated by the surface reflection of shrinkage cracks in underlaying layers, which are treated with hydraulic binders and are often used in mixed-type Spanish pavements.
   These cracks provide an easy inlet for solid particles and water and the resulting deterioration can

Reflective Cracking in Pavements. Edited by J.M. Rigo, R. Degeimbre and L. Francken.
© 1993 RILEM. Published by E & FN Spon, 2–6 Boundary Row, London SE1 8HN. ISBN 0 419 18220 9.

affect, not only the even surface of the pavement but, also, the structural capacity itself, a fact which is more significant and may result in a noticeable reduction of the service life of the compromised pavement.

Therefore, the development of these cracks in the wearing courses, has to be, either prevented or delayed. One of the available alternatives, that forms part of the well-known thin layers technology, the fibers-reinforced microsurfacings is currently applied in Spain, usually as surface membranes (SAM membrane).

2 System description

The membranes, consisting of microsurfacings reinforced with fibers, are a modern development of the slurries SEAL, although the pure bitumen emulsions have been replaced by modified bitumen emulsions. This change has lead to advantages from the viewpoint of thermal susceptibility, resiliency and flexibility at low temperatures, etc.

Since more bitumen is bound, due to the addition of fibers, a rather high mechanical stability is achieved, free of any possible bleeding, and the mixes are characterized by a rather high flexibility, traction resistance and excellent waterproof characteristics. The microsurfacings are usually applied in one or two layers, as required by the extent of the pavement cracks.

Microsurfacings are usually applied in double layers. Fine aggregates (0/3 mm.) are used in the first layer and coarser aggregates up to 6/8 mm. are used in the second layer.

3 Components

3.1 Aggregates

The aggregates applied for this type of membranes have to clean and of variable size grades as required for their use in SAMI or SAM membranes. In the first type, the aggregates have to fit the Spanish LB-4 or LB-3 grading envelope, although a high content of finer fractions is recommended.

In the second layer of SAM membranes, since these are in direct contact with wheel tyres, the aggregates have to be hard, highly resistant to polishing and according to the LB-2 grading envelope of the Spanish Specifications.

3.2 Fibers

They are plastic type and usually meet the following characteristics:

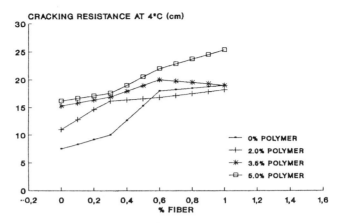

Fig. 2.
Variation of cracking resistance with fiber and polymer content using 12 % of residual bitumen

In the Cracking Test, the test pieces are subject to flexural strength. The ends of the metallic holder, in which the test piece is set, are bent until a crosswise crack can be seen in the surface. In this way, the microsurfacings flexibility can be measured. The Test is carried out at 4ºC (Figure nº 3).

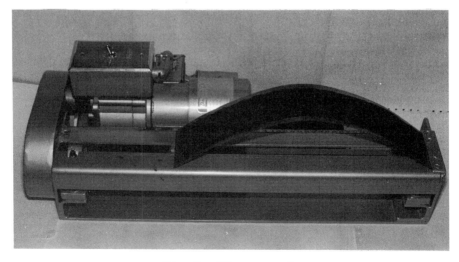

Fig.3. Flexurometer

The usually recommended emulsion rates range from 17/25% in weight, in proportion to the aggregates.

- Break elongation exceeding 40%
- Melting point over 250º
- Water absorption below 1%
- Tensile strength higher than 5,000 kg/cm

3.3 Emulsion

It is a cationic emulsion of bitumen modified by elastomeric products of the SBR types. The residual binder resulting from an evaporation process is characterized by a rather low thermal susceptibility (penetration index larger than 1.5), high plasticity interval (interval between FRAAS brittleness temperatures, and a ring and ball softening temperature which exceeds 75ºC), average resiliciency measured by the Elastic Recovery Test (above 80%), and high toughness (in excess of 20 kg.cm).

The optimal contents, both of the elastomeric products in the emulsion and of fibers in the microsurfacing have to be determined, since those values are a basic requirement for the performance of the system.

Both components are measured by the so called Flexibility Test which uses a flexurometer to define the minimum content of polymers in the residual bitumen and the fiber content which the system needs for to obtain a good performance (Figs. 1 and 2). Those values are 5% polymers in the residual binder and 0.5% of fibers referred to the aggregates, although the emulsion and the extent of cracking for these values can range up to 1.2%.

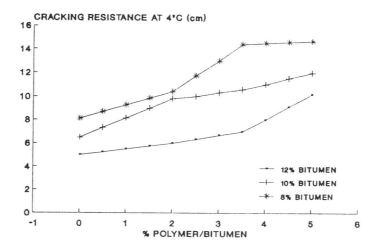

Fig. 1.
Variation of cracking resistance with binder and polymer content

4 Application

The mixing and spreading equipment to be applied in this type of membrane is similar to the one used for standard slurries, although the fitting of mechanical devices is required for the addition of fibers. These can be added either dry or, else, through a wet process, and, in that case, the fibers have to be previously dispersed in water.

If the latter procedure is favored, the use of tyre rollers is recommended to help the outflow of the breaking water. This allows for easier curing and increased compaction of the mix. The amount applied by square meter ranges from 6 to 8 kg per layer.

5 System applied in the National Highway IV

The National Highway IV, which links Madrid to Andalucia, in Southern Spain, is subject to heavy traffic, which in the area close to Madrid, reaches 110,000 vehicles/day of which, around 18%, are heavies. The pavement of this highway was widened from 2 to 3 lanes in each direction. The project required to extend on the E-2 type subgrade, built according to Spanish specifications, a 20 cm layer of soil-cement as sub-base, a 25 cm dry rolled concrete as the road base layer, and 15 cm of asphaltic mixes applied in two layers. In the rolled concrete base, 8 cm deep joints were cut every 15 meters and a 0.85 m wide geogrid was applied over the lengthwise and crosswise joints of both the old and new pavements, between the bituminous base course and the wearing course.

After three years in service, a number of crosswise cracks have developed all over the stretch of road. The cracks are spaced about 7.5 meters and the joints cut in the dry rolled concrete (every 15 meters), as well as other intermediate joints, are reflected in the pavement due to the concrete failure. The cracks, which extend the full width of the lane, are continued, many times, in the outer shoulder of the road and are usually clustered in families, although some isolated cracks can also be found. Lengthwise cracks, all along the stretch of road, although less frequent, are also noticeable.

The high traffic of heavy vehicles which run over the mentioned lane, combined with the low temperatures and the rains, have resulted in the progressive deterioration of the pavement. The situation called for immmediate steps to prevent the extension of the cracks and joints throughout the bituminous pavement.

Two types of actions were defined: one which could be described as "rehabilitation", consisting in finding procedures, at the lowest possible cost, that would guarantee the good performance of the pavement for a 2-3 year period, at the end of which a final solution would be applied. In the second case, which might be called "reconstruction", a number of crack preventing procedures would be attempted to achieve the final repair of the pavement. The latter is planned to be implemented during August 1.992, and we think it will be the subject of reports in future Conferences.

From the different technologies which were proposed to be tried as Rehabilitation Techniques two approaches were selected, both based in the SAM membrane. The first approach consisted in a surfacing treatment of modified bitumen binders with elastomeric products at the rate of: 7 kg/m^2, in two layers, and alternating the binders at the rate of 1.6 and 1.1 kgm^2 and spraying respectively, 12/18 and 6/12 aggregates, which were previously lacquered. The second system consisted in applying a microsurfacing, reinforced with fibers. The first system was not applied due to prevailing low ambient temperatures.

6 Works development

The membrane was laid in two layers. The first was applied in November, while temperature ranged from 2 to 14ºC. The curing was slow and the amount applied averaged 6.6 kg/m^2. Before they were applied, a prior priming was sprayed in amounts of about 300 gr/m^2 of 50% emulsion.

The composition of the membrane applied for this layer was as follows:

- Silica sand 0/6 mm 100
- Modified cationic emulsion (63%) 15
- Water 10
- Fiber 0.3
- Cement and additives 1

The residual binder of the emulsion (produced by evaporation) showed the following characteristics:

- Softening point 64ºC
- Penetration . . . . 62 dmm
- FRAAS brittlenes 16ºC
- Plasticity interval 80ºC
- Elastic recovery 85%

As the winter arrived, low temperatures and heavy rains, prevented the completion of the second layer, which had to be postponed until next May.

The composition of this second layer was as follows:

- Silica sand 0/5 66
- Porphyric fine gravel 34
- Modified cationic emulsion (63%) 17
- Water 9
- Fiber 0.6
- Cement and additives 1

The application was carried out at ambient temperatures ranging from 8 to 22ºC. The curing times were short and the average amount applied was 7.5 kg/m^2.

The resulting membrane shows a mechanical performance defined by the following tests:

- Losses during the Wheel Track
  Abrasion Test . . . . 90 Gr./m^2
- Flexibility at 4ºC
  (measured by flexurometer) . . . . 160 mm

Fig.4. Microsurfacing first layer

7   Conclusions

Due to its length (30 Km), this job represents the largest surface covered by a SAM membrane consisting of a cold microsurfacing, reinforced with fibers. The brief time elapsed from the time this report was drafted to the completion of the works, did not allow us to submit data and findings which are adequate to prove the sucess of the procedure.

Fig. 5. Microsurfacing second layer

However, we would like to highlight that, although only one layer has been applied, under very harsh winter conditions and subject to heavy traffic, the performance of the membrane is quite good since the number of cracks detected has been minimum, restoring the watertightness of the pavement and preventing further deterioration.

On the other hand, in the areas which were not treated, the combined action of traffic and rains has significantly deteriorated the pavement causing an increase in the maintenance of the works.

Once the second layer is applied, we feel the planned target will be achieved, i.e., to extend the service life of the pavement, without any maintenance, for an additional 2/3 years term. At the end of this term, this stretch of road will have to be rebuilt.

25 APPLICATION OF GEOSYNTHETICS TO OVERLAYS IN CRACOW REGION OF POLAND

W. GRZYBOWSKA, J. WOJTOWICZ and L. FONFERKO
Cracow University of Technology, Cracow, Poland

Abstract

The first, preliminary attempts at applying geosynthetics to pavement overlays were undertaken in Poland at the Cracow University of Technology in 1978, in a form of laboratory research using the Dutch geogrid Structofors. The samples of asphalt concrete with Structofors were submitted to static, bending loads at chosen temperatures. It was found, that the modulus of Structofors was lower than the modulus of the asphalt mixture, so the presence of the geogrid in the samples had no influence neither on the moment of crack under the static load nor on the critical load value. Works concerning this problem were continued in 1989 when possibilities appeared to produce Polish nonwoven geotextiles to overlays. Laboratory and field examinations were preceded by preliminary theoretical analysis, based on the elastic theory. In laboratory tests, samples of asphalt concrete with Polish polyester nonwoven geotextile were submitted to static bending, repeated dynamic loads and to shearing stresses. It was found, that number of cycles to crack is nearly doubled, but the adhesion between layers is worsened, which can be advatageuos to relieve the stress concentration around the crack in the lower layer. On the basic of the laboratory test, in 1991 a series of experimental sections 1300 m long was constructed, on the international road E 40 (Calais-Koln-Wrocław-Kraków-Lvov-Kiev).
<u>Keywords:</u> Geotextiles, Overlays, Asphalt Concretes, Reflective Crackings, Sheer Properties Tests, Surface Treatment.

1 Introduction

The present state of road pavements in Poland, and the continuous increase of the number of heavy trucks with excessive axle load, requires the implementation of new technologies, prolonging the longevity of road pavements, but not engaging undue expenditure, resource and time. On the basis of experiences in Western countries using geosynthetics to pavement overlays the first, preliminary attempts in aplying this technology in Poland were undertaken as far back as 1978, on a laboratory scale, at Cracow University of Technology. The asphalt concrete samples, strengthened with the Dutch geogrid Structofors, were submitted to static, bending loads in various temperatures. Analysing the results it was found, that one of the main conditions which the reinforcement should fulfill, namely that the modulus of elasticity of reinforcement should be greater than the matrix modulus, for the synthetic geogrid was not met. The sample improved with the geogrid cracked after exceeding the critical elongation of nearly the same value as sample without the geogrid, and also at nearly the same critical load. So the conclusion was, that geogrid had no influence on the moment of crack and critical load of asphalt mixtures in conditions of static loads. These observations were confirmed by theoretical analyses, performed 10 years later on the basis of the elastic theory, with the use of the Finite Element Method (Grzybowska, Salamon et al. 1989). As was found, the efficiency of geosynthetics in overlay structure under a single load, depends on the relation between the modulus of geosynthetics and the modulus of the existing pavement (Fig.1). Because the available geosynthetics have a modulus of elasticity lower than the modulus of bituminous mixtures, especially at low temperatures, considering the synthetic grid as a typical reinforcement working in the range of elastic strain is doubtful.

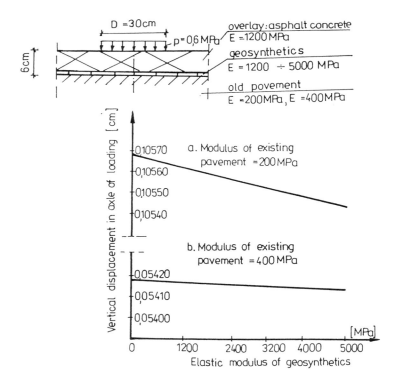

Fig. 1. The effect of elastic modulus values on the vertical displacement of geosynthetics layer under loading (Grzybowska, Salamon et al. 1989)

Recently the possibility of another approach to the application of geosynthetics in pavement overlays has appeared, due to initiating in Poland of the production of nonwoven, spunbonded polyester geotextiles. These, after saturation with asphalt, could act in the overlays as the insulation against the penetration of water inside the structure as well as the interlayer preventing the asphalt bleeding onto the surface and retarding the propagation of cracks from the lower layer upwards. Some properties of this geotextile are as follows:
chemical type of material - pure polyester, material structure - nonwoven, spunbonded geotextile, density 220 g/m^2, rupture force along fibers in temp. 145°C - 9.7 kN/m, elongation in temp. 145°C - 58%,, transversely the same characteristics are respectively: 7.4 kN/m and 83 %.

In this paper some of the findings obtained recently on the basis of laboratory tests and observations on constructed experimental sections (using of Polish nonwoven geotextile) are presented.

2 Laboratory tests

2.1 Bending under a static load

This test was performed to study the character of co-action of the geotextile and asphalt layers in conditions of increasing load and to check the adhesion of the whole structure. The measurements were performed on the beams, whose structure and dimensions are presented in Fig. 2. As an asphalt mixture a medium grained asphalt concrete of a dense structure was used for both layers. All samples were compacted in a static manner.

The beams were simply supported at both ends and loaded uniaxially at a rate of 12 mm/min. The temperature of the test was 20°C.

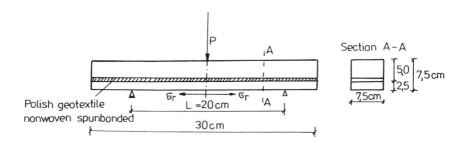

Fig. 2. Test of bending under a static load.

The results are presented in Fig. 3.

As was observed, the bending strength value and the moment of crack of bituminous mixtures was not affected substantially by the presence of the geotextile. The influence of the geotextile

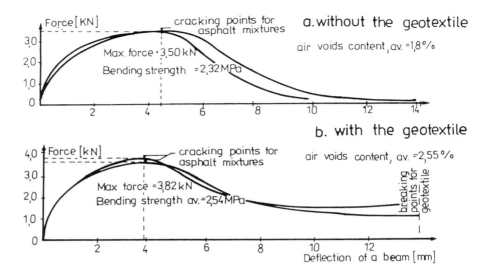

Fig.3. Results of the bending test for asphalt concrete beams with and without geotextile, in the temperature 20°C.

on the behaviour of the samples after cracking of the asphalt mixture manifested itself only in some prolonging of the integrity of the beams commensurate with the elongation behaviour of the geotextile.

For the whole time, adhesion between the layers as estimated visually was very good.

2.2 Bending under the repeated load

To check the co-action of the geotextile with asphalt layers in conditions closer to a more realictic pattern of loading on a road, some fatigue examinations were conducted with using the MTS hydraulic press. Tests were performed on the samples of the same structures and dimensions as in paragraph 2.1. Samples were supported at both ends and submitted to the uniaxial repeated loading according to the function of haversine invert with frequency 5 Hz and amplitude 2 kN, that is a bending stress equal to 1.33 MPa, at the temperature of 20°C. Measurements were conducted to the moment of cracking of the bituminous mixtures.

The results obtained are presented in Table 1.

In conditions of the cyclic loading, samples with the geotextile exhibit the resistance to a crack substantial greater than samples without the geotextile. Due to the presence of the geotextile interlayer, asphalt concrete sustains nearly twice the number of loading cycles than without it.

2.3 Shearing tests

The study of the role of a geotextile in overlays as a stress-relieving medium diminishing the stress concentrations above the crack in the lower part of a pavement structure,

Table 1. Results of the test in repeated load conditions

No of sample	Type of structure	Number of cycles until to crack	Average St.dev. Var.coeff.
1.A	Without geotextile	890	952.5
2.A	" "	990	56.8
3.A	" "	1010	0.06
4.A	" "	920	
1.B	With geotextile	2100	1810.0
2.B	" "	2740	751.6
3.B	" "	1170	0.42
4.B	" "	1230	

requires the examination and quantitative estimation of adhesion between the geotextile and asphalt layers.

To realize this aim, a method of direct shear test has been developed at Cracow University of Technology. The samples, in the form of beams as in the paragraph 2.1 were cut to pieces of length d = 68-80 mm and submitted the shear stresses, as presented in Fig. 4.

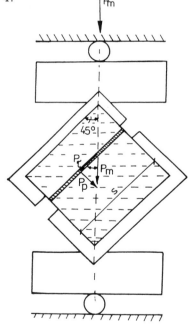

Fig.4. Shear test for asphalt concrete samples with geotextile interlayer

The shear stresses were calculated as follows:

$$P_\tau = \frac{P_m}{2^{1/2}} \qquad (1)$$

$$\sigma_\tau = \frac{P_\tau}{F_\tau} \qquad (2)$$

$$F = d \times s \qquad (3)$$

where:

P_m - maximum force acting on a sample.
P_τ - maximum shearing force.
σ_τ - shear stress.
Rate of load application: 1 mm/min, temperature: 30°C.

The results of measurements are presented in Table 2.

Table 2. Results of the measurement of shear stresses

No of sample	Type of structure	Max. shear force (N)	Max. shear stress (MN/m^2)
1.C	Without geotextile	1237.4	0.21
2.C	" "	1202.1	0.21
3.C	" "	1661.7	0.28
4.C	" "	2262.7	0.37
5.C	" "	2086.0	0.37
6.C	" "	1909.2	0.33
Average:		1726.5	0.30
Standard deviation:		401.8	0.068
Variance coefficient:		0.233	0.230
1.D	With geotextile	1173.8	0.20
2.D	" "	735.4	0.13
3.D	" "	827.3	0.15
4.D	" "	834.4	0.14
5.D	" "	586.9	0.11
6.C	" "	756.6	0.13
Average:		819.1	0.14
Standard deviation:		178.4	0.029
Variance coefficient:		0.218	0.198

As is shown, the presence of a geotextile interlayer diminishes by more than 2 times the adhesion between asphalt layers, for instance, the profiling and wearing layers. This phenomenon is undoubtedly advantageous for the prevention of reflecting cracking because the geotextile absorbs the part of crack energy from the lower cracked layers and does not transfer it upwards. At the same time, however, the bending stress in the upper layers increases, and also the susceptibility to horizontal forces (acceleration, deceleration, braking and centrifugal forces) becomes greater, which can decrease the stability of an overlay structure.

2.4 Conclusions from the laboratory tests

The presented results of the preliminary laboratory tests show that nonwoven, spunbonded geotextile has a positive influence on the fatigue properties of asphalt samples at a temp. $20^{\circ}C$. Some possibilities of diminishing of stress concentration around the crack in the lower layer is also noticed, due to decreasing adhesion between the layers. At the same time, however, the worsening of adhesion is disadvantageous, because of increasing bending stresses in the upper layer, and worsening of the resistance to the horizontal stresses.

3 Experimental sections

The experimental sections were located on the international road E 40 (Calais-Koln-Wrocław-Kraków-Lwow-Kijów), at Km 411+237-412+500 along the longitudinal joint between the bituminous shoulder and the bituminous carriageway. The aim of these works was the renovation of the road, because of insufficient bearing capacity and much damage on the shoulders. Applied polyester geotextile had the parameters given in page 1. The geotextile was laid on the damaged shoulder, covering also the lane-shoulder joint, which had separated over its whole length. The applied structure of overlay is presented in Fig.5.

The surface was covered with 1 kg/m^2 of asphalt D 70 and overlaid manually with a geotextile to a width of 2.5 m. To assure a good saturation of geotextile with asphalt, a rubber roller was applied immediately after laying. The asphalt concrete was then mechanically laid and compacted. The works were done under good atmospheric conditions. It was observed, that the adhesion between the bituminous layers and the geotextile was good. The evenness, in spite of the manual laying of the geotextile was sufficient, and the resistance to thermic influence was very good.

For comparative purposes, the geotextile was laid: a) on the left shoulder on the profiling layer, and b) on the right shoulder indirectly on the repaired old pavement.

Control surveys included:

boring the samples from upper layers to check the adhesion
examination of evenness
comparative deflection tests using the Benkelman beam
(before, 1 month after and 8 months after renovation)

It was found that the influence of the geotextile on deflection values was not noticeable.

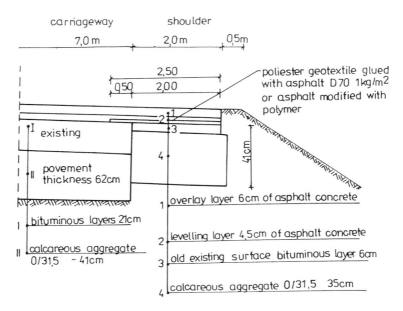

Fig.5. Structure of experimental section, with shoulders improved with geotextile.

Observations carried out recently have showed that sections improved with the geotextile, as well as control sections, are in very good condition. Too short a period of operation does not allow to formulate any more detailed conclusions.

5 Prospective works

Based on works hitherto, the programme for future implementation in 1992-1994 has been developed in collaboration with Swiss specialists from the Swiss Branch of International Geotextile Society. This programmme includes the implementation of double surface treatement in the renovation of roads in current use, applying the Polish and Swiss geotextile, as well as overlays with the geotextile and the geogrid.

6 References

Colombier, G. (1989) Fissuration des chaussees: nature et origine des fissures; moyers pour maitriseer leur remonte, in **Proceedings of Conference on Reflective Cracking in Pavements** (eds J.M. Rigo and R. Degeimbre), State University of Liege, Liege, pp. 3-22

Grzybowska, W., Wojtowicz, J., Salamon, W., Fonferko, L., et al. (1989) Nowa metoda wzmacniania nawierzchni bitumicznych przy zastosowaniu geotekstyliów. **Report prepared at Cracow University of Technology for Ministry of Education.**

Haas, R., Ponniah, E.J., Design oriented evaluation of alternatives for reflection cracking through pavement overlays, in **Proceedings of Conference on Reflective Cracking in Pavements** (eds J.M. Rigo and R. Degeimbre), State University of Liege, Liege, pp 23-46.

Jaecklin, F., en collaboration avec la TKSVG (1991) **Le Manuel des Geotextiles.** Association Suisse des Professionèls de Geotextiles, St. Gall.

26 MINIMIZATION OF REFLECTION CRACKING THROUGH THE USE OF GEOGRID AND BETTER COMPACTION

A.O. ABD EL HALIM and A.G. RAZAQPUR
Center for Geosynthetic Research, Information and Development, Carleton University, Ottawa, Canada

Abstract

Results from finite element analyses and laboratory tests are presented which suggest that in the presence of construction induced cracks in new asphalt overlays, geogrid reinforcement will be ineffective in preventing reflection cracking. As is customary, the reinforcement layer was placed at the interface of the overlay and the existing pavement surface. However, when through improved compaction, by means of a newly developed compactor, called AMIR, construction induced cracks were minimized, the same reinforcement proved to be effective in controlling reflective cracking.

Keywords: Compaction, Finite Element, Geogrid Reinforcement, Overlay, Pavement, Reflection Cracking, Experiment.

1 Introduction

Reflection cracking involves the cracking of a flexible overlaying above underlying cracks or joints. The cracks are thought to occur as a result of traffic loads and thermal stresses (Coetzee and Monismith 1980). Research on reflection cracking is extensive (Sherman 1982). Most of the existing theories dealing with the problem are based on the assumption that the existing cracks in the underlying layer are the major contributor to what is observed on the top of the new asphalt overlay (Sherman 1982).

In spite of the extensive investigation of reflection cracking, to date no effective solution has been found (Degeimber and Rigo 1989). Results of recent research has shown that existing compaction equipment will induce hairline cracks during the compaction of the new asphalt layer. These construction cracks will produce a structural layer that has significantly different properties and characteristics than the one used in the analytical modelling of the problem of reflection cracking. As a result, solutions recommended and developed are based on models which fail to solve the real problem. For example, the use of reinforcing elements at the interface of the old and new pavement layers has met with little success in preventing reflection cracks from propagating to the new surface.

This paper presents analytical and experimental data which show why the reinforcement fails to control reflection cracking.

2 The AMIR Compactor

In practice construction induced cracks, known as "checking", have been generally ignored. Pavement engineers and researchers usually attribute these cracks to the properties of the asphalt mix, temperature during compaction, weak support conditions or lack of skilled operators. Halim et al. (1990) showed that the widely used conventional steel wheeled rollers induce hairline cracks during compaction. As subsequently demonstrated, these construction induced cracks, which occur at the surface of the newly compacted asphalt layer, can have a detrimental effect on the minimization of reflective cracks.

Halim (1984) used the concept of relative rigidity to explain how construction cracks occur. It was shown that cracks are formed due to the mismatch between the geometry and rigidity of the steel roller and the asphalt mix during compaction. This conclusion led to the development of a new compactor called Asphalt Multi-Integrated Roller (AMIR). The AMIR compactor replaces the cylinder shape with a flat plate, and provides a flexible material (rubber belt) at the asphalt/compactor interface. Svec and Halim (1991) showed that the use of the AMIR compactor can provide crack-free asphalt pavement. Using finite element analyses, it will be shown next that construction induced cracks render reinforcement ineffective in controlling reflection cracking.

3 Finite element modelling

3.1 General
Consider the pavement system in Figure 1, which illustrates an existing cement pavement and an overlay, with a reinforcement layer between them. The figure also shows the presence of a vertical crack in the existing concrete pavement and two potential construction induced cracks in the overlay. All cracks are assumed to be 2 mm wide, and while the reflective crack is a through crack, the construction cracks are assumed to have penetrated only halfway through the thickness of the overlay. The dimensions, loading and material properties used in the analysis are shown in the same figure.

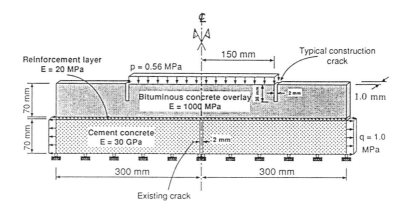

Fig. 1: Pavement system modeled by finite element

Due to symmetry only half the system was discretized by four node plane strain elements. A total of 289 elements were used to model the problem. Systems with and without construction cracks were analyzed. Cracks were modelled by gaps between elements. Since the purpose of this study was not determination of crack propagation and other fracture processes, crack tip elements were not employed. In all the analyses, full bond between the various layers and linear elasticity were assumed. The applied load consisted of vertical loads (traffic loads) or of horizontal loads (thermal loads).

3.2 Finite element results and discussion
The purpose of the reinforcement is to resist the tensile stresses caused by the applied loads and to prevent the reflection cracking. The cracks will not propagate if the reinforcement reduces the tensile stresses in the overlay to a level below the tensile strength of the asphalt cement. Thus to check the effectiveness of the reinforcement, we plot the principal tensile stresses along a vertical section which passes through the construction crack in the overlay. The same stress will be plotted for pavements with and without a reinforcement interlayer. Similarly, pavements with and without construction cracks will be considered.

Figures 2 and 3 show the tensile stresses along section C-C due to applied horizontal and vertical loads, respectively. Although these figures show the stresses in both the existing pavement and the new overlay, we are mainly interested in the overlay stresses. Thus, Figure 2(a) shows that with or without construction cracks, the reinforcement layer actually increases the tensile stresses in the overlay. However, when construction cracks are

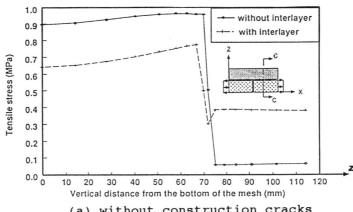

(a) without construction cracks

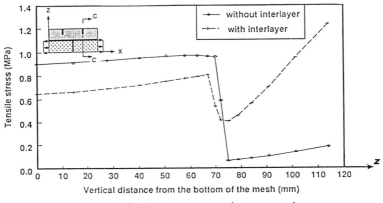

(b) with construction cracks

Fig. 2: Principal tensile stresses along Section C-C due to horizontal loads

present, as in Figure 2(b), the increase in stresses caused by the presence of reinforcement is more then 600 percent. The lowest stresses in the overlay are produced when construction cracks are absent and there is no reinforcement. It should be emphasized that it is the relative values of
stresses that we are concerned with in the present discussion.

Considering the stresses under vertical loads, Figures 3(a) and 3(b) show the principal tensile stresses along section C-C for systems without and with construction

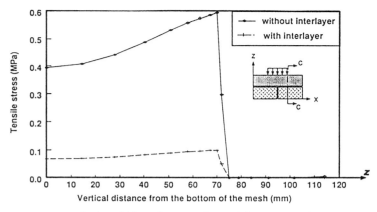

(a) without construction cracks

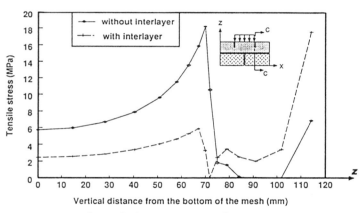

(b) with construction cracks

Fig. 3: Principal tensile stresses along Section C-C due to vertical loads

cracks, respectively. Figure 3(a) shows that in the absence of construction cracks, the overlay stresses are basically unaffected by the presence of the reinforcement, except in the vicinity of the reinforcement. In the latter region, the overlay tensile stresses are substantially reduced by the reinforcement. When construction cracks are present, as in Figure 3(b), we notice that the overlay stresses increase substantially due to the presence of the reinforcement, and the reinforcement becomes harmful to the overlay.

From the above results, we conclude that the placement of the reinforcement at the existing pavement interface

with the overlay does not prevent reflective cracking when
construction cracks are present, but on the contrary
increases the probability of crack formation in the
overlay. If construction cracks are present, then under
both traffic loads and thermal loads, the reinforcement
will increase the tensile stresses in the overlay, leading
to early initiation of cracks. It should be pointed out
that the stresses in other sections of the overlay, that
are not plotted for the sake of brevity, show similar
trends.

4 Experimental investigation

4.1 Experimental program
To verify the above findings, a large scale field trial was
designed and carried out in Ottawa in 1990. Two types of
polymer grids, a polypropylene and a polyester, were used
in the investigation. The objective of using two different
types of grids was to ensure that the construction cracks,
mentioned earlier, do not depend on the type of grid used.
Subsequently, for each grid type one section was compacted
using vibratory and rubber rollers while the other section
was compacted using the new AMIR roller.

Two weeks after the construction of the test sections,
large size slabs were cut and recovered from the field.
Each slab was 600 mm long by 300 mm wide. The thickness of
the slabs ranged from 50 mm to 75 mm. A total of 64 slabs,
reinforced and unreinforced, were recovered from the field
and carefully transported to the laboratory for testing.

The laboratory testing was performed using a rig which
simulates the horizontal loading conditions in the finite
element model. It consists of a rigid table with two
horizontal steel plates, one fixed the other moveable. The
test can be preformed by gluing the bottom of the slab to
the surface of the plates. Horizontal displacement is
induced by the movement of one plate. The load versus
displacement relation is recorded automatically.

5 Observations and test results

The test results described herein relate to the horizontal
loading case referred to in the finite element analysis.
When asphalt slabs, reinforced or unreinforced, compacted
by means of current rollers, were tested, two different
types of cracks were observed. The first type of cracks
was initiated at the bottom of the specimen at the slab-
plate interface. At the same time, surface cracks
propagated downward until they met the cracks at the
bottom, Fig. 4. When the AMIR compacted slabs were tested,
only the cracks at the slab-plate interface were observed.

These cracks propagated upward until they reached the asphalt/reinforcement interface. As more displacement was applied, the cracks at the asphalt/reinforcement interface changed direction and propagated along the plane of the reinforcement as can be seen in Figure 5. Those observations were supported by the test results. Figure 6 shows four typical stress/displacement relationships obtained from the testing program. As can be seen in the figure, the AMIR compacted samples show higher strength than the steel compacted ones.

Fig. 4: Crack pattern at failure

Fig. 5: Crack pattern at failure

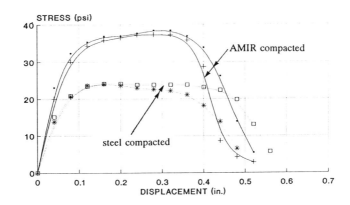

Fig. 6: Tensile stress-displacement relations

6 Conclusions

1. Construction induced cracks can adversely affect the performance of reinforced asphalt pavements.
2. Asphalt pavements, reinforced or unreinforced, must be analyzed as constructed.
3. Reinforced pavements can provide an effective solution to the problem of reflection cracking if the newly constructed asphalt layer is crack-free.

7 References

Abd El Halim, A.O. 1984, **The Myth of Reflection Cracking,** Proceedings of the Canadian Technical Asphalt Association, Vol. XXIX, pp. 202 to 214.

Abd El Halim, A.O. and O.J. Svec. 1990, **Influence of Compaction Techniques on the Properties of Asphalt Pavements,** Proceedings of the Canadian Technical Asphalt Association, Vol. 35, pp. 18 to 33.

Coetzee, N.F. and C.L. Monismith. 1980, **Reflection Cracking: Analysis, Laboratory Studies, and Design Considerations,** Proceedings of the Association of Asphalt Paving Technologists, Vol. 49.

Degeimbre, R. and Rigo, J.M. 1989, "**Forward**", Proceedings of First International Conference on Reflective Cracking in Pavements, Liege, Belgium.

Sherman, G. 1982, **Minimizing Reflection Cracking of Pavement Overlays,** National Cooperative Highway Research Program Synthesis of Highway Practice 92.

Svec, O.J. and Abd El Halim, A.O. 1991, **Field Verification of a New Asphalt Compactor, AMIR,** Canadian Journal of Civil Engineering, Vol. 18, pp. 465 to 471.

27 NEW SYSTEM FOR PREVENTING REFLECTIVE CRACKING: MEMBRANE USING REINFORCEMENT MANUFACTURED ON SITE (MURMOS)

J. SAMANOS
Screg Routes, France
H. TESSONNEAU
Screg Sud-Est, France

The first experiments on the totally new process discussed in this paper were conducted in France in 1988. Since then, almost 2.500.000 m^2 have been laid. The technique involves in situ creation of a reinforced membrane composed of a layer of elastomer binder onto which continous threads are sprayed immediately after it is laid. The threads interweave to create the reinforcement. This paper describes the process, gives the mix designs used, and introduces the machine that was especially designed to manufacture and lay the reinforced membrane. The results of laboratory tests conducted by one of the laboratories of the French administration and the Belgian Road Transport Research Center are also presented and discussed. These tests show that this process is one of the most efficient processes for preventing reflective cracking. Some examples of its application are mentioned. Its principal advantages over processes using shop-manufactured materials (geotextiles or geogrids) are outlined.

DESCRIPTION OF PROCEDURE

Principle

The principle (see Figure 1) consists of a composite system formed by simultaneously spraying a bituminous binder and continuous organic threads that interweave to create reinforcement, and then spreading chipping onto this membrane before applying the wearing course.

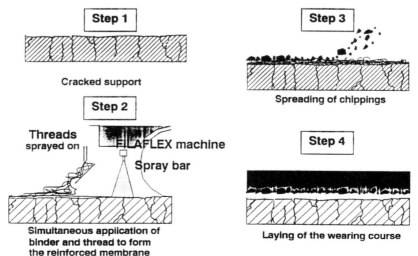

FIGURE 1 Steps in MURMOS application

Reflective Cracking in Pavements. Edited by J.M. Rigo, R. Degeimbre and L. Francken.
© 1993 RILEM. Published by E & FN Spon, 2–6 Boundary Row, London SE1 8HN. ISBN 0 419 18220 9.

Formula

The binder is an elastomeric bituminous emulsion applied at a rate of 0.9 to 1.8 kg/m^2 (equivalent to a residual binder content of 0.6 to 1.2 kg/m^2).
The polyester threads are sprayed on at high velocity. The multistrand threads ranging in size from 140 to 500 decitex (1 decitex is the weight in grams of 10.000 m of thread) are generally spread to a density of 100 g/m^2, with variation vetween 40 and 120 g/m^2 possible.
Chipping (6/10 or 10/14 mm) coverage is about 5 to 10 L/m^2.
There are several options for the wearing course : e.g., Glg-type chip seal (course chippings, binder interlay, and fine chippings) or conventional asphalt concrete, but the most frequent wearing course is fiber-based asphalt, which, because of its high binder content, further improves the performance of the membrane.

MACHINE FOR APPLYING MURMOS

This new MURMOS process required the development of a special machine. Once all the parameters for application of this concept had been defined with a sprayer-drawn prototype that sprayed threads under pressure in a 1-m-wide strip, two high-throughput machines were built. They were in fact slightly different, but the following equipment was common to them both :

* A roadgoing trailer with two steering axles that carries all the subunits required to make the membrane,

    (1) A power source for the hydraulics units and (2) the pump feeding the thread injectors,

* Two independent thread-spraying units, each including :

    (3) A 900-kg reel of thread on a 1.5-m-long, telescopic arm that can be extended overboard to obtain a total width of application of 3.70 m (with both reels),

    (4) A tensioning system that unwinds the 900 threads from the drum, and

    (5) A thread spray bar (seven injectors on one machine, nine on the other) placed immediately behind the binder nozzles and connected to the pump by flexible hoses. As for the binder spray bar, the thread spray bar can be extended laterally to vary the width of application anywhere between 1.25 and 3.70 m.

    (6) A 6.000-L emulsion tank (on one machine only),

    (7) A built-in binder spray bar (extendable from 1.25 to 3.70 m) for application of a uniform membrane at a constant rate, onto which the threads are sprayed immediately before spreading of chippings and laying of the wearing course,

The layout of the Filaflex machine is shown in Figure 2.

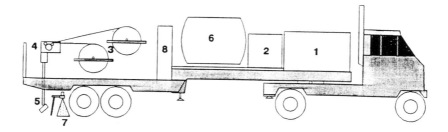

FIGURE 2 Description of machine

LABORATORY ASSESSMENT

Different structures were tested on a variety of devices to identify the most suitable systems.

Tests conducted at the Regional Laboratory of Ponts et Chaussées in Autun, France (Ref. 1, 2, 3)

Principle behind the test

The tests involved examining the speed at which a crack rises through a composite pavement including a system intended to delay the reflection of cracks and a wearing course.
Each test sample was representative of the system and was submitted to the following two simultaneous loads at a constant temperature (5°C) :

* Slow, sustained, longitudinal tensile stress to simulate thermal contraction ; and

* Cyclic vertical bending at 1 Hz to simulate traffic.

The progress of the crack was monitored by means of a serie of electrical conductors.
The test apparatus is shown on Figure 3. The test made it possible to assess various characteristics linked to the efficiency of the system studied (reflection of crack, speed of development, and time for full cracking of the system).

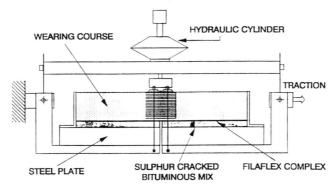

FIGURE 3 Contraction and bending test apparatus of the regional laboratory at Autun

Results

Five MURMOS systems were studied. The differences between them concerned:

* Binder content (per square meter),
* Type of binder, and
* Thread content (per square meter).

The compositions of the system are presented in Table 1, along with the experimental results.

Table 1 : Results of Laboratory Testing of Murmos Samples of Various Composition

TEST	BINDER	BINDER CONTENT (kg/m^2)	THREAD CONTENT (g/m^2)	CRACK STARTING TIME (minutes)	RUPTURE TIME (minutes)
1	Biphase elastomer bituminous emulsion	1,2	50	110	450
2	"	1,2	100	160	570
3	"	1,6	50	200	520
4	"	1,6	100	260	630
5	"	1,0	80	270	540

In all cases, the wearing course was 4-cm thick fiber-based asphalt with the following characteristics :
* Continuous grading curve : 0 to 10 mm
* Binder content : 7.2 percent of 60/70 pen
* Fiber content : 1.5 percent.

This table, as well as information on other existing solutions assessed with the same test procedure, indicates that the best performance was obtained with the system containing 1.6 kg/m^2 of biphase elastomeric bituminous emulsion and 100 g/m^2 of thread (Figure 4).

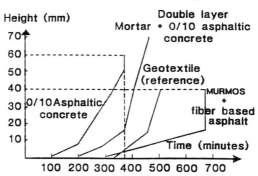

FIGURE 4 : **Results of contraction and bending tests on various systems for preventing relfective cracking**

Tests conducted by the Belgian Road Transport Research Center

The Belgian Road Transport Research Center has developed a tensile and compressive test device that simulates the movement of a cracked concrete support as a result of thermal stresses (Ref. 4). This apparatus was described at the RILEM Congress in Liège in March 1989.

Principle of the test

The outstanding feature of this test device is that it stimulates crack opening and closing at a speed close to that observed in situ as a result of daily temperature variation, i.e., several 10ths of a milimeter per hour.

With this apparatus, deformation is induced by the expansion or contraction of the frame to which the sample is attached. Such slow and perfectly controllable deformation is obtained by periodically changing the temperature of a liquid circulating in the bars of the frame. The sample is kept at a constant -5°C or -10°C throughout the test. The roadbase lies on a bed of steel balls that allow free and practically frictionless movement. The apparatus is fitted with sensors for measuring the stresses applied, the opening of the crack, and the displacement of the pavement relative to the cracked support. Visual observation of the phenomenon is facilited by a video system : filming is controlled by the data acquisition and verification system.
The system to be tested is glued to the road base, a block of precracked concrete. It is then covered with a wearing course. The test samples measure 60 cm in length, and 7 cm in width, and the concrete is 7 cm thick. The speed at which cracks open and close is 15 m/min, i.e. 1 mm/hr.
The test is computer-controlled, with continuous measurement of the opening of the crack in the support and in the asphalt overlay, the displacement of the wearing course relative to the support, and the applied stress. The progress of the crack is video recorded.

The maximum strain (tensile stress) applied at the edges of the crack inducer is about 25 percent (i.e., 1 mm of opening during a 2-hr cycle for a starting gap of 4 mm).
The results were summarized by C. Clauwaert and L. Franken : "In general, the interface systems with the highest tensile strengths crack rapidly whereas those comprising emulsions of low-viscosity bitumen can withstand high relative displacement between the overlay and the cracked support without cracking or separation of the various layers of the structure".

Results

Given the results of the MURMOS tests performed at the Regional Laboratory in Autun, only systems containing 100 g/m^2 of thread and 1.6 kg/m^2 of emulsion were tested. Table 2 presents descriptions of the structures and the results obtained with the test at -10°C.

It appears that none of the MURMOS systems tested cracked aftre 20 cycles at -5°C. The same result was obtained with the test at -10°C.
At -5°C, the tested systems using geotextiles and geogrids indicated forces in concrete three or four times higher than that obtained with test sample RTE 3.1. At the time this paper was written, the results of testing at -10°C indicated that no system other than that described in Table 2 withstood cracking. This excellent performance was ascribed to a decoupling effect caused by the thread reinforcement in the membrane that considerably reduced the maximum stress induced in the asphalt overlay when the crack opened.

To sum up the results of the laboratory tests performed by the French Public Service in Autun and by the Belgian Road Transport Research Center, it appears that the composite overlay presented in this paper is one of the best of the crack-inhibiting courses known to date.

Table 2 : Tests performed at the Belgian Road Transport Research Center

TEST SAMPLE	R.T.E. 1	R.T.E. 2	R.T.E. 3-1	R.T.E. 3-2
System for preventing reflective cracking	M.U.R.M.O.S. PMB emulsion 1.6 kg/m2	M.U.R.M.O.S. PMB emulsion 1.6 kg/m2	M.U.R.M.O.S. PMP emulsion 1.6 kg/m2	M.U.R.M.O.S. PMB emulsion 1.6 kg/m2
Wearing course	Fiber based asphalt	Type 1 overlay (7 cm)	Type 1 overlay (7 cm)	Type 1 overlay (7 cm)
Test temperature (°C)	-5°C	-5°C	-5°C	-10°C
OBSERVATIONS				
Cycles	20	26	20	20
Cracking	None	None	None	None
Maximum elongation of support (mm)	1,012	1,021	1.10	1,014
MEASUREMENTS (cycle)	1st	1st	1st	1st
Max force in concrete (N)	2,745	4,595	1,501	3,346
Max elongation in overlay (mm)	0,180	0,180	0,051	0,084
Max stress in overlay (MPa)	0,89	0,97	0,30	0,68
Max stress in interface system (MPa)	0,125	0,210	0,069	0,159
MEASUREMENTS (cycle)	20th	20th	20th	20th
Max force in concrete (N)	530	2,216	504	1,452
Max elongation in overlay (mm)	0,098	0,179	0,025	0,039
Max stress in overlay (MPa)	0,17	0,47	0,10	0,30

PMB = Polymer Modified Bitumen Emulsion

PRACTICAL EXPERIENCE

As of mid-1992, almost 2.5 million m² of MURMOS systems had been applied in France, in Belgium and Spain.

The first test sections were laid at the end of 1988 and early 1989. These sections enabled definition of the various application parameters, which led to the development of the MURMOS machine, and they also gave the opportunity for long-term testing of several types of composite systems under different conditions.

These test sections have been followed up. Some details of actual applications are given in the following paragraphs as well as in the paper presented by Y. DECOENE (Ref. 5).

RN 86 in the Ardèche

A 20.000 m² stretch of national highway RN 86 in the center of France (Ardèche) was upgraded in 1988. The MURMOS innovative technique was selected as part of the contest organized by the regional administration.
The technique consisted of two processes :
* Preventing cracks rising to the surface of the old semirigid pavement. The solution proposed replaced the basic solution, which consisted of 12 cm of conventional asphalt concrete.
* Preventing cracks rising to the surface of the new pavement or the existing pavement reinforced with a cement-treated layer.

The wearing course is a 4-cm-thick fiber-based asphalt concrete.

A55 Highway

The 155 highway is a highly cracked thick flexible pavement. A 30.000 m² stretch was upgraded in August 1989. The T0+ traffic corresponds to 1.200 trucks per day in each direction.
The wearing course is a 2-cm-thick fiber-based asphalt concrete.

A6 Highway

A 100.000 m² stretch of highway was upgraded in May 1990. Traffic there is also more than 1.200 trucks per day in each direction.
The wearing course is a 4-cm thick fiber-based porous asphalt. In this case, the process was applied to three types of support :
* Continuous reinforced cement concrete,
* Californian cement slabs stabilized with connectors (in order to prevent relative vertical movement), and
* Thick cement concrete slabs.

DESIGN, FIELDS AND LIMITS OF APPLICATION

Design

On the basis of the laboratory tests outlined above and the field experience acquired since 1988, the following standard design has been adopted :

Binder	:	Elastomer-modified bitumen emulsion	1.6 kg/m²
		Residual bitumen content	1.0 kg/m²
Threads	:	Polyester	100 g/m²
Chippings	:	6/10	5 to 6 l/m²
Wearing course:		Pure-bitumen asphaltic concrete	Max size 10 mm
			Thickness 6 to 7 cm
	or	Fibre-reinforced asphaltic concrete	Compoflex Thickness 4cm
			Mediflex
	or	*GLg chipseal	4/6 - 10/14

GLg chipseal	4/6 aggregate	:	7 to 8 l/m²
	Binder	:	1.6 kg/m²
	10/14 aggregate	:	8 to 9 l/m²
	Fibres	:	100 g/m²
	Binder	:	1.2 kg/m²

Of course this design is to be adapted to the type and condition of the road surface and traffic conditions.

On absorbant surfaces (aged or porous) or surfaces with old cracks, the emulsion content should be increased to as much as 2 kg/m^2 in some cases, if not more. At the same time the thread content should be increased to as much as 120 g/m^2.

In some special cases, a second layer of emulsion will be laid at the same time as the wearing course (hot mix), using a self priming paver.

Fields and limits of application

Experience shows that MURMOS can be laid on cracked pavements bearing all kinds of traffic, e.g.
- flexible pavements with cracks due to fatigue or ageing
- semi-rigid pavements with shrinkage cracks
- cement concrete pavements (CRCC or slabbed).

In the case of highly deteriorated pavements, MURMOS should not be applied with excessively thin wearing courses.

Furthermore, if the road surface has cracks or joints with relative vertical movements greater than 30/100 mm (approximately), cracks will quickly reflect up to the surface of a thin course. In this case, the following procedures should be applied :
- either a thick overlay which may contain MURMOS, or
- in the case of cement concrete pavements, reduce relative vertical movement by installing connectors before applying MURMOS and an asphaltic concrete : this was done most successfully on the A6 motorway in 1990.

CONCLUSIONS

First experimented with in 1988, a new system for preventing reflective cracking has been developed. MURMOS consists of manufacturing a reinforced membrane in situ by simultaneously spraying a layer of generally modified binder and continuous threads that interweave to form the reinforcement. This composite system is then covered with a wearing course.

After 3.5 years of experiments, the main conclusions are as follows :

* Laboratory tests indicate that the performance of the MURMOS system is among that of the best,
* With the specially designed machine, daily laying rates are high ; they can reach 30.000 m^2 per day,
* In distinction to systems using shop-manufactured reinforcement, the MURMOS process presents no problem with repsect to bends or road width (no folds in curves or problem of overlapping),
* Although relatively recent, the upgrading work performed with the MURMOS process up to mid-1992 amounting the almost 2.500.000 m^2 on a wide variety of roads (cracked flexible pavements, semirigid pavements, and cement concrete roads), demonstrates the excellent performance of the process in the field, and,
* The MURMOS process is felt to be an economic solution for cracking problems on all types of road (flexible, semirigid, and rigid).

---
*CRCC : Continuously Reinforced Cement Concrete

REFERENCES

1. C. Le Noir and M. Bailie. Etude de Système anti Remontée de Fissures. Procédé FILAFLEX. French Ministry of National Public Works. Housing, National Development, and Transport. Road Transport Divisions. CETE. Ponts et Chaussées Regional Laboratory at Autun, 1990.
2. Fissuration de Retrait des Chaussées à Assises Traitées aux Liants Hydrauliques. Ponts et Chaussées Laboratory Bulletin Nos. 156 and 157. June and Aug. 1988
3. T.H. Vecoven. Méthode d'Essai de Systèmes Limitant la Remontée des Fissures dans les Chaussées. Ponts et Chaussées Regional Laboratory in Autun. RILEM Congress on Reflective Cracking in Pavements : Assessment and Control, Liège, Belgium. March 8-10, 1989.
4. C. Clauwaert and L. Franken. Etude et Observation de la Fissuration Réflective au Centre de Recherches Routières Belge. Belgian Road Transport Research Center. RILEM Congress on Reflective Cracking in Pavements : Assessment and Control, Liège, Belgium, March 8-10? 1989.
5. R. Dumont and Y. Decoene. Application of a Geotextile Manufactured on the Jobsite of the Belgian Motorway Mons-Tournai

28 THE USE OF A POLYPROPYLENE BITUMINOUS COMPOSITE OVERLAY TO RETARD REFLECTIVE CRACKING ON THE SURFACE OF A HIGHWAY

A.R. WOODSIDE, B. CURRIE and
W.D.H. WOODWARD
Department of Civil Engineering and Transport, University of
Ulster, Jordanstown, Northern Ireland

Abstract
For many years reflective cracking has been a problem in the maintenance of a road infrastructure. Various options have been tried in the past. However, these systems have necessitated other ancillary works. Consequently, the authors have developed a composite material consisting of a polypropylene mat sandwiched in a thick slurry bituminous matrix. A laboratory testing programme and road trials are outlined, indicating how the life of the road pavement mat be enhanced and the reflective cracking deferred. The paper discusses the additional advantages of such a system and concludes by indicating how this composite mat may affect the "life-time" costing of a pavement.
Keywords: Highway Surface, Polypropylene, Bitumen, Overlay

1 Introduction

Surface cracking and reflective cracking cause considerable maintenance problems on much of the world's infrastructure. This is particularly evident on the heavily trafficked routes. The simplest solution available has been to scarify the old bituminous surface, treat the joints and then resurface with a superior material such as Hot Rolled Asphalt or Asphaltic Concrete. However, this method frequently is often only a temporary solution with the old problem reappearing with time.

An alternative solution adopted by several road authorities has been the use of a structural layer as the most viable solution. This method, however, necessitates alteration of pavement surface ironwork such as gullies and manhole covers, and the raising of kerbs which in many cases causes surface water drainage problems. One can argue in favour of either system but the authors would suggest that there are valid reasons against their selection, the main one being the cost of the replacement/overlay and the associated remedial work.

The cheapest form of remedial action carried out in the United Kingdom is to simply surface dress or "chip-seal" the road surface and trust that its life may be extended for a number of years before the cracking/crazing becomes visible once more. This method has proved most successful when a polymer binder has been used as the binding agent. For example, the use of bitumen with Ethylene-Vinyl-Acetate or

Styrene-Butadiene-Styrene co-polymers provide a degree of elasticity to overcome the movement due to surface cracking.

However, if an overlay was provided which instead consisted of a thin structural layer and possessing a tensile strength characteristic it might prove to be a feasible solution to this problem. The authors decided to examine the use of woven membranes laid in conjunction with a bituminous material could enhance its properties and so provide a viable, cheap and long-term solution to the problem of surface cracking.

In an attempt to itemise what properties would be required from such a surfacing, the authors list the following as being necessary:

(a) absorb crack movement by the elasticity of the new surfacing material.
(b) provide an adequate bond to the existing road surface.
(c) compatible bonding of individual materials.
(d) cost effective.
(e) simple to lay.
(f) regulate the original surface.
(g) non-susceptible to temperature changes.
(h) conserve energy.
(i) environmentally acceptable.
(j) high skid resistant surface.

Consequently, the authors have developed a composite material known as Ralumat. This consists of a stress absorbing polypropylene mat sandwiched in the micro asphalt surfacing known as Ralumac. This new process was developed by the Highway Engineering Research Team at the University of Ulster's Technology Unit in conjunction with Colas (Northern Ireland) Limited.

2 Ralumac/Ralumat

Ralumac is a polymer modified, cold applied micro asphalt which is German in origin and has been used extensively in the United Kingdom and in Ireland by Colas Limited. Based on a two layer thick slurry it has proven to be a cost efficient and environmentally acceptable material which is successfully used to fill ruts, restore profiles, treat cracked and porous surfaces, and provide a stable and long-lasting skid-resistant surface.

Because of these reasons it was decided by the authors to further develop this already successful surfacing by incorporating a stress absorbing membrane or mat to see if its performance could be enhanced.

This involved a range of comparative performance test methods to be devised at the University of Ulster.

3 Testing programme

The following groups of tests were carried out:

(a) tensile testing of membranes
(b) component material compatibility assessment

(c)   skid resistance assessment
   (d)   comparative point load testing
   (e)   tensile testing
   (f)   retardation of reflective cracking assessment

Each of these will now be briefly discussed and some of the main findings outlined.

3.1 Tensile testing of membranes

A wide range of potentially suitable membranes were initially considered. Of these, a large number were deemed unsuitable or not viable due to its intended use and financial constraints. However, polypropylene matting appeared to be a feasible option as it was found to bond to a bituminous mixture and have the ability to stretch or be elongated for a reasonable amount without breaking. Ideally this should be a distance greater than the sum of the widths of the cracks over which it is to be laid.

To assess the suitability of possible types a large number were tested using an Instron machine under direct tension with samples of 50mm length assessed with a cross-head separation speed of 20mm/min.

3.2 Component material compatibility

It is essential that the individual components of any bituminous mix are compatible and behave in a prescribed manner. Laboratory testing showed that polypropylene membranes provided an excellent bond with all types of bituminous mixtures providing the binder content was adequate. However, not all aggregate and binder combinations were successful. For example, binder was observed to strip from quartzite aggregates. Consequently, tests were carried out to assess the compatibility of the chosen materials. This had been carried out by Colas (Northern Ireland) Limited and indicated which of a number of sources of aggregate could combine with the bituminous content to give the required properties of Marshal Stability >10kN and Flow Value of <4mm.

To assess performance, comparative wheel tracking trials were carried out which compared the original micro-asphalt with its reinforced polypropylene mat derivative, and to traditional surfacing materials such as Hot Rolled Asphalt and Close Graded Macadam.

The polypropylene membrane still requires an overlay to seal the surface, provide skid resistance and prevent the ingress of water. Further testing was carried out on surface dressed/membrane and Ralumac/membrane combinations. These were wheel-tracked using a modified wheel-tracking machine with a locked wheel on every return passage. This test quickly caused the surface dressed/membrane combination to fail whilst the original micro-asphalt did not.

3.3 Skid resistance assessment

Typically, thin overlays do not have high levels of skid-resistance due to the type and size of aggregate used. But, because of the high skid resistant coarse aggregate (<8mm) used in Ralumac, the resulting skid resistance values are comparable to other skid-resistant types of surfacing.

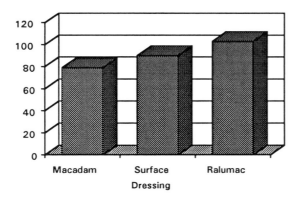

Fig.1. Skid resistance results

Figure 1 shows results obtained using the British Pendulum Tester which indicate that the micro-asphalt has an average Skid Resistance Value of 103 compared with 90 for surface dressing and 79 for close graded macadam. In this test a higher result indicates a more skid resistant surface.

3.4 Comparative point load testing

A point load test was developed to assess the loading characteristics if the reinforced micro-asphalt material. This consisted of placing prepared slabs between two supports 200mm apart and applying a gradually increasing central point load. Deflection of the sample was measured as the sample was tested to destruction. All testing was carried out at a standardised temperature of 22^0C. For comparative purposes a number of different types of bituminous mix were tested. Figure 2 is a summary of some of the results produced.

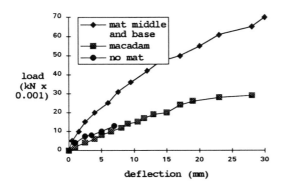

Fig.2. Load v deflection for different surfacing treatments

It can be seen that the commonly used 14mm Close Graded Macadam mix failed under a load of 0.028kN, the micro-asphalt surfacing at 0.014kN and 0.070kN for the polypropylene reinforced micro-asphalt.

3.5 Tensile testing

Tensile testing was carried out using an Instron machine to compare the following:

(i)   polypropylene mat on its own (17 strands).
(ii) micro-asphalt on its own.
(iii) micro-asphalt reinforced with 17 strands of polypropylene

Some of the results are shown in Figure 3. From the test results obtained, average values for failure were as follows:

(i)   mat on its own   1.72kN
(ii) micro-asphalt   0.14kN
(iii) reinforced micro-asphalt (mesh in bottom)   1.95kN
(iv) reinforced micro-asphalt (mesh middle and bottom)   3.85kN

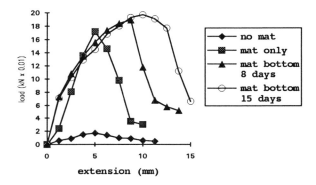

Fig.3. Tensile test - load v extension

The results indicate how the mat has contributed to the tensile strength of the overall material. The micro-asphalt on its own had a relatively small value of 0.14kN. This compares to 1.95kN when the mat is used. This would indicate that the warp in the mesh is being used to increase the tensile strength of the material by assisting in bonding the polypropylene to the micro-asphalt. Furthermore, it is evident that two layers of mat on their own - 3.44kN, is less than two layers used in conjunction with Ralumac - 3.85kN. Again, this suggests that the two materials are bonding to form a composite which is of higher tensile strength than the sum of the individual members. It was also noted that the composite material failed at a definite peak which was reached in a reasonably uniform manner showing no indication of slippage (bond-breaking) between the two components. Thus the point of failure can easily be predicted.

3.6 Retardation of reflective cracking assessment

Modifications to a wheel-tracking machine enabled reflective cracking to be simulated in the laboratory. Various combinations of reinforced and non-reinforced micro-asphalt were assessed under differing test conditions such as loading and width of crack separation. Some of the results are shown in Figure 4 where the number of passes required to produce a cracked test specimen are shown for different crack widths and degree of reinforcement.

The results show how the polypropylene reinforced material can delay the formation of reflective cracking.

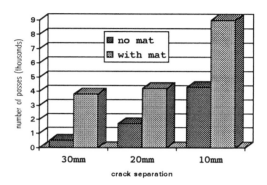

Fig.4. Reflective cracking - number of passes to form a crack

3.7 Road trials

A small scale road trial was carried out on a lightly trafficked residential road. Areas were selected which displayed signs of reflective cracking and using small sections of fabric these areas were reinforced while adjacent areas were surfaced in the normal manner. The result, after two years, is that the reflective cracking has reappeared in the surrounding areas adjacent to the reinforced. areas. Thus the life of the reinforced overlay appears to be much greater than that of the unreinforced.

However, problems were experienced in the laying process and research is now underway to design a mechanism whereby the fabric may be laid together as an integrated part of the mix.

Furthermore, a two stage operation, ie fabric laying followed by slurry laying increases the overall cost of the operation and thus makes it unattractive when compared with a hot mix overlay such as Hot Rolled Asphalt.

The approximate cost of the polypropylene reinforced micro-asphalt system would appear to be approximately £1.55 compared to £1.85 for a Hot Rolled Asphalt.

The authors believe that this process has greatest potential in some of the hot climate developing countries where labour is relatively inexpensive.

4 Conclusions

The simplicity of a cold applied micro asphalt system is an attraction in itself especially to the experienced contractor. The surfacing provides the same element of waterproofing and restoration of skid resistance as surface dressing without the need to worry about binder spread rates, chipping size, rates of spread, or the adverse affects of temperature and rainfall.

However, neither surface dressing nor surfacings such as Ralumac offer a satisfactory long-term solution to the problem of reflective cracking.

The results obtained from the authors' investigation using polypropylene matting to enhance the performance of this micro asphalt would indicate that additional tensile strength and overall performance could be achieved by the inclusion of such a membrane and appears to be a step closer to overcoming the problem of reflective cracking.

5 Acknowledgements

The authors wish to acknowledge the help and co-operation they have received in the implementation of this research by The Department of Civil Engineering & Transport at The University of Ulster and Colas (NI) Ltd.

29 FABRIC REINFORCED CHIP SEAL SURFACING AND RESURFACING

C.J. SPRAGUE
Sprague and Sprague Consulting Engineers, Greenville, USA
D.M. CAMPBELL
Hoechst Celanese Corporation, Spartanburg, USA
P.J. KUCK
Hoechst AG, F+E Spunbond, Bobingen, Germany

Abstract

A cost-effective surfacing and resurfacing technique has emerged world-wide which combines a geotextile and a traditional low cost bituminous surfacing method to produce a durable, weatherproof road surface for a fraction of the price of typical flexible pavement structures. This system has become known as <u>fabric reinforced chip seal</u> or <u>fabric reinforced surface dressing</u>. Trial installations have been installed and monitored in North and South America, Europe and Australia. In each case the fabric reinforced chip seal has successfully waterproofed the road structure, arrested disintegration and provided a durable skid-resistant surface. Various installations around the world are summarized to substantiate the effectiveness of the fabric reinforcement. Additionally, detailed installation costs and procedures as well as material and equipment recommendations are made based on the referenced installations.
Keywords : Chip seal, Surface dressing, Waterproof structure, Skid resistance, Resurfacing.

1 Introduction

There are thousands of miles of low volume and rural roads in North America which must be maintained or improved to faciliate safe access to and commerce within and between modestly populated areas. Frequently these roads were initially constructed to minimize capital outlay while still providing acceptable performance under existing environmental and traffic conditions. Yet, over time traffic volumes and wheel loads increase, accelerating the deterioration of road structures which are already subject to the influences of climatic cycles, extreme weather events and lack of maintenance.
   Regardless of whether the initial road construction was the result of private development or public funding, it is common for long-term maintenance of these roads to be the responsibility of the state, province, or municipality in

Reflective Cracking in Pavements. Edited by J.M. Rigo, R. Degeimbre and L. Francken.
© 1993 RILEM. Published by E & FN Spon, 2–6 Boundary Row, London SE1 8HN. ISBN 0 419 18220 9.

which the road is located. This seemingly unavoidable responsibility and the financial burden that goes with it has led engineers to seek out cost-effective means for maintaining, i.e. surfacing and resurfacing, these facilities.

One such cost-effective alternative has emerged which combines a geotextile and a traditional low cost bituminious surfacing technique to produce a durable, weatherproof road surface for a fraction of the price of typical flexible pavement structures. This system has become known as <u>fabric reinforced chip seal</u> or <u>fabric reinforced surface dressing</u>.

2 Fabric reinforced chip seal

A chip seal is a low cost road surfacing technique that involves the spraying of a "tack coat" of asphalt, or bitumen, on the road surface followed by placing a thin layer of uniform size aggregate into the tack coat creating a thin "bound" surface layer. This procedure can be repeated to create a thicker, more durable surface.

A fabric reinforced chip seal is a modification of the conventional chip seal which involves laying a fabric on a specially proportioned tack coat which is first sprayed on the road surface, then spraying asphalt, placing aggregate and compacting a chip seal layer on top of the fabric. The fabric binds to the road surface, becomes saturated with asphalt and provides greater integrity to the chip seal.

3 Initial road condition

Commonly a low volume road becomes a candidate for maintenance when it's surface clearly indicates that the road is deteriorating. This deterioration may be in the form of excessive ruts, bumps, or holes in the case of "dirt" roads, or it may be in the form of various forms of cracking and pot-holing in the case of paved roads. In either case, the surface is often an indication that overall road structure failure has occured and reconstruction is likely to be the only long-term solution.

Yet, an intermediate or partial solution implemented immediately may arrest the deterioration at its current level and lead to a lower cost permanent solution. Specifically, if the cause of the deterioration, commonly water, can be controlled and the road structure can thus be stabilized at its present level, the present road may very well prove suitable as a base upon which to increase the road structure in the future. This has been the approach taken by engineers in using fabric reinforced chip seals.

4 Expectations

The use of chip seals, or surface dressings as they are sometimes called, for the surfacing and resurfacing of roads is not new. They have been used to extend the life of road structures in the following ways:
1. Waterproofing the road structure
2. Arresting disintegration
3. Providing a skid-resistant surface

Yet, chip seals have traditionally been very short-term solutions, usually requiring resurfacing in 2 to 4 years as a result of continued road deterioration and chip loss.

Fabric reinforced chip seals on the other hand present a significantly longer-term enhancement to the road structure. There are three basic functions which the fabric reinforcement in the chip seal performs [1].
1. Seal and bridge cracks to prevent any further infiltration of water.
2. Arrest disintegration by holding cracks together and stabilizing underlying aggregate or pavement.
3. "Normalize" the surface by providing a uniform substrate on which to apply the chip seal.

Various installations around the world have substantiated the effectiveness of the fabric reinforcement [1, 2, 3, 4, 5].

5 Materials

<u>Tack coat and binder.</u> Usually a medium to rapid curing cutback asphalt or a rapid setting emulsified asphalt is used. Though waiting time may be required to allow for sufficient curing, or evaporation of the liquefying agent, these "liquefied" asphalts are much easier to work with than straight asphalt cement which must be heated and kept recirculating. Special consideration regarding the proportion of residual asphalt contained in the tack coat is necessary when working on steep grades which may cause the cutback or emulsion to run if too liquid. A curing time ranging from 30 to 90 minutes should be expected depending on the residual asphalt content, spray rate and temperature.

<u>Fabric reinforcement.</u> Relatively light-weight nonwoven needlepunched geotextiles have proven suitable. Recommended fabrics have area weights ranging from 120 to 200 g/m^2 and are composed of either polyester or polypropylene. Some shrinkage has been experienced when using polypropylene fabrics though it is not clear that this adversely effects performance [1]. Additionally, heat treating one side of the fabric has been proposed as a benefit in reducing the likelyhood of fabric adhering to sticky tires which may traverse its surface. This benefit has not been validated quantitatively though the additional fabric stiffness resulting from heat treating does make the fabric more sensitive to wrinkling during lay-down on corners.

Stone chips. Commonly used stone, or chip, is relatively uniform in gradation and ranges in size from 4 to 18 mm in largest dimension. Two layers of stone are often used with the second layer being either the same size as the first or slightly smaller to effectively embed in the first layer when compacted.

Distributor truck. A tank truck with a horizontal full-width spray bar is required for spraying asphalt cutbacks or emulsions. The truck must be equipped with heating devices and recirculation capability if straight asphalt cement is to be used. The truck should include a hand-held spray wand for localized applications (necessary when fixing wrinkles, preparing overlaps or filling in streaks from nonuniform spray).

Fabric lay-down device. Rolls of fabric are both bulky and heavy, therefore to lay the fabric without excessive wrinkling over large distances it is important to use a fabric lay-down device. Acceptable devices include frames mounted to the bucket of a front end loader, specially modified tractors, or hand held "braking" devices. Regardless of the specific device chosen, it is important to brush the fabric into the tack coat and to have an absolute minimum amount of play in the steering to control any "wandering" which can lead to wrinkles during lay-down.

Chip spreader. Either a self-propelled hopper/spreader or a back-dumping dump truck with a spreader attachment is required to uniformly spread stone chips. Chips must be discharged in such a way that the spreader only travels on just-spread chips, never on fresh binder.

Compaction equipment. Compaction equipment is necessary to properly "seat" the stone in the binder. Also, if the tack coat has overly stiffened, which can happen in cooler weather, the fabric may be rolled prior to chip sealing to insure bonding to the pavement surface. Though a steel wheeled roller provides a large compaction effort, the large drum tends to bridge over the irregularities which are so common in a deteriorating pavement. Therefore, to achieve sufficient "seating" of the stone pneumatic tire compaction equipment is required.

Other equipment and supplies.
Road sweeper/vacuum for preparing the old pavement surface and for removing excess chip from the new surface.

Gasoline and rags for cleaning residual asphalt from people and equipment.

Push brooms are useful for assuring fabric contact with the pavement surface and eliminating small wrinkles.

Knives or cutters are needed to cut out excessively large wrinkles (larger than 2,5 cm or 1 inch) or to discontinue fabric laying prior to the end of a roll.

6 Installation procedures

Road surface preparation.
   Subgrades: Subgrade should be graded, properly compacted, and crowned.
   Cracked pavement surfaces: Sweep paved surfaces and fill potholes.
   Spray tack coat. A tack coat should be sprayed onto the existing prepared subgrade or cracked pavement surface. The tack coat should be a cutback or emulsified asphalt proportioned to provide the required amount of residual asphalt. Straight asphalt cement is certainly acceptable if a uniform spray pattern can be maintained. Uniform, even spraying of the tack coat at the appropriate application rate is very important to complete saturation of the fabric and therefore the ability of the system to waterproof the road.
   Fabric installation. The fabric should be installed using an appropriate installation device but not until the tack coat has sufficiently cured. All end joints should be overlapped (3 to 15 cm or 1" to 6") in the direction of paving and sprayed with additional tack coat to insure adequate saturation. Adjacent rolls (side by side) should be placed with a positive overlap (3 to 15 cm or 1" to 6"). Sufficient tack coat should be over-sprayed in order to insure proper bonding between fabric layers. When a cutback or emulsified tack coat is used, caution should be exercised to insure that adjustments are made to the application rates so that the residual asphalt contents are maintained and that the tack coat has fully cured (volatiles or water totally evaporated) prior to any fabric placement.
   Spray binder. Spray the surface of the fabric with the quantity of cutback or emulsion required when placing a conventional chip seal. Inadequate binder application has been found to directly correlate to excessive chip loss [1, 4], so attention should be given to assuring that the binder is sprayed at no less than the specified application rate. The quantity required will be based on type and size of chips as well as the spread rate. No additional binder is required by the fabric. (Note: This step is sometimes skipped.) Instead tack coat in excess of that needed to saturate the fabric and impregnate the underlayer is sprayed. The excess is then available to bond the chip from beneath.
   Chip spreading. Spread chip uniformly being careful not to allow wheels to track in the binder. This will require back-dumping or the use of a self-propelled hopper/spreader. The chip may be spread immediately after the binder is sprayed to facilitate trafficking.
   Compaction. Compact the stone chip surface with a steel drum roller followed by a pneumatic tire roller. Numerous passes of the pneumatic tire roller should be made. It is this "kneading" action which is so important in properly seating the chip [1]. Excess stone may be removed by sweeping after compaction.

Multiple treatments. Additional repetitions of binder spraying, chip spreading, and compaction may be appropriate. A double treatment using a slightly smaller chip gradation on the second repetition is a common surfacing technique. It is important to allow the previous layer, or treatment, to completely cure before placing the next binder/chip layer. Trafficking of intermediate binder/chip layers may be allowed, and even desirable, since automobile and truck traffic will provide additional pneumatic compaction.

7 Review of chip seal experience

Winnipeg, Manitoba, Canada. Maintenance of gravel roads in Manitoba generally includes treating them approximately three times yearly with a calcium lignosulfonate dust suppressor and once a year (in the spring) with an application of well graded gravel at approximately 130 metric tons per lane kilometer (= 43 kg/m²).

The fabric reinforced chip seal is seen as a possible method of eliminating the annual gravel treatment as well as reducing the need for dust suppression. Additionally, it is anticipated that the road surface will be smoother and retain chip more effectively while still maintaining an acceptable appearance. In July 1992 the Manitoba Department of Highways and Transportation installed a 400 m (1/4 mile) fabric reinforced chip seal trial section on the Heritage Highway (Manitoba 238).

Heritage Highway details. The Heritage Highway is historically significant and must be maintained with a historical appearance. All materials used on the trial section are outlined in Table 1. The installation proceeded very smoothly using special hand "brakes" to assist in the manual installation of the fabric and conventional Department equipment for placement of the chip seal.

Within a few weeks of the surfacing, during which there were several rainy days, overloaded trucks used the roadway. The trucks caused modest rutting and extensive cracking in the "control" sections while the sections surfaced with fabric reinforced chip seal exhibited no visible distress. The damage to the control sections was extensive enough to require resealing. This short-term superior performance of the trial section is likely a result of the continous weatherproofing provided by the asphalt impregnated fabric reinforcement. The Department will continue to monitor the installation's performance.

Greenville County, South Carolina, USA. In late 1986, the Greenville County Engineering Department began investigating the use of paving fabrics to reduce maintenance costs associated with low volume county roads. The county had traditionally stretched its maintenance dollar by resurfacing low volume roads using split seal bituminous surfacing

TABLE 1
REINFORCED CHIP SEAL EXPERIENCE

CHIP SEAL ELEMENT	MANITOBA	GREEN-VILLE	SCDHPT	UK	FRENCH	AUSTRALIA
Existing surface	Subgrade	Pavement	Pavement	Pavement	Subgrade	Subgrade
Tack coat	MC500 Cutback @2.0 l/m²	CRS-2 Emulsion @1.5 l/m²	CRS-2 Emulsion @1.6 l/m²	200 sec Cutback @0.8 l/m²	Cutback/ Emulsion @1.1 l/m²	70 pen Cutback @1.1 l/m²
Geotextile	140 g/m² CF,NP,NW PET	120 g/m² CF,NP,NW PET	140 g/m² ST,NP,NW PP	140 g/m² CF,NP,NW PP	>130 g/m² CF,NP,NW PET	>180 g/m² CF,NP,NW PET
Binder	MC500 Cutback @1.7 l/m²	CRS-2 Emulsion @1.1 l/m²	(a)	(a)	Cutback/ Emulsion @ N/A	(a)
Chip	4 - 12 mm @27 kg/m²	6 - 14 mm @16 kg/m²	Sand @4 kg/m²	14 mm @15 kg/m²	6 - 14 mm @19 kg/m²	5 - 7 mm @ N/A
Binder	MC500 Cutback @2.0 l/m²	CRS-2 Emulsion @1.1 l/m²	CRS-2 Emulsion @1.5 l/m²	200 sec Cutback @1.6 l/m²	Cutback/ Emulsion @ N/A	70 pen Asphalt @1.5 l/m²
Chip	-	6 - 10 mm @11 kg/m²	8 - 18 mm @13 kg/m²	14 mm @15 kg/m²	4 - 6 mm @16 kg/m²	10 mm @ N/A
Cost/m²	$C N/A vs $C1.00	$1.77 vs $1.23	$1.50 vs $ N/A	N/A	N/A	$A6.75 vs $A4.00

KEY: CF = continous filament
    ST = staple fiber
    NP, NW = needlepunched nonwoven
    PET = polyester
    PP = polypropylene
    (a) = binder was not sprayed directly on the fabric, rather chips or sand were spread directly on and rolled into the fabric.
CONVERSIONS: 1 l/m² = 0.22 gal/yd²
               1 kg/m² = 1,84 lbs/yd²
               1 g/m² = 0,0294 oz/yd²
               1 m² = 1.197 yd²
               1 $C = $US 0.84 (8/21/92)
               1$A = $US 0.72 (8/21/92)

(double treatment of chip seal) using the department's own equipment.

The department was well aware of the widespread successful use of paving fabrics in asphalt overlays but was concerned that many of the roads needing to be resurfaced were in such poor condition that a typical 2,5 - 5 cm (1 - 2 inch) thick overlay would prematurely degrade. Looking to achieve the benefits of incorporating fabrics into the road structure without wasting money on expensive asphalt surfacing of structurally inadequate roads, the county engineer decided to try a 2-step approach. First, low volume roads would be given a fabric reinforced chip seal to stabilize the current surface, prevent further infiltration, and permit chip to fill irregularities. Secondly, 2 - 3 years later, after the road structure - pavement and subgrade - had stabilized, proceed with an overlay.

To-date this approach has been taken on several county roads. The first road to receive a fabric reinforced chip seal, Shagbark Circle in the summer of 1987, has recently been over-laid. The performance of the chip seal on this road surpassed all expectations and could have continued to provide a very adequate riding surface for some time but increased development in the area increased public pressure to upgrade road appearance.

Shagbark Circle details. Large sections of Shagbark Circle had deteriorated to the point of requiring complete reconstruction. Alligator cracking, block cracking and pot-holing were widespread. In lieu of the costly task of ripping out the pavement down to the subgrade it was decided that this pavement would provide a severe test for a fabric reinforced chip seal. The chip seal, if successful, would stabilize the pavement structure and arrest further deterioration.

The road averaged 6,0 m (18 feet) wide and had localized areas of poor roadside drainage which corresponded to some of the areas of greatest pavement deterioration.

The materials used for constructing the chip seal are outlined in Table 1. All materials are those typically used by the department in conventional chip seals. Fabric was supplied in both 3,80 m (12,5 ft) and 1,90 m (6,25 ft) widths to cover the road in 2 passes without excessive overlap or waste using an installation frame attached to the bucket of a front-end loader. Conventional county-owned equipment was used to construct the chip seal. The installation procedures used were in accordance with those outlined earlier.

South Carolina Department of Highways and Public Transportation experience. As part of an on-going program to evaluate the effectiveness of geotextiles in pavements, the SCDHPT constructed two trial sections using fabric reinforced chip seals in 1986. One trial was done on SC 215 in Fairfield County [2] and the other was done on Oakdale Street in Rock Hill, SC [3]. The SCDHPT uses chip seals extensively to maintain rural and low volume roads and was interested in seeing if the benefits of paving fabric in overlays - waterproofing and reducing reflection cracking - could be extended to chip seals. The SCDHPT is monitoring performance of these trial sections.

SC 215 and Oakdale St. details. These 2-mile test sections of 2-lane road used 23.400 m² (28,000 sy) of geotextile each. The fabric was installed using a tractor specially modified to efficiently install large fabric rolls. The materials used in the chip seal are detailed in Table 1. The Oakdale Street project used a three layer system, placing three chip seal treatments after the initial fabric installation. This much heavier surface is not detailed herein. A unique aspect of these installations is the application of sand immediately over the fabric and the subsequent trafficking for 6 - 7 days before proceeding with the placement of the chip seal.

U.K. experience [1]. Research in the U.K. has focused on the use of fabric reinforced chip seals over distressed pavements. Several trial sections have been installed and monitored. Various combinations of fabrics and tack/binder coats have been used. The procedure used in the U.K. involves an application of tack coat on the road surface, then laying of the fabric followed by chip application. This minimizes trafficking directly on the fabric and any resulting wrinkling. Table 1 gives the materials used to construct a typical trial section.

Monitoring of performance produced the following observations:
* Areas of insufficient tack coat or binder correlated with excessive chip loss.
* Manual unrolling of fabric was clumsy and produced wrinkling.
* Pneumatic tire compactors do the best job of compacting the numerous local irregularities.
* Crack bridging and waterproofing effectiveness was apparant.
* Chip retention is enhanced by pneumatic tire traffic.
* Effects of lateral forces, i.e. braking or turning, on the system should be further investigated.

French experience [4]. The French have conducted research into fabric reinforced chip seals in Europe as well as in South America. The research has focused on surfacing existing "dirt" roads. Though the French have generally followed the installation procedures outlined earlier, they have also utilized the U.K. and Australian approach of not spraying binder directly on the fabric when conditions raise concerns about wrinkling. Typical materials used in the research and field trials are detailed in Table 1.

Performance evaluations of the field trials have indicated the following:
* Short-term performance is similar to conventional system.

* Single layer chip seal is more durable than double layer.
* Distinct correlation between insufficient binder and deterioration.
* Binder application rate = conventional rate + (1.1 x fabric saturation rate).
* Longer life and slower rate of deterioration.
* 20 - 30 % higher cost.
* Does not make up for structural inadequacy of pavement.

<u>Australian experience</u> [5]. Trial installations have been reported which focus on providing a surface to "dirt" roads to protect expensive subgrades using conventional surfacing as well as fabric and polymer modified bitumen enhanced surfacings. Many low volume, rural roads are constructed of clay which expands and contracts when wetted and dried, respectively. Additionally, the surface becomes very slippery when wet. The fabric reinforced chip seal surface has the potential to waterproof the subgrade, i.e. minimize fluctuations in moisture content, bridge cracks and provide a safe all-weather surface.

Over 50 kilometers (31 miles) of pavement have been sealed using this system. The installation procedure used is similar to that used in the U.K., i.e. binder is not sprayed directly on the fabric.

Peformance evaluations of the field trials have furnished the following observations:
* Conventional chip seal surfaces suffered substantial loss of stone and severe cracking.
* Fabric reinforced chip seals were holding up with only some wheel path deformation after 4 1/2 years.
* Shoulders are vulnerable.
* Equally effective over clay, silt and sand subgrades, as well as lime stabilized subgrades.
* Polymer modified bitumen tack coat requires greater application rate (+ 10 %) and still needs fabric.
* Higher tack coat application rates are required for silty and sandy subgrades because of their high surface areas.
* The binder application rate can be adjusted if the subgrade take-up is too great.

8 Cost effectiveness

As shown in Table 1, costs associated with fabric reinforced chip seal applications have been reported in a few cases. These reported costs clearly indicate that there is an increased cost associated with incorporating a fabric and additional tack coat into the surfacing or resurfacing of a road. Yet, there is evidence that this added cost may be easily offset by longer life and reduced maintenance.

Continued monitoring and reporting on the performance of the installations referenced will be necessary to definitively determine cost effectiveness.

9 Conclusions

The suitability and practicality of using fabric reinforced chip seals has been demonstrated world-wide. The technical merits of the system have been documented in a wide variety of climates and in association with both prepared soil subgrades and existing distressed pavements [1, 2, 3, 4, 5]. Long-term evaluation of existing installations with respect to conventional practices will be necessary to assess overall cost effectiveness. Still, short-term indications are very promising.

10 References

[1] Walsh, I. D. (1986), **Geotextiles in Highway Surface Dressing**, Proceedings of the Third International Conference on Geotextiles, Vienna, pp. 135-140.
[2] Craven, A. (1986), **Reinforcement Objective: Chip Seal and Geotextile Combined In Road Paving Job**, Geotechnical Fabrics Report, July/Aug, IFAI, p.12.
[3] Craven, A. (1986), **Renovation in Rock Hill**, Dixie Contractor Magazine, December 5, p. 22.
[4] Morel, G., H. Perrier, M. Require and J. Perfetti (1990), **Geotextile Reinforced Surface Dressings**, Proceedings of the Fourth International Conference on Geotextiles, Geomembranes and Related Products, the Hague, pp. 203 - 208.
[5] Sutherland, M. and P. Phillips (1990), **Geotextile Reinforced Sprayed Seal Roads in Rural Australia**, Proceedings of the Fourth International Conference on Geotextiles, Geomembranes and Related Products, The Hague, pp. 209 - 212.

30 JUTE FIBRE FOR PRODUCTION OF NON-WOVEN GEOTEXTILES TO PREVENT REFLECTIVE CRACKING IN PAVEMENTS

S.N. PANDEY and A.K. MAJUMDAR
J.T.R.L., (I.C.A.R.), Calcutta, India

Abstract
Jute, also known as golden fibre has been primarily used for packaging in past. The fibre contains lignin and cellulose including hemicellulose besides waxes, sugars, minerals etc. Due to stiff competition from synthetic fibres, its uses in packaging has been considerably reduced. In view of this, it has become necessary to develop diversified jute products for new uses. Recently the natural fibre has a special appeal in the field of geotextile engineering. The cost of jute fibre products are expected to be price competitive. The advantages of jute material is it's strength, excellent absorbency, better drapability, environmental compatibility, biodegradability and annual-renewability.

This paper discusses the feasibility of using jute geotextile for geo-technical application based on the laboratory investigations. The results indicate that though bio-degradable, it holds a good promise as a potential geo-textile for stabilising soft subgrades resulting in better geo-technical performance. Attempt is made to study to develop different types of geo-jute fabrics in the various geo-technical areas to protect Reflective Cracking such as soil erosion, soil conservation in pavements slopes of canals, dykes, road sides, railway track slopes, mountain slopes and mountain roads.

Suggestions have been made to improve the properties of jute fibres and to make it rot proof to provide durability. Various advantages of jute fibres and its blends with other fibres as geo-textiles and its application in civil engineering has been discussed.

JTRL has developed a PP/Jute needle punched Non-woven sandwich blended fabric which can be used as geo-textiles. Jute was sandwiched between polypropylene and soil burial test was carried out with the aid of culture of fungi developed from cowdung and other waste materials.

JTRL has also commercialised some non-woven fabrics with the technology of Mechanical, Thermical, Chemical bonding method which can be used in the area of geo-textiles and filtrations and to protect reflective cracking in pavements. Cyclic loading, strip strength of the burial and unburied fabrics were determined. It was observed that polypropylene covering jute from both sides help to prevent the jute degradation in the fabric. C.B.R. and pore size distribution were carried out. It was also observed that the stress-strain characteristics of soil are better with the jute fibre than without it.

Reflective Cracking in Pavements. Edited by J.M. Rigo, R. Degeimbre and L. Francken.
© 1993 RILEM. Published by E & FN Spon, 2–6 Boundary Row, London SE1 8HN. ISBN 0 419 18220 9.

Keywords: Jute Geotextiles, Environment Friendly, Sandwiched/ Betumenised Jute Nonwovens.

1 Introduction

Geotextiles made of natural fibre were used thousands of year before the development of Geosynthetics. However, the arrival of cheap and durable polymers after world war, it has provided the basis for today's geotextile industry. Natural fibres and their qualities have been over-shadowed in the market by the perceived advantages of greater strength and durability that synthetics can offer. Natural jute-fibre materials have their own distinct advantages and provide cost effective geo-textile forms in several applications.

Slope erosion protection.
Chanal liners.
For separation, filtration and reinforcement.
Improving roads pavements/drainage and waterproofing of cracked roadway.

Jute is one of the cheapest natural fibre, available in abundance in India, Bangladesh, China, Thailand, Nepal, Indonesia and several other countries. It has excellent physico-chemical properties which make it most suitable fibre at the lowest price for use in Geotextile.
Besides, several other items are fabricated from Geo-textiles for specific end uses. For Road Engineering, the items include are Jute non-woven, woven, bitumen coated, synthetic coated, chemical and thermically bonded non-wovens.
Jute Geotextiles are used,[Pandey and Majumdar (1989); Pandey and Majumdar (1990a)]for drainage and waterproofing of cracked roadway :

(a) Impermeable sheet (Extrusion coated jute non-wovens) to prevent seepage, behind structural defence formed of concrete slabs stones, cement morter etc.
(b) Permeable fitter cloth (Needle punched jute non-wovens) to permit seepage but prevent loss of soil, behind structural defence formed of concrete slopes, stone, gabions etc.
(c) Reinforcement (Jute combination fabrics) to strengthen soil (Reinforced) to compact or consolidation of surface of road.

The overall strength of road surface may be increased in two ways, [Pandey and Majumdar (1990b); Pandey and Majumdar (1990c); Pandey and Majumdar (1992)].

(a) Injection (Bitumen impregnated jute based or laying of Geo-textiles on surface of non-wovens).
(b) Earth reinforcement (With PP/Jute needle punched sandwich blended fabric).

Cement grout or resin/bitumen treated Jute woven/non-wovens and hardener may be injected to protect surface cracking of road.

2 Improving Rigid Roads Pavements

The natural fabrics can reduce the pumping hazards by introducing between pavement and subrads. It can retard reflecting cracking by introducing between wearing course and concrete pavement.
In reinforced soil supporting structures, the natural material can be used as reinforcing elements. The natural fabrics with protective covers used surfacing elements.

3 Improving Flexible Road Pavements

The natural fibres can be used as a separator between base course of granular material and subgrade. In addition to that the thickness of road pavement can be reduced with introduced with the introduction of the composite type jute fabrics reinforcing elements.

4 Experimental Procedure

Preparation of mechanical bonded needle punched jute based non-woven geo-textiles :
Fire sets of 100% jute based non-woven geotextiles, 300 G/M^2, were prepared with different procedure at DILO (Germany) non-woven line recently installed at JTRL Pilot Plant, which are given in Table-1.

Table 1. List of different types of jute based geotextiles which have been prepared for observations

Type of Jute based Geo-textiles	Fabric weight GM/M$_2$	Fabric Quantity (100% Jute)
A	300	Jute Nonwovens
B	300	Jute web is sandwiched between PP fibres
C	300	Jute non-wovens impregnated by bitumens
D	300	Chemical (Acrylic) bonded Jute non-wovens
E	300	Thermically bonded jute nonwovens using PP as thermosetting fibre

5 Result and Discussions

Five types of jute based Geo-textiles mentioned in Table-1 were subjected to the soil for different periods of soil incubation. The samples were collected from soil at the intervals of 6, 12 and 18 months. Table-2 indicates Tenacity values/Abrasion resistance of sample No. B, C, D and E are also better than sample A.

Strength of sample A detoriated when incubation time was increased. The tenacity value of samples no. B, C, D and E are better than sample A when subjected to soil incubation. This improvement might be due to protection provided by coating with polypropylene, acrylic and bitumen to jute fibre.

Table-3 illustrates the effect of jute non-wovens on C.B.R. values. The laboratory test results conclusively shows that the strength characteristics of soil are better with, jute based geo-fabric than without it. The sample E shows better performance than other four, as per requirement of end uses of the five jute based non-wovens.

Table-4 shows the other mechanical and hydraulic properties of jute based non-wovens. In this Table analytical values of area density g/M^2, Thickness (mm), Water permeability litre/M^2/Sec, Air permeability M^3/M^2/hr, Ball bursting strength (kg), Cone (60°) bursting strength (kg), Pore Size (M) +, have been mentioned.

Table 2. Tenacity (g/tex) of Jute based non-wovens before and after soil burial test

Sample number	Tenacity (g/tex)				Loss in Mgm abrasion resistance of normal fabrics
	0 month	6 month	12 month	18 month	
A	0.90	0.76	0.60	0.50	70
B	1.20	1.10	1.00	0.90	30
C	1.80	1.75	1.70	1.60	15
D	1.60	1.50	1.40	1.30	20
E	2.40	2.30	2.20	2.10	10

Table 3. Effect of jute based non-wovens geo-textiles (300 g/m^2) on C.B.R. value

Type of jute based nonwoven geo-textiles	A	B	C	D	E
C.B.R. value at 20% moisture content					
Without fabric	07.0	07.0	07.0	07.0	07.0
With fabric	10.0	14.0	18.0	16.0	22.0

6 Thickness Measurement

Fabric thickness plays an important role in allowing drainage water to pass through the fabric when it acts as filter media of permanent road structure. At the same area density sandwiched blended jute non-wovens give highest thickness than other geotextiles. Thermally bonded geo-textiles give lowest thickness as because the same jute non-wovens (mixed with 3% polypropylene fibre) fabric were passed through a Thermos Galinder (Benz) Pilot plant, at a temperature of 225°C, 40 kg/cm pressure and 1 m/min speed.

Table 4. Mechanical and hydraulic properties of Non-woven Jute based Geo-textiles

Varieties of Jute based Geotextiles	A	B	C	D	E
Area density, gm/m²	300	300	300	300	300
Thickness (mm)	10	12	8	7	6
Cyclic loading plastic deformation	5	6	7	5.5	8
Water permeability Litre/m²/Sec (5 mm wt)	97	94	55	67	40
Air permeability m³/m²/hr (5 mm wt)	2150	1800	1200	1000	866
Ball bursting strength kg	120	146	227	190	249
Cone (60°) Bursting strength (kg)	32	49	51	35	60
Pore size (μ) +	115	100	79	75	63

7 Water Permeability

This test was carried out to determine the capacity of jute based geo-fabric to conduct or to allow the water flow across or through fabric. Thickness of particular quality and construction of Geo-fabric plays an important role. It was found out that permeability constant varies significantly with the weight of the fabric.

In order to determine the permeability coefficient of the Jute Geo-fabrics, permeameter was used. Flow through fabric was measured and permeability constant was obtained from the Darcy's Law.

8 Machine Parameters

Test pressure = 0.3 meters = 30 cms
Area of cross section = 4.0 cm² = 4×10^{-4} m²
Formula used $K = \dfrac{Qh}{A.L}$

K = Permeability coefficient (m/s) at 293 K
A = Cross section of flow (m²)
L = Fabric thickness (cm)
Q = Flow through fabric m³
H = Hydrostatic pressure head (cm)

Actual area of cross section of flow was 19.56 cm², but to make flow meter reading compatible with maximum flow meter readings, capacity of the machine, cross section of flow was reduced to 4.0 cm² with the help of resin cloth.

From Table-4 it has been observed water permeability of sample A is 97 (highest) and E is 40 (lowest). It can be concluded that if non-woven fabric treated with some treatment or procedure it looses its permeability capacities.

9 Air Permeability ($m^3/m^2/hr$)

It has been noticed from Table-4 that air permeability of samples A, B, C, D and E are 2150, 1800, 1200, 1000 and 866 respectively. Like water permeability, same trend of results are observed, when needle punched Geotextile fabrics are treated with resin or bitumen its air permeability reduced.

10 Pore Size Determination

Filtration is one of the main functions of geo-textiles used for road construction. It is while retaining suspended soil particles. In order to evaluate the performance of a geotextile with respect to some of its functions such as drainage and filtration, it is necessary to obtain its hydraulic properties. The experiment was performed as follows :

4 g of bentomite clay mixed in 300 cc of sodium oxalate as the dispensing agent, was kept for 24 hours for conditioning. After 24 hours this suspension was passed through sieve having opening size 75μ. So as to confirm that suspension contain particles having maximum diameter 75 or less than that.

This suspension was then passed through fabric the filtrated solution was collected in the cylinder the total volume of suspension in the cylinder was made on litre by adding water to suspension. Hydrometer was used to calculate the density of soil suspension, velocity of particles and diameter of particles after specified time intervals at a depth 'HR' which increases as particles settle with the increase in the time interval. Readings were taken at 1/2, 1, 2, 4, 8, 15 and 30 minutes and 1, 2, 4 and 24 hours etc.

The fabric piece weight was measured before filtration and after filtration. After filtration wet fabric piece was kept in order to calculate fabric's oven dry weight, so as to determine the quantity of bentonite clay blocked or trapped in the fabric piece.

Bentonite clay grain size distribution was determined by hydrometer analysis using 4 g of bentonite clay with 300 cc of sodium oxalate as dispersing agent. Standard grannulary curve for bentonite clay was obtained. The pore size distribution of jute based geo-fabrics have been determined.

The Jute non-wovens (A), Sandwiched non-wovens (B), Bitumenized fabrics (C), Chemical bonding fabrics (D), Thermically bonded fabric (E) are having pore size of 115μ, 100μ, 74μ, 75μ and 63μ respectively. It is obvious diameter of pore are gradually decreasing when fabrics are treated with heat and chemical treatments.

11 Cyclic Loading Test

This test reveals the fact that how much number of cycles were required to cause mechanical damage of jute based geo-fabrics. This cycle is repeated until failure occures. The fatigue resistance of jute based geo-fabrics is the ability of fabric to withstand repeated loading before undergoing catastrophic failure. The method basically involves loading the fabric specimen longitudinal to a known length and the releasing back to zero. From Table-4, it is observed that thermically bonded jute based geo-fabric has shown better performance in case of cyclic loading test than other four types of fabrics.

12 Cone (60°) Bursting Strength

This resistance of geotextiles to damage during installation due to dropping of sharp edged or sharp pointed stones on to geo-textiles is evaluated through cone drop test. In this test a 250 mm diameter specimen is clamped to yield a clear diameter of 150 mm and a 60° brass cone weighing 1 kg is dropped through a height of 500 mm. The diameter of the resultant hole is measured with a conical device. The results of the tests provide a convenient means of qualitative comparison of geotextile when used in combinations with other direct tensile test results. In this test also thermically bonded jute based geo-fabric shown highest performance.

13 Ball Bursting Strength (kg)

This test is an index test and is widely used for quality control. It is also used for design of fabric when used for separation purpose of road construction. The simulated situation is shown the most common test is the mullen burst test. A 100 mm diameter specimen is held against an inflatable rubber membrane by an angular clamp. The fabric (31 mm dia) is hydraulically distended in to the shape of a hemisphere with the application of pressure the fabric is deformed and when no further deformation is possible, bursting of the fabric occurs. The maximum pressure recorded is given as bursting strength. According to Table-4 Ball Bursting strength of five developed fabrics as follows :

    (a) Jute based non-wovens (needle punched) 120 kg
    (b) Sandwiched Geo-fabric   146 kg
    (c) Bitumenised fabrics   227 kg
    (d) Chemical bonded non-wovens   190 kg
    (e) Thermically bonded non-wovens   249 kg

Here, it is observed thermically bonded non-wovens shows highest performance.

14 Laboratory Testing for Simulation of Full Scale Applications

The speed of propagation of cracks was monitored within a test set-up consisting of jute based geo-fabrics and a concrete wearing course for comparison of the effectiveness (appearing of cracks, speed of propagation, time for total rupture) of the different systems tested. Test results were compared with a reference test I, carried out in the same test set up, represented by two types.

(a) One layer only of asphalt concrete
(b) Combination of two centimeters of bituminous sand plus one layer of asphalt concrete.

Four different systems using all the jute based geo-fabrics according to Table-1 were tested. The comparison therefore focused at Fig.1 on the quality of the total system.

Analysis of initial propagation speed and the time it takes for the system to fully crack show a similar performance for the four systems with geotextile as for the test with bituminous sand within the geotextile group and advantage was observed for the three tests using thermically bonded jute based geotextile.

A very important conclusion that can be drawn is that these jute based geo-fabrics are easily applied on a wide range of pavements with different wearing courses and climates.

Also it is important to observe according to Fig.1 that the thermically bonded jute based non-wovens remains intact, even after complete propagation of the cracks. This greatly minimises penetration of the surface water into the under-lying support layers.

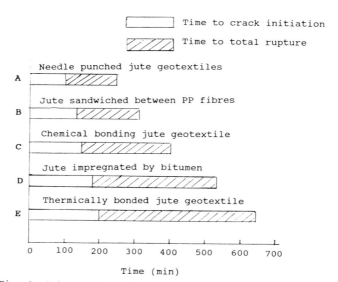

Fig. 1 Laboratory testing for simulating actual field condition to see performance of jute based non-woven geotextiles to retard reflective cracking in pavements

15 Conclusion

From these studies it could be concluded that jute based Geo-fabrics might be useful for road construction to protect reflective cracking.

The research into the functioning of fabrics in motorway and major highway construction to reinforce the subgrade/sub-base interface has suggested that the fabrics play short/long term role in road performance and to check reflective cracking. Once the road has been fully constructed and is in use, the fabric becomes superfluous and hence the biodegradability of jute would not pose problems for this end use. These studies reveal that, to protect reflective cracking in road construction, jute based geo-fabrics are found to function in a three fold way :

(a) It separates the subgrade from sub-base thus preventing the punching of the base material into the subgrade and at the same time the fines form the subgrade are also prevented from gaining entry into the road structures.
(b) It acts as drainage layer to remove excess water from softening the subgrade.
(c) It helps to improve the bearing capacity and settlement behaviour of the subgrade by virtue of its action as a fabric reinforcement.

16 References

Pandey, S.N. and Majumdar, A.K. (1989) Manufacturing, Testing and Application of Jute as Geo-textiles. Proceedings of International workshop on Geo-textile at Bangalore, India.

Pandey, S.N. and Majumdar, A.K. (1990a) Jute and Allied fibres as Non-woven Geo-jute and their application in civil engineering. International reinforced soil conference, University of Strathclyde, Glasgow, U.K.

Pandey, S.N. and Majumdar, A.K. (1990b) Application of Jute and Allied fibres in non-woven geo-textiles. Proceedings of technical textile conference held at Princeton, New Jersey, USA.

Pandey, S.N. and Majumdar, A.K. (1990c) Design of a fixture determining the puncture strength of geo-fabric. Proceedings of 4th International Conference on geo-textiles and geo-membrane at Hague, Netherland.

Pandey, S.N. and Majumdar, A.K. (1992) An attempt to select geo-synthetics for water resource project. Proceedings of National Workshop Role of Geosynthetics in water resources projects, at New Delhi.

31 BITUMINOUS PRE-COATED GEOTEXTILE FELTS FOR RETARDING REFLECTION CRACKS

M. LIVNEH, I. ISHAI and O. KIEF
Transportation Research Institute, Technion-Israel Institute of Technology, Haifa, Israel

Abstract

Geotextile felt technology has recently come into use in Israel in order to retard the process of reflection cracking in new asphalt overlays. Thus, in an effort to establish design criteria for this technology the tracking wheel model was suggested for predicting the bituminous felt behavior. A number of tests were conducted within this model for which the efficiency of various felts in comparison with the tack-coat interface was evaluated. The performance of the 3/250 bituminous felt was found to be very impressive.
Keywords: Reflection Cracking, Laboratory Testing, Fatigue, Bituminous Felts, Overlay.

1 Introduction

There are many methods intended to retard reflection cracking in in newly overlayed asphalt pavements. In most cases, these methods yield unsatisfactory results, and the search for an efficient and economic method is thus very important, as described recently by Caltabiabiano and Burnton (1991). Beginning from about the mid-seventies, the use of geotextile technology, began to slowly spread over the world, in order to supply solid engineering solutions to various performance problems, including the propagation of reflection cracking. According to Lytton (1989), the application of geotextile felts in asphaltic overlays is based on their ability to perform one or more of the three following activities: (a) Reinforcement - the ability to absorb the tensile stresses which develop in the asphaltic overaly, (b) Strain release - the ability to spread the strains created through a controlled "sliding" action, and (c) Sealing - the ability to block the penetration of water from the surface of the cracked pavement to the lower layers.

Even after 20 years of practical use and monitoring of the performance of the geotextile felts, the mechanism by which reflection cracking is retarded is yet to be fully understood. As a consequence, there are considerable differences, as reported by Maurer and Malashesku (1989), between the predictions derived from laboratory standard tests and the actual field performance. Obviously, these differences can also be attributed to the partial knowledge regarding the qualities required of the felt in order to obtain appropriate performance. This is witnessed by the conflicting variety of criteria and recommendations intended to determine the required properties of these felts, as reported by Koerner (1990).

Thus, it seems that one cannot accurately predict the performance of a specific felt without conducting advance fatigue experiments in a simple laboratory wheel tracking device. In light of the above, the objectives of this paper were to develop a simple laboratory fatigue test which will simulate field conditions as closely as possible on the one hand, and will differentiate between the efficiency levels of bituminous pre-coated geotextile felts on the other hand. For these objectives, 3 kinds of bituminous Polypaz felts, 3M, 3/250 and 4/180, were used, showing that their basic characteristics do not always indicate their final behavior.

2 The preliminary test results

In order to examine the efficiency of the bituminous felts alone, it was decided to limit the scope of the laboratory investigation to one type of sand-asphalt mixture made of dolomite aggregate (where 100% passing sieve #10, 55% sieve #40 and 15% sieve #200) and 8% of 60/70 bitumen content (derived from the Marshall test as the optimum content).

As stated above, preliminary tests were conducted in order to examine the possible correlation between the properties of the various felts and their retarding capability. These tests included the conventional standard tests and also the "Narrow Strip" tensile test. The characteristics of two out of the three membranes mentioned before are given in Table 1. In addition to this table, the characteristics of the modified bitumen used for pre-coating the felts, are: penetration: 30-90 1/10mm; softening point: >120°C; and ductility: >50cm. Fig. 1 presents the results of the "Narrow Strip" tensile test for all three membranes examined in this work. Table 1 and Fig. 1 indicate that a 3M type bituminous membrane (made of fiber-glass) has a far lower strain resistance than the two other membranes, as compared with the closely identical tensile strain capability of the two 3/250 and 4/180 membranes (made of polyester).

In addition to the above standard tests, a direct shear test at the interface was also used in order to find out the differences in the possible performances of the various tested felts. The results of these tests are presented in Fig. 2, where the testing procedure was according to Uzan et al. (1978). This figure indicates that there is no difference in the magnitude of the shear patterns and values of the three types of membranes. These derived shear values at the interface were lower, as expected, that those of a tack-coat applied at the interface. Despite this, the behavior characteristics of these three membranes are actually equal under the tracking wheel, as shall be demonstrated in the following sections. Moreover, taking all of the above tests into account, indicates that there might be some possibilities for similarity between the behavior of the 3/250 and the 4/180 felts, while the behavior of the 3M felt is dissimilar. Again, this will be discussed in the following sections.

3 Laboratory wheel-tracking device

The comparison of the efficiency of the different interface treatments was carried out with a laboratory wheel tracking device. This device creates a linear motion by means of a metal plate which moves back and forth between two rails. An elastic base made of rubber is placed above the plate and the sand-asphalt beam is placed above it. On top of the beam is a constantly loaded boat trailer wheel comprised of a hoop and tire. A schematic diagram of this device is presented in Fig. 3. The length of the cracking along both sides of the asphalt beam was determined visually by means of a sophisticated magnifying glass. The wheel-tracking device was located in a closed room with thermal insulation and temperature control for a constant 25°C. The dimensions of the sand-asphalt beams examined in the moving wheel tests were length, width and height dimensions of 70x10x10cm respectively. Three different types of beams were prepared for all tests, as shown in Fig. 4.

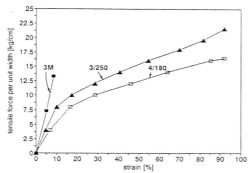

Fig. 1. Narrow strip tensile tests for three types of felts.

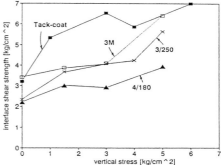

Fig. 2. Direct shear tests for four interface treatments.

Table 1. Manufacturer's data regarding the characteristics of the different membranes tested.

Property	Units	Standard	3/250*	3M*
Raw Material			Polyester	Fiberglass
Production Method		All membranes are nonwoven		
Membrane thickness	mm.		3	3
Felt weight	[gr/m^2]		250	105
Tearing force in tension	[kg/cm^2]	DIN52123 DIN52133	20/17	15/15
Elongation in tension	%	DIN52123	50/50	3/3
Stability under high temperature	[°C]	DIN52133	+115	+115
Stability under low temperature	[°C]	DIN52133	-20	-20
Type of polymer			SBS**	SBS**

* 3/250 - 3mm thick membrane with 250 gr/m^2 polyester fabric reinforcement ** SBS - Styrene - Butadiene - Styrene

In all these beams, an artificial groove, 30-50 mm. deep and 4mm. wide, was made at their center, along their entire bottom width. The edge of the groove was rounded. The center edges of the beams were painted white in order to facilitate the identification of crack formation and to monitor their rate of development. As mentioned before, the sand asphalt beams where placed on two 34.5x10x15 cm. rubber beams. A gap of once cm. was left between the rubber beams, in direct continuation of the groove in the beam above it. A U profile was installed along the breadth of the beams and their entire height, thus preventing any horizontal or longitudinal movement of the beams, without interfering with their ability to bend under the wheel load. Table 2 summarizes the results of the fatigue longevity obtained from the tests conducted on various beams by means of this wheel-tracking device.

In addition to the findings of Fig. 8, it is important to note that a phenomenon of dual cracking was observed in some of the tests. This phenomenon is characterized by the propagation of a crack which begins at the edge of the artificial groove and the propagation of a crack which begins on the upper surface. Germann and Lytton (1977) who found the same phenomenon in their laboratory tests, noted that it was also observed in in-situ tests segments carried out in the State of Florida. Here, this phenomenon was observed in half of all beams which did not contain membranes, and took place in <u>all beams</u> which included membranes of all of the various types. In order to demonstrate this phenomenon, Figs. 5 and 6 present the visual results of cracking propagation over time as derived from two fatigue experiments conducted on sand-asphalt beams with 3/250 bituminous pre-coated geotextile felts.

The term "UP" denotes the distance which the crack advanced from the edge of the artificial groove at the center of the beam upwards, and whose initial length was given at the beginning of the test. The term "DOWN" denotes the distance advanced by an additional crack which began at the upper part of the beam and propagated downward. A graphic description of the relationship between the total length of the crack (the sum of the crack from "below" and the crack from "above") and the time elapsed since the beginning of the test, is presented in Fig. 6.

The average time was multiplied by 2880 in order to obtain the number of tracking-wheel repetitions per hour, as described in Fig. 7. Finally, it should be mentioned that there are various reasons for the crack's propagation along the surface of the sample downward, however, it is beyond the scope of this paper to detail them.

4 Analysis of results

The results of the fatigue tests presented in Fig. 8, indicate that only the fatigue life associated with a 3/250 type membrane is longer than that associated with the tack-coat interface, and that this advantage is highly significant. The fatigue life ratio for this membrane is 3.8 - a far higher value than for the monolithic beam. Again, the explanation for this phenomenon of the fatigue life ratio is beyond the scope of this paper and is presented by Kief et. al. (1992). As expected, the fatigue life of a monolithic beam is higher than that of the control sample (with a tack-coat interface) but not at the same impressive rate as with the 3/250 bituminous felt.

Koerner (1990), provides a number of recommendations regarding the characteristics of asphaltic layer membranes in order to obtain maximal performances. Minimum requirements were posited for membrane strength, elongation, bitumen penetration and softening point. As can be seen in Fig. 2 which summarizes the results of the "Narrow Strip" tensile tests conducted on the membranes, and in Table 1 which details membrane characteristics, all membranes meet the minimum requirements, except for the 3M membrane whose strain capacity was far lower than required. Accordingly, the fatigue life ratio obtained from the use of this membrane is very low - <u>0.60</u>. The qualities of the 4/180 membrane fulfill all Koerner's (1990) requirements. Fig. 1

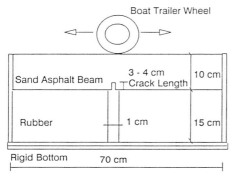

Fig. 3. The laboratory wheel-tracking device.

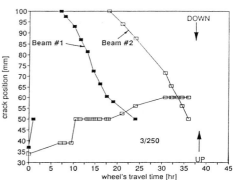

Fig. 5. Crack position vs. wheel's travel time.

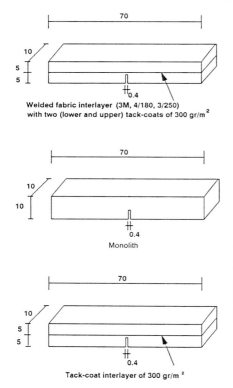

Fig. 4. Flexural fatigue beam specimen.

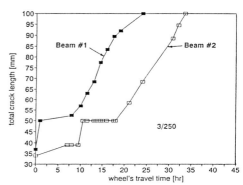

Fig. 6. Total crack length vs. wheel's travel time.

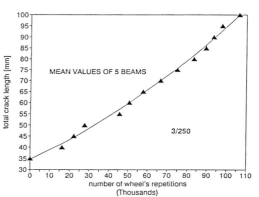

Fig. 7. Total crack length vs. average number of wheel repetitions.

348

also shows that the tensile strain capacity of the 4/180 membrane is almost identical to that of the 3/250 membrane. The behavior of these membranes under shear (Fig. 2) is also almost identical, and thus the difference in the fatigue life between the two membranes may stem from the difference in the thickness of the bituminous coats. However, it is more conclusive to state that one cannot predict the quality of any membrane's performance without conducting advance fatigue tests. As shown before, the laboratory simulation indicates that the use of membranes which meet accepted criteria can lead to significant retardation of reflection cracking (the 3/250 membrane), but also to earlier fatigue failure than expected in a conventional 5 cm. layer (4/180 type membrane).

5 Conclusions

The main conclusions are: (a) As stated, one cannot determine in advance the quality of a certain membrane's performance, without conducting advance fatigue tests. (b) Despite its limitations, it seems that the testing device adopted and improved in this work, can serve as a reliable method for diagnosing the efficiency of the various treatments for reflection cracking retardation, such as bituminous felts, geogrids, steel meshes, SAMI, etc. (c) The fatigue life of an asphaltic overlay is influenced by the various properties of the bituminous felts; including its

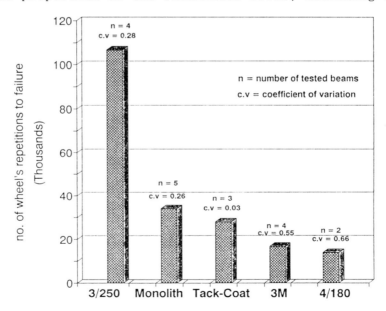

Fig. 8. Summary of fatigue test results.

modulus of elasticity, (i.e., a membrane with a low strain capacity significantly decreases the fatigue life) and its thickness, (i.e., the thicker the membrane the shorter its fatigue life) for two membranes with identical strain capacities.

Finally, it should be emphasized that the 3/250 bituminous felts exhibited remarkable performance in the laboratory tests. This performance is obviously granted if the qualities of the modified bitumen and geotextiles are strictly kept.

6 Acknowledgements

The authors would like to express their sincere thanks to PAZKAR Ltd., for sponsoring this research. Thanks are also due to Messrs. M. Ron and N. Livnat, and G. Svechinsky for their useful advice

7 References

Caltabiabiano, M.A. and Brunton, J.M. (1991) Reflection Cracking in Asphalt Overlays. **Asphalt Paving Technology**, Vol. 60

Germann, F.P. and Lytton, R.L. (1977) Methodology for Predicting the Reflection Cracking Life of Asphalt Concrete Overlays. **Research Report No. 207-5, Flexible Pavement Evaluation and Rehabilitation,** Research Project 2-8-75-207, March.

Kief, O., Livneh, M. and Ishai, I. (1992) Use of Bituminous Felts for Retarding Reflection Cracking. **Transporation Research Institute, Technion - Israel Institute of Technology,** Research report No. 92-179 (in Hebrew & summary in English).

Koerner, R.M. (1990) **Designing with Geosynthetics**, Second Edition.

Lytton, R.L. (1989) Use of Geotextiles for Reinforcement and Strain Relief in Asphalt Concrete. **Geotextiles and Geomembranes**, Vol. 8, No. 3.

Maurer, D.A., Malasheskie, G.J. (1989) Field Performance of Fabrics and Fibers to Retard Reflective Cracking". **Geotextiles and Geomembranes**, Vol. 8.

Uzan, J., Livneh, M. and Eshed, Y. (1978) An Investigation of Adhesion Properties Between Asphaltic-Concrete Layers. **Asphalt Paving Technologists**, Vol. 47.

PART FIVE
CASE HISTORIES

32 COMPARATIVE SECTIONS OF REFLECTIVE CRACK-PREVENTING SYSTEMS: FOUR YEARS EVALUATION

G. LAURENT
C.E.T.E. de l'Ouest, Nantes, France
J.P. SERFASS
Screg Routes, Guyancourt, France

Abstract
The purpose of this paper is to present the results of the monitoring of six full-scale sections, laid in 1988 on National Highway 23 (Sarthe). Four sections comprise a crack-retarding interlayer; two reference sections were also placed.
The main features of the experiment are described : heavy traffic, existing pavement showing typical shrinkage cracks.
The six sections, the behaviour of which is analyzed, are the following :
- N° 1 (reference) : thin asphalt concrete overlay
- N° 2 (reference) : crack sealing prior to application of the same overlay
- N° 3 : sand asphalt interlayer plus thin asphalt concrete overlay (plain bitumen)
- N° 4 : geotextile interlayer plus thin AC overlay
- N° 5 : sand asphalt interlayer plus thin AC overlay (SBS-modified)
- N° 6 : fiber and SBS-modified sand asphalt interlayer plus thin fiber-modified AC overlay

The various mix formulas and characteristics are presented and discussed. Manufacturing and laying operations are also described.
The position of the cracks in the existing pavement were recorded accurately. Since the overlay was completed, the sections have been systematically monitored, enabling the crack reflection rate to be established for each system.
After four years of service, including two significant winters, this comparative experiment shows the effectiveness of the system combining sand asphalt and thin asphalt concrete, both fiber-modified.
The follow-up of these sections will be continued for several years, in order to draw all possible conclusions.
Keywords : Crack-Retarding Interlayer, Asphalt-Concrete, Sand-Asphalt, Geotextile, Fiber-Modification, SBS-Modified

1 Introduction

In the last few years, the laboratory evaluation of anti-cracking systems has greatly improved; new methods and machines have been developed. However, at present, no laboratory testing can be considered as fully representative of what is happening in a real pavement. The construction and monitoring of full-scale trial sections is therefore necessary, first to adjust testing methods and design models, then to evaluate and compare various anti-cracking systems.
The Highway 23 trial sections, laid in 1988, were aimed at comparing the retarding effect of several "anti-cracking" techniques on the reflection of cracks caused by the shrinkage and thermal movements of hydraulic binder-treated pavement layers. To enable a meaningful assessment to be made, reference-sections without any crack-retarding interlayer were also laid.

Reflective Cracking in Pavements. Edited by J.M. Rigo, R. Degeimbre and L. Francken.
© 1993 RILEM. Published by E & FN Spon, 2–6 Boundary Row, London SE1 8HN. ISBN 0 419 18220 9.

2 The field sections

The comparative sections were placed on the National Highway 23 (Paris-Nantes) in the Department of Sarthe.

2.1 Existing pavement

The old, flexible pavement was overlaid in 1972 with 200 mm of blast furnace slag-treated graded crushed aggregate (20 mm maximum size) and 60 mm of asphalt-concrete (14 mm standard mix).

The first maintenance operation took place in 1976 : it consisted in a 40 mm course of asphalt concrete (14 mm "thin" type mix). The transverse cracks originating from the slag-treated base reflected rapidly through this maintenance course. By the end of 1978, all cracks had reflected; the intervals between cracks range from 7 to 10 m.

The cracks were sealed in 1984. It quickly appeared that this type of maintenance was not sufficient and it was decided to apply a new maintenance overlay, consisting in 40 mm of asphalt concrete.

It was also decided to seize this opportunity to lay trial sections of crack-retarding systems.

2.2 Other characteristics

The traffic is heavy. In 1990, the average daily number of commercial vehicles (over 5 tonnes of payload) amounted to 800 per lane (T0 class) and the average daily number of all vehicles to 12,000 (on the two lanes).

The deflections were measured shortly before overlaying. They range between 15 and 20/100 mm, which shows that, apart from the transverse cracks, the existing structure is in good condition

2.3 The techniques compared

Each trial section is 500 m long. Two reference sections were placed. Both include the same "thin" maintenance asphalt mix.

Section N°	Techniques. Reference sections	Thickness (mm)
1	Thin asphalt concrete, Administration-type - 14 mm maximum size, gap-graded 2-6 mm	40
2	Ditto, with prior crack sealing	40

Four sections with a crack reflection-preventing interlayer were laid.

Section N°	Techniques. Crack-retarding sections	Thickness (mm)
3	Sand asphalt (3 mm max size) interlayer, plus thin asphalt-concrete (as above), both with plain bitumen	20 + 40
4	Modified binder and geotextile interlayer, plus thin asphalt concrete (as above)	40
5	Sand asphalt interlayer, plus thin asphalt concrete both with SBS-modified bitumen	15 + 30
6	Fiber and SBS-modified sand asphalt, plus thin fiber-modified asphalt concrete	20 + 35

All the sections were placed in the fall of 1988 (October and early November), which is a little too late in season for these thin layers. It should be noted, in particular, that section N° 4 was laid in humid weather, which locally caused poor adhesion to the support.

2.4 Sections monitoring

On each 500 m section, all cracks were recorded accurately, prior to the application of the overlays. Each transverse crack was located and drawn on a specific chart, an example of which is given in Figure 1.

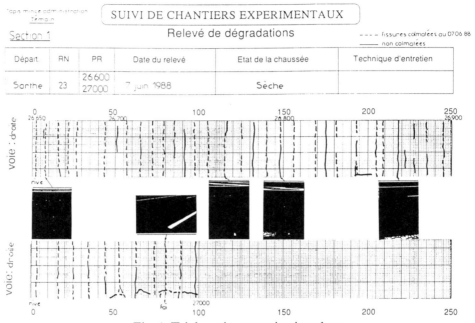

Fig. 1. Trial sections monitoring chart

Photographs were taken to illustrate each type of distress.

During the construction of the sections, all materials and finished layers were carefully controlled.

3 References and trial sections characteristics

3.1 First reference : 40 mm asphalt concrete

The "basic" reference consists of 40 mm of asphalt concrete. The mix, known as "Administration thin AC" is a gap-graded type (maximum size of aggregate : 14 mm, gap in gradation between 2 and 6 mm). The binder is a 60/70 plain bitumen. The binder content amounts to 5.2 ppc and the total proportion of fines to 7%.

This formula is commonly used for maintenance in the region. This mix has been applied on sections 1, 2, 3 and 4.

3.2 Second reference : crack sealing, then thin asphalt concrete
The cracks were sealed with a SBS-modified asphaltic crack sealant a few weeks before the application of the "thin" (40 mm) asphalt concrete described above.

3.3 Plain bitumen sand asphalt interlayer, plus thin asphalt concrete
The formula of the stress-relieving sand asphalt is the following :
- 2 mm crushed sand   78%
- 3 mm natural (rounded) sand   20%
- Additional filler   2%
- 80/100 pen bitumen   10 ppc

The total fines content reaches 15%. The design degrees of compaction, ranging between 95 and 97%, have been easily obtained on site. The thin asphalt concrete wearing course was placed directly on the sand asphalt interlayer (no tack-coat is necessary).

3.4 Modified binder and geotextile interlayer, plus thin asphalt concrete
The fabric used is a non-woven polyester geotextile, needled-punched and point-calendered. Its mass per unit area is 200 g/m2. Its vertical compressibility is very small. Its lower side is fluffy, in order to enhance binder adhesion. Its upper side is, on the contrary, smooth, to improve trafficability.

The modified binder has two functions : i) stick the fabric to the existing surface - ii) saturate the geotextile. It is a SBS-modified bitumen, the main characteristics of which are :
- Ring and Ball Softening Temperature   72°C
- Fraass Breaking Point   -24°C
- Viscosity at 170°C   270 cP

The initial binder spray rate was 0.8 kg/m2; this rate rapidly appeared too low for the existing surface, which was leaned and porous. On the second half of the section, the spray rate was increased to 1.2 kg/m2.

Immediately after the binder had been sprayed, the fabric was applied with a special taping machine. Adequate bonding was obtained and the traffic of paver and supplying trucks caused no problem. However, a few wrinkles occured; they were eliminated by cutting or heating.

The thin AC wearing course described above was then placed. The heat of the mix softens the interlayer binder, which comes up through the textile, saturates it and provides tacking for the AC overlay.

3.5 SBS-modified sand asphalt interlayer, plus thin SBS-modified AC
The sand asphalt is made with crushed fractions (maximum size : 4 mm). The SBS-modified binder content is 10.5 ppc. Its average thickness is 15 mm. A tack coat (300 g/m2 of cationic emulsion) was sprayed before its application.

The thin asphalt concrete is a 10 mm max. size, 2-6 mm gap-graded formula :
- 6-10 mm crushed aggregate   66%
- 2 mm crushed sand   33%
- additional filler   1%
- SBS-modified bitumen   5,8 ppc

The thickness of this wearing course is 30 mm.

3.6 Fiber - and SBS-modified sand asphalt interlayer, plus thin fiber-modified AC
The sand asphalt (20 mm thick) has been formulated as follows :
- 2-4 mm crushed aggregate   37%

- 2mm crushed sand   62,4%
- mineral fibers   0,6%
- SBS-modified bitumen   8,8 ppc

The wearing course AC (35 mm thick) corresponds to the following composition:
- 6-14 mm crushed aggregate   48%
- 2-6 mm crushed aggregate   20%
- 2 mm crushed sand   22%
- additional filler + fibers   10%
- 60/70 plain bitumen   7,4 ppc

As expected from the lab study, low voids contents were obtained in the field. The job was completed in two days : the sand asphalt interlayer in the mornings, the AC wearing course in the afternoons. Because of the high binder contents, no tack coat was necessary

4 Monitoring and 4 years-performance

Visual inspection has been conducted twice a year, by the end of the winter and of the summer. The surface condition has been accurately observed. All cracks or beginning of cracks have been reproduced on control charts. On each section, the rate of crack reflection can therefore be established.

The following table summarizes the extent of cracking of each section, before and after overlaying. The results are plotted in Figure 2.

Section		Number of cracks				
		Before overlay	After overlay			
N°	Characteristics		% of cracks reflected			Condition in March 92
			04/90	02/91	03/92	
1	Reference : 40 mm AC	83	0	10	23	Cracks tend to split
2	Reference : Crack sealing + 40 mm AC	110	1	24	34	Ditto
3	20mm sand asphalt + 40 mm AC	113	0	6	18	Fine cracks
4	Geotextile + 40 mm AC	92	0	11	16	Fine cracks and crazing
5	15 mm SBS-mod. sand asphalt + 30mm SBS-mod.AC	99	0	3	19	Fine cracks
6	20 mm fiber mod.					

HIGHWAY 23 - TRIAL SECTIONS

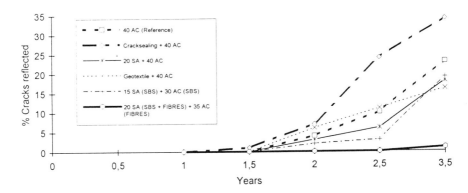

Fig. 2. Crack reflection rate

After four winters, cracks reflection reaches significant levels. In June, 1992, the performance of the different overlays can be analyzed as follows :
- The second reference (Section N° 2) has the poorest performance. The high rate of crack reflection may be attributed to the bonding effect of the crack sealant.
- The various crack reflection-retarding systems tested perform better than the reference 40 mm AC, but the gain, in terms of percentage of cracks reflected, is limited, except for Section N° 6.
- The state of the reflected cracks is significantly better in the sections comprising a stress-relieving interlayer than in the reference sections : the cracks are finer and do not tend to deteriorate by splitting and spalling.
- The reflection rate is approximately the same for sections 3 (sand asphalt), 5 (SBS-modified sand asphalt) and 4 (geotextile). But, on the geotextile section, fine crazing has appeared, as a consequence of a localized lack of bonding to the support. This poor adhesion can be attributed, firstly to an insufficient rate of binder (the existing surface was lean and porous), secondly to the humidity of the fabric, resulting from adverse climatic conditions.
- The best results are clearly those obtained on section N° 6, in which only one fine crack can be observed. This section differs from sections 3 and 5 essentially by the type of wearing course : that of section 6 consists in a fiber-modified, high-binder-content asphalt concrete.
- It should be noted that no rutting has occurred on any of the sections.

5 Provisional conclusions

This field experiment shows that, until now, the retarding effect of a sand asphalt or a geotextile stress-relieving interlayer, is comparatively limited. They have, however, a marked beneficial influence on the state and the **predictable evolution** of cracked areas. It can be expected that, with this types of interlayer, the surface **impermeability will be preserved** and that no particular maintenance operation shall be required for a period of time still to be determined.

Increasing the sand asphalt thickness is a tempting way, but higher thicknesses would lead to rutting problems.

As regards geotextile interlayers, **wrinkles** are practically inevitable and may initiate cracks in the AC wearing course. Moreover, to obtain the expected stress-relieving effect and proper bonding of the overlying AC, a **high spray rate is necessary for the interlayer binder**.

The **most interesting issue** in this experiment is **the influence of the asphalt concrete overlay type and formulation**. The best result has been achieved with an **asphalt concrete containing a high proportion of binder-rich coating mastic**.

This type of mix can readily be obtained by adding **fibers**, that fix a high quantity of binder, without impairing the resistance to creeping and rutting.

On the basis of the present assessment, reliable crack-preventing solutions can be selected. New experiments are scheduled, to develop even more effective systems.

33 ASSESSMENT OF METHODS TO PREVENT REFLECTION CRACKING

M.E. NUNN and J.F. POTTER
Transport Research Laboratory, Crowthorne, UK

Abstract
This paper describes a series of road trials that are being carried out to assess the performance of a variety of maintenance treatments to inhibit reflection cracking in new overlays. The performance of each treatment will be related to the severity of the cracks being treated, the overlay thickness and the intensity of the traffic.
Extensive observations have shown that in the UK reflection cracks in flexible composite roads normally initiate at the surface and that the cracks develop mainly during the winter months. The early indications from the present trials suggest that crack initiation is strongly influenced by the thickness of bituminous overlay and the different treatments examined have less influence.
Key words Reflection Cracking, Surface Initiated Cracking, Modified Binder, Geogrid, Geotextile, Stress Absorbing Membrane, Yield Strain.

1 Introduction

Lean concrete laid continuously without joints is used widely as a roadbase under a bituminous surfacing in the United Kingdom. However, like most cement or hydraulically bound materials, it suffers from thermally induced transverse cracks and after a few years of service the cracks often begin to appear in the bituminous surfacing. In their early stages of development the transverse surface cracks are not considered to be a structural problem but infiltration of water can weaken the foundation layers and fines can be pumped to the surface creating voids under the roadbase. This allows heavy traffic to rock the slabs which induces further cracking, spalling and general deterioration. The time when the cracks first appear in the surfacing depends on the thermal properties of the concrete, spacing between transverse cracks, thickness of the bituminous layers, properties of the surface layer and traffic loading.

Reflective Cracking in Pavements. Edited by J.M. Rigo, R. Degeimbre and L. Francken.
© 1993 RILEM. Published by E & FN Spon, 2–6 Boundary Row, London SE1 8HN. ISBN 0 419 18220 9.

Reflection cracks can also occur in bituminous overlays on jointed concrete roads above the original joints between the concrete slabs. The load transfer across joints and their ability to respond to thermal movements are the major controlling factors in the formation of reflection cracks.

The road maintenance engineer needs to know what is the most appropriate and cost-effective method for treating reflection cracking. In this paper, trials to assess a variety of maintenance methods to inhibit reflection cracks on roads with lean concrete roadbases and on overlaid jointed concrete roads are described and their performance to date is reported.

2 Mechanisms of crack propagation

Extensive field studies by Nunn (1989) into the mechanisms of propagation of reflection cracks show that cracks generally start at the surface of the road and propagate downwards to meet an existing crack or joint in the underlying concrete layer. Clearly this has important implications for the design of maintenance treatments to prevent the occurrence of reflection cracks in overlays on cracked roads. This work showed that surface initiated cracking depended on the thickness of the bituminous layer, the resistance of the binder to age hardening and the temperature regime at the site; high summer temperatures cause hardening and low winter temperatures cause embrittlement.

3 Methods of minimising reflection cracking

The following methods of inhibiting reflection cracking in new overlays on cracked roads are being assessed in road trials:

  Stress absorbing membrane interlayer (SAMI)
  Reinforcement of the bituminous overlay
  Increasing the thickness of bituminous overlay
  Modifying the bituminous materials
  Treating the existing bituminous materials
  Treating the concrete

Stress absorbing membrane interlays (SAMI's) are designed to act as slip layers to prevent crack movements in the cement bound roadbase from being transferred to the overlay. SAMIs can be laid as continuous membranes like geotextiles or they can be specialised treatments applied over individual cracks.

Asphalt reinforcement materials are generally geogrids or high tensile strength glass or steel fibres added to the bituminous mix. Geogrids are usually manufactured from materials like polypropylene, polyester or glass fibres. Installation procedures generally recommend that the geogrids are placed over the cracked road or on a regulating layer before the structural overlay is applied. Consequently they may prove ineffective for cracks that initiate at the road surface.

Modifying bituminous wearing course materials has the potential to deter cracks which initiate at the road surface. Polymer additives have been mixed with the bituminous binder in the surface layer to make it more ductile so that it can better accommodate the thermal movement that occurs at the underlying cracks. Softer binders also have the potential to deter the onset of cracking but it is important to balance the likelihood of cracking against the reduced resistance to deformation which can result from the use of too soft a binder.

There are also three methods of treating the existing road before overlaying that have the potential to reduce reflection cracking. The first requires removal of the existing bituminous layers and subjecting the exposed cement-bound roadbase to controlled fracturing at regular intervals. The cracked roadbase is then seated with a heavy roller, prior to the application of the new overlay. This method of pavement rehabilitation known as 'crack and seat', has been used on jointed concrete pavements in the United States and in Europe, but not on roads with cement bound roadbases and bituminous surfacings. The second treatment is to remove the bituminous material to about half a metre either side of the cracked roadbase. This excavation is then reinstated before overlaying with new bituminous material possibly containing a modified binder. The third treatment is to rout out the crack on the existing surface using a thermal lance and apply a bituminous seal before overlaying.

4 Road trials

Trials were constructed on cracked roads to evaluate different methods of inhibiting reflection cracking in new overlays. The performance of the treatments in each trial is being assessed in relation to the performance of control sections with conventional overlays. The trials include test sections with thin overlays to produce an early indication of performance as well as sections with more conventional thicknesses of overlay.

The performance of these treatments depends on the condition of the crack being treated. It is important therefore that each treatment is applied to cracks having a range of severities to produce a comprehensive comparison of performance. In trials located on multi-lane roads the effect of traffic loading is being studied. Table 1 gives the general details of the trials and Table 2 lists the treatments studied.

Table 1. Road trial sites in UK

Site location	Date built	Length (m)	Treatments	Overlay thickness (mm)	Number of cracks
1. A30(east) Launceston	1987	1120	2 geogrids 2 SAMIs 1 inlay	40 - 80	170
2. A30(west) Launceston	1989	1400	8 bit. mixes 1 geogrid 1 geotextile	40 - 150	194
3. M20 Swanley	1987	2400	4 bit. mixes	50	180
4. A303 Amesbury	1989	920	2 bit. mixes	40 & 100	179
5. A30 Bodmin	1990	2700	3 geotextiles proprietary materials crack and seal excavate and replace	40 & 80	218
6. M2 Kent	1990	1000	2 bit. mixes 1 geotextile	75 - 200	52 joints
7. A45 Suffolk	1990	450	2 bit. mixes	100	90 joints
8. A30 Exeter	1991	1600	crack and seat	115 & 175	289
9. A40 Whitchurch	1992	1914	crack and seat	175 & 350	164
10. M5 Taunton	1992	3600	crack and seat 1 SAMI 1 geogrid	100 & 150	306 joints

Table 2   Treatments studied

Geogrid	Geotextile	SAMI	Bituminous mix	Treatment to surface	Treatment to concrete
Tensar AR1 Hatelit Glasgrid	Polyfelt PMG14 Lowtrack 45/45 Amopave Typar BM41	Fibredec Surface dressing Colas preformed PavePrep Fibrescreed	HRA + EVA HRA + SBS HRA + SBR 50 pen HRA 100pen HRA 200pen HRA	HRA inlay HRA overlay Fosroc Nitroseal 999 Fibrescreed	Crack and seat

4.1 Assessment of road condition before treatment

To assist the assessment of performance, techniques have been developed in the UK to assess the severity of cracks requiring treatment. Detailed visual condition surveys were carried out on the trial sections and the location, extent and visual severity of the cracks recorded. The degree of load transfer across a crack will depend on the degree of aggregate interlock. The Falling Weight Deflectometer (FWD) with the geophone configuration shown in Fig 1 was used to give a measure of load transfer and hence of crack severity. A good load transfer is characterised by the ratio D_7/D_2 being close to unity. The measurements were made in the winter when the crack was wide. The FWD also gave a measure of the structural support provided by the foundation beneath the crack. Details of the technique are described by Potter (1991). These data were also used to select the location of the test sections on each trial.

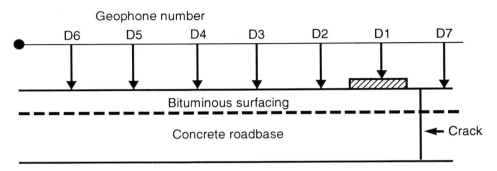

Fig. 1 FWD geophone positions across crack

The opening and closing of a crack is caused by temperature changes in the concrete and the amount of movement depends on the length of the concrete slabs each side of the crack, the thermal properties of the concrete and the frictional restraint between roadbase and sub-base. For trials constructed on jointed concrete roads, the horizontal thermal movement at the joints was measured directly at different road temperatures throughout the year by measuring the change in separation between metal studs inserted either side of the joint. It is not, however, practical to measure this movement directly in flexible composite roads and therefore it is proposed to use the ratio L_i/L_m as a relative measure of the horizontal thermal movement at the crack for a particular site where, L_1 is the average distance between the cracks on each side of the crack being considered and L_m is the mean crack spacing for the site. The greater this ratio the greater will be the thermal movement at the crack and the strain induced in the overlay. Consequently it is necessary to take this into account when assessing the performance of different treatments.

4.2 Monitoring performance

The performance of the treatments is being monitored by visual inspection surveys carried out at 3 to 6 monthly intervals to record crack growth and to identify the time of year when crack growth occurs. The structural condition is being assessed by measuring the change in longitudinal profile, development of wheel-track rutting and by deflection measurements.

For the test sections in which the wearing course mixtures were modified, core samples were taken at the time of laying and annually thereafter to determine the elastic stiffness, ductility and binder properties of the wearing course materials. Cores were also cut through some cracks shortly after they appeared in order to establish whether or not the cracks had initiated at the surface.

5 Performance of treatments

The trials described have been constructed during the last 5 years. At present, because the test and control sections in the more recently constructed trials are not showing much reflection cracking, only tentative conclusions can be drawn. As more data become available, each site will be analyzed separately to determine the effectiveness of the treatments by relating the time for reflection cracks to appear to the thickness of overlay and to the severity of the original cracks assessed in terms of their horizontal and vertical movement before treatment.

Figure 2 shows the general development of cracking in lane 1 of the test sections with a 40mm wearing course overlay at the trial on A30 Launceston bypass eastbound. The reflection cracks initiated and developed during the winter months when the bituminous overlay was most brittle and least able to withstand the strains induced by the opening of the crack in the concrete due to thermal contraction. Vertical movements caused by the passage of traffic also increase during cold weather. Measurements at various times of the year using the FWD have shown that vertical movements of the slabs either side of the crack can be up to 50 per cent greater for a fall in temperature of 10°C. The development of cracking in the less heavily trafficked lane 2 follows a similar pattern but with a lower incidence of cracking.

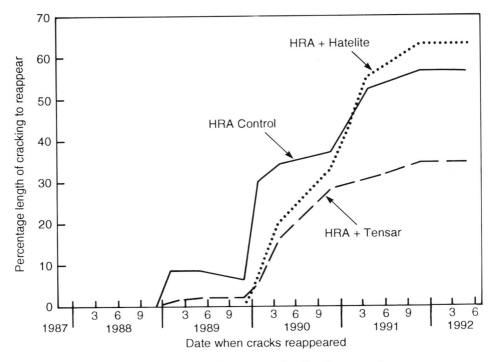

Fig. 2 Development of reflection cracks

Figure 2 also shows that the geogrids delayed the initiation of cracking by about 1 year and at the end of the third year the two geogrids had preformed in a similar manner to the control. Subsequently the rate of development of cracks in the Tensar section has slowed. On the same trial site with a thicker 80mm overlay reflection cracks have appeared in the control section but none are visible in the sections containing the geogrids.

Cores cut through cracks that have recently appeared showed that in the overlay the cracks initiated at the surface and propagated downwards to meet the crack in the underlying layer. The same behaviour was reported by Nunn (1989) for new roads with lean concrete roadbases. A tensile creep test has been developed to determine the strain that the wearing course can withstand before it cracks. Table 3 gives values of the yield strain determined by this test shortly after the materials were laid in the road and again at a later date.

Table 3. Properties of wearing course materials laid in the road trials.

Site (section)	Material	Penetration of recovered binder (0.1 mm)	Yield strain (%) Initial	Last tested	Age (yrs)
A30(L)	HRA 50pen + SBR	-	1.46	1.10	2
A30(J)	HRA 50pen control	48	1.40	1.08	2
A30(H)	HRA 100pen	74	3.01	1.95	2
A30(F)	HRA 50pen	42	2.72	1.32	2
A30(D)	HRA 50pen control	52	1.50	1.17	2
A30(I)	HRA 200pen	122	2.24	1.60	2
A30(G)	HRA 50pen control	47	1.52	1.14	2
A30(E)	HRA 100pen	58	1.21	0.96	2
A30(C)	HRA 50pen	44	1.22	1.14	2
A303(A)	HRA 100pen + SBS	-	3.27	2.16	1
A303(B)	HRA 100pen + SBS	-	2.80	2.32	1
A303(C)	HRA 50pen control	52	1.19	1.19	1
A303(D)	HRA 50pen control	52	1.48	1.21	1
M20(A)	HRA 100pen + SBS	-	2.16	1.05	3
M20(B)	HRA 50pen + SBR	-	1.63	0.30	3
M20(C)	HRA 50pen + EVA	-	1.56	1.04	3
M20(D)	HRA 50pen control	50	1.77	1.10	3

Table 3 shows that wearing course materials become more brittle with time and therefore they are less able to resist tensile strain. In an earlier study by Nunn (1989) the ductility of the wearing course was found to correlate well with the incidence of reflection cracking. If this is universally true then the most promising materials would be those that use softer binders with possibly a polymer additive to counter potential deformation problems. The indications from the trials are that more ductile wearing courses and thicker overlays inhibit the onset of reflective cracking. This is illustrated in Table 4 which gives the percentage of overlaid cracks above which reflection cracks have initiated after 3 years of service in the trial on A30 Launceston bypass westbound.

Table 4. Crack initiation in overlays with wearing courses of different ductility

Yield strain of wearing course (per cent)	Percentage of cracks initiated Thickness of overlay (mm)				
	<60	60-79	80-99	100-119	>120
< 1.49	58	25	0	25	0
1.50 - 1.99	50	11	0	0	0
2.00 - 2.49	0	0	0	0	0
>2.50	0	0	0	0	0

The trials are showing that the thickness of overlay has a strong influence on the time when reflection cracks re-appear. Table 5 shows the percentage of reflection cracks that have re-appeared in the overlay during the first 3 years of the trial on the westbound carriageway at Launceston bypass.

Table 5. Crack initiation in overlays of different thickness

Thickness of overlay (mm)	Percentage of cracks initiating after		
	Year 1	Year 2	Year 3
40 - 49	0	34	41
50 - 59	0	12	25
60 - 69	0	0	12
70 - 79	0	0	6
80 - 99	0	0	0
100 - 119	0	0	6
120 - 139	0	0	0

Although it is well known that a thicker overlay delays the appearance of reflective cracking at the surface, the quantification of the relationship between overlay thickness and delay will give a basis against which to compare other preventive measures. This relationship is being developed to take account of the initial severity and thermal behaviour of the cracks. All treatments incur a cost which could be expressed in terms of buying time before cracks re-appear. When performance and cost data are available for the various methods of inhibiting reflection cracking, the cost effectiveness of the different treatments can be readily compared, using the bought time approach, with the cost of a conventional overlay of equivalent performance. Furthermore the confirmation that cracking and therefore irreversible damage starts at the surface will lead to more direct methods of preventing the initiation of reflective cracking in new overlays.

6 Conclusions and recommendations

The trials have demonstrated that wearing course materials become increasingly brittle with age and that cores taken through new reflection cracks in a strengthening overlay show that the cracks initiate at the road surface and propagate downwards. This helps to interpret the early indications from the trials that more ductile materials as well as thicker overlays are better at inhibiting reflection cracking. A better understanding of how reflection cracks form will lead to more rational methods of treatment.

At present, only a small proportion of the original transverse cracks have reappeared in the new overlays. The trials, therefore, require further monitoring to allow sufficient performance data to be collected for statistical relationships to be established between the onset of reflection cracking, horizontal and vertical movements at the cracks being treated, overlay thickness and applied traffic for each treatment being studied.

7 Acknowledgements

The work described in this paper forms part of the programme of the Transport Research Laboratory and the paper is published by permission of the Chief Executive.

8 References

Nunn, M.E. (1989) An investigation of reflection cracking in composite pavements in the United Kingdom. Proceedings of the Conference on Reflective Cracking in Pavements, Liege, pp. 146-153.

Potter, J.F. (1991) Long term pavement performance trials in the UK. Proc. Int. Conf. on SHRP and Traffic and Safety in Two Continents. Gothenburg, Sweden.

Crown Copyright 1992. The views expressed in this paper are not necessarily those of the Department of Transport. Extracts from the text may be produced, except for commercial purposes, provided the source is acknowledged.

34 EXPERIENCE OF DU PONT DE NEMOURS IN REFLECTIVE CRACKING: SITE FOLLOW UP

G. KARAM
Du Pont de Nemours, Luxembourg

Abstract :
Reflection cracking of roads which have been repaved is a significant problem which can be alleviated through the use of thermally bonded non-woven polypropylene fabrics. This paper describes the advantages of Typar* BM-41 made by Du Pont de Nemours, and presents several case histories with varying degrees of success. Based on these results recommendations regarding the selection of thermally bonded non-woven polypropylene fabrics and conditions for use are presented.

Keywords :
Thermally bonded non-woven polypropylene fabric (T.B.N-W. PP fabric), Bitumen & Emulsion, flexible Road, Rigid Road, Semi-Rigid Road.

INTRODUCTION

The first repaving trials of Du Pont de Nemours (Luxembourg) S.A. with Typar* Geotextile, a T.B.N-W. PP fabric started in the early eighties. Since then a new product named BM41 has been specially developed and installed for this application on over 60 sites all over Europe. A multitude of laboratory research work has been performed, this includes the determination of the exact amount of binder to saturate the fabric and to guarantee optimum adhesion or the effect of the fabric in retarding reflective cracking and in providing a moisture barrier. Several laboratories (LRPC Autun (F), CRR Bruxelles (B), Delft (NL) and Liège University (B)), have confirmed the efficiency of the system but experiences on site will really tell if a system performs adequately. We will present 5 sites where a T.B.N-W. PP fabric was installed and subsequally inspected in July 1992. The 2 first cases were already presented in the first RILEM conference held in Liège on March, 1989.

---

* Du Pont's registered trademark for its spunbonded polypropylene

Reflective Cracking in Pavements. Edited by J.M. Rigo, R. Degeimbre and L. Francken.
© 1993 RILEM. Published by E & FN Spon, 2–6 Boundary Row, London SE1 8HN. ISBN 0 419 18220 9.

Case History Nr. 1

Site :
RN 427, Meefe (Wasseige/Belgium) the road was built in 1937. Due to increasing trafic and partially non-stabilized subsoil, this road was heavily dammaged and many cracks appeared on the surface.
Repaving date : June 1985
An asphalt concrete of 2 X 4 cm was laid. The local engineer asked for a testing area where a T.B.N-W. PP Geotextile was laid on one side of the road.
Surface : 4000 m²
Road structure : "Rigid" concrete road of 25 cm thick
Repaving procedure :

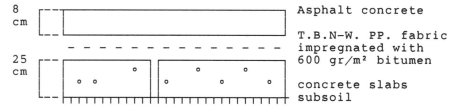

RN 427 : Reflective cracking system

Before the repaving, the joints were sealed and the potholes filled with concrete. 1.1 l/m² of emulsion at 55 % was applied by mechanical spraying at a temperature of 40 to 60°C. Then the Geotextile was laid manually and 2 layers of asphalt concrete of 4 cm each were installed at a temperature of 150°C.

Inspection July 1992 : 7 years later.
The part of the road covered with a fabric (the left part going to Wasseige) shows very few cracks compared with the right side where no Geotextile was installed

FIG. 1
RN427 MEEFE(WASSEIGE)/Belgium, the pavement in the village half with Typar* and half without.

Case History Nr. 2

Site : Motorway A64, near Zaragoza (Spain). The pavement was cracked due to the shrinkage of the aggregate cement subbase and to thermal stresses. The road was constructed in 1975
Repaving date : beginning 1988
Surface : 15.000 m²
Road structure : "Semi-rigid"

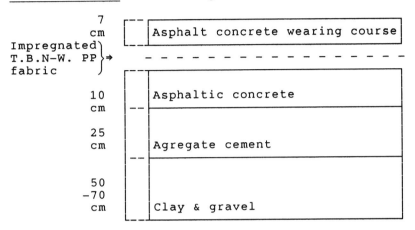

Repaving procedure : The repair was done, only where large cracks were observed. The pavement was milled off and a first sticking layer was sprayed.
An asphalitic binder of 10 cm was laid, then 1000 gr/m² of emulsion at 65 % was sprayed followed by manual laying of T.B.N-W. PP fabric. Then asphalt wearing course was installed.
Inspection July 92 : 4 years later.
The local motorway maintenance authorities were present during the inspection.
- It was clearly visible that the system including the T.B. N-W. PP fabric stopped the reflective cracking.
- The cracks which appear on the section without T.B.N-W. PP fabric stop at the edge of the section where the system was installed.
  These cracks appear about 3 years after the repaving.

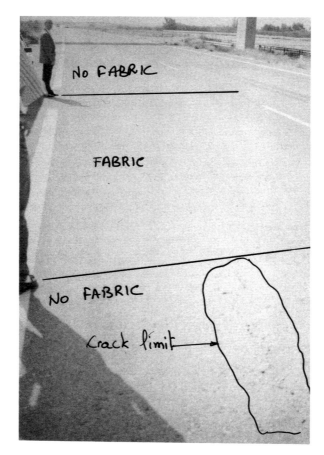

FIG. 2
A64 Motorway near ZARAGOZA/Spain
Cracks stop at the fabric limit.

Case History Nr. 3 :

Site : Highway E42 Namur to Seraing (Belgium)
this road had already been repaved.
Repaving date : October 1988
Surface : 800 m²
Road structure : "Semi-Rigid"
Repaving procedure :
- the damaged overlay was milled off down to the old pavement
- spraying of 1000 gr/m² of emulsion at 65 %

- unrolling of Typar* BM 41 and 2 preimpregnated BM41 strips over a large longitudinal crack
- installation of 4 cm Asphalt concrete layer

Inspection July 1992 : 4 years later.
No cracks were observed in the sections where the BM 41 or the preimpregnated BM 41 were laid. Some cracks appear on the untreated sections.

FIG. 3
E42 highway NAMUR to SERAING/Belgium
No cracks reappear on the left sides where a T.B.N-W. PP fabric was laid.

Case History Nr. 4 :

Site: B268 - Losheim (Germany)
Repaving date : September 1989
Surface : 25.000 m²
Road structure : "flexible" pavement
               fatigue cracks
Repaving procedure :
- filling of potholes and sealing of large cracks
- spraying of 800 gr/m² bitumen (180/200 pen)
- manual installation of the T.B.N-W. PP fabric
- installation of 4 to 5 cm Asphalt concrete "Splittmastik".

Inspection July 1992 : 2.8 years later
No cracks were observed.

FIG. 4
B268 near LOSHEIM (Germany)
no cracks reappear on the
section where T.B.N-W. PP
fabric was installed.

FIG. 5
B268 near LOSHEIM
section where no fabric
was installed

Case History Nr. 5 :

Site: RN 91 at Namur (Belgium)
Repaving date : March 1987
Surface : 5.000 m²
Road structure : "flexible"
            fatigue cracks, due to very heavy
            traffic
Repaving procedure : was following,

4-5 cm new overlay

- - - - - - - - - - - - - - -   Impregnated T.B.N-W. PP
                                  fabric

old asphalt layer
base

- milling off 2-3 cm of the old asphalt overlay
- spraying of 1200 gr/m² emulsion at 65 %
- installation of T.B.N-W. PP fabric
- overlaying of 7 cm Asphalt concrete

On a milled surface, the emulsion tend to concentrate in the lower part of the corrugations. Consequently the fabric adhesion is reduced. Unsufficient adhesion between the layers on one section of the road (left side going to Namur from the red lights near "Saint Luc Hospital") has resulted in almost immediate appearance of cracks during compaction. On that section the fabric and the new pavement were removed and a new overlay was reinstalled without fabric.

Inspection July 1992 : 6 years later
No cracks appear on the section with T.B.N-W.PP fabric. Large cracks started on the section without T.B.N-W.PP fabric, these cracks stop at the edge of the section where Typar* was installed.

FIG. 6
RN91 NAMUR/Belgium
On the left side where a thermally T.B.N-W. PP fabric was installed, no cracks reappear.

CONCLUSIONS :

Correct installation is a key to the performance of the reflective cracking control system. For systems using a thermally bonded non-woven polypropylene fabric, here are our recommendations :

a) the supporting surface has to be as clean as possible. Dust or dirt could reduce the adhesion. The potholes and the road irregularities must be filled and cracks more than 5 mm width must be sealed.

In case of rigid pavements, the road base must be stabilized to prevent vertical movement of concrete slabs. In case of too uneven surfaces, a levelling course is recommended.
b) the binder, (Pure Bitumen or Bitumen Emulsion) has to be applied on a dry surface. We recommend the spraying of 600 to 800 gr/m² pure bitumen. If emulsion is used, quantity of residual bitumen shall be similar. Wait until emulsion has broken and water has evaporated completely before laying down the Fabric. Do not use cutback. Specific site conditions may request different quantities of bitumen.
c) Unroll mechanically or manually the fabric under tension so to prevent wrinkles or folds. If these occur, cut, open and overlap excess fabric. Use 300 gr/m² bitumen to give the correct adhesion on overlaps. Overlaps should be 10 to 15 cm and always in the direction of overlay installation.
d) Avoid breaking and unnecessary maneuvering of the vehicles on the fabric. Install the new overlay, ensuring that the temperature does not exceed 200°C and is not inferior to 130°C.

Field experiences and laboratory tests have demonstrated that a thermally bonded non-woven polypropylene fabric is fully adapted to delay reflective cracking.
Following properties can be outlined :
- does not absorb rainwater, the water can be easily brushed away and the fabric does not need to be removed (peeled off) in case of rain during installation
- with 600 gr/m² of bitumen absorbtion, the fabric provides a moisture barrier for normal traffic conditions
- the fabric is almost incompressible. The amount of bitumen absorbtion is constant, properties remain the same and less fatigue stresses are developped
- the fabric can be easily laid manually as well as with a laydown machine
- the fabric rolls can be easily cut to desired width and handled on site. This is very useful for curves or road enlargements
- the system provide a more uniform bonding of the pavement layers and a good resistance to the trafic load stresses which result in less degradation of the overlay
- BM41 mechanical properties are suitable for this application.

35 SOFT PAVEMENT WIDE INTERLAYER IN JAPAN

K. INOUE
Japan Seal Industries Co. Ltd, Osaka, Japan

Abstract
Prof. Fukuoka's studies have been already specified in the joint reflection crack field(ACC/PCC) in Japan.
Till now many sheets have been used for soft pavement(ACC/ACC). I would like to talk about this soft pavement(asphalt concrete overlay-ACC/interlayer/damaged road-ACC) and that wide applications.
Materials are made by an American manufacturer;polypropylene staple fiber, needle punched and heat treated non-woven fabrics. I would like to say this is quite different from spunbonded non-woven fabrics.
In 1978, our fabrics were initially used in Kumamoto Pref. and the quantities have reached to 200,000m^2 there.
In Japan, more important techniques are milling and overlay, or preferably immediate overlay after milling. These were started by Osaka municipal engineers and later in Hyogo Pref.
I myself have watched Osaka prefectural operation for many years. This operation was done as the prior operation to the following 2 larger ones.
I thought this operation was important, because it was soft pavement wide interlayer and that heavy traffic was suspected although this was not milling and overlay.
Recent operations are also announced now.
Keywords: Soft Pavement, Wide Interlayer, ACC, PCC, Milling and Overlay.

1 Introduction

First of all, I would like to explain that the word "soft pavement" means asphalt cement concrete(ACC) overlay on elastic pavement (ACC). Of course, I will talk about interlayer laydown and hot ACC overlay.
As I talked with discussion members at Tokyo, my opinion was that, in order to retard reflective cracking in pavements, we had to discuss separately as follows: (1) overlay by hot ACC on Portland cement concrete (PCC). (2) overlay by hot ACC on ACC narrow interlayer, say less than 1 meter. (3) overlay by hot ACC on ACC with wide interlayer, which is my theme. (4) overlay by cold ACC on ACC. (5) other new method. I think each application needs proper material i.e. interlayer materials. (Fig. 1)
This speech is specially focussed on wide interlayer. Because wide application needs a large quantity of interlayer materials. So they

Reflective Cracking in Pavements. Edited by J.M. Rigo, R. Degeimbre and L. Francken.
© 1993 RILEM. Published by E & FN Spon, 2–6 Boundary Row, London SE1 8HN. ISBN 0 419 18220 9.

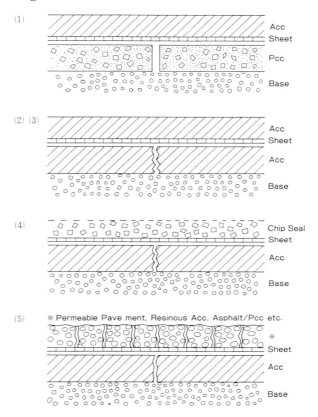

Fig 1

must be cheap and that quick application is also required. In Japan a fair quantity of interlayer has been used as category (2) narrow fabrics overlay.

Also many kinds of sheets have been used for soft pavement(ACC/ACC). For example, in U.S.A. Please refer to NEEP 10 Table 1(*1). Many materials are also tested in many countries.

The materials which I have used are made by an American manufacturer ;polypropylene staple fiber, needle punched and heat treated non-woven fabrics. I would like to say this is quite different from spunbonded non-woven fabrics. Spunbonded non-woven fabric has a serious problem. That is "peeling" when traffic is released before overlay. And I think it also has less elongation characteristics.

2 Time to time story

At the 1st. RILEM here in 1989, Prof. Fukuoka introduced his 1971-74 studies in Japan which were published in 1975(*2).
I have entered into this field since 1976, because I was called as a specialist of market developers of staple fiber needle punched non-woven fabrics. And I found that reflective cracking preventive field was easy for me to enter into. Of course, actually it is not easy.
I have learnt much from Prof. Fukuoka's reports. In Japan, these studies are specified as one of techniques to retard reflective cracking to joint cracking(ACC/PCC). This is the category (1) overlay by hot ACC on PCC, which I talked just before. I would like to introduce Japanese technical hand book, say "Manual for road maintenance" page 111, 1978 Japan, which we call "a yellow book".(*3)
Some operations of my customers failed, but most ones were successful, depending on how to evaluate. Please refer to Mr. Raymond A. Forsyth's comment.(*4) These comments were expressed at International Geotextile Seminar, Nov. 1989, Osaka,Japan.
I think that careful watching is necessary to evaluate. That is same in U.S.A. and Japan. And also the same in other countries. Road conditions and traffic characteristics are always different, and time difference makes different traffic characteristics, especially in rapidly changing areas. And we need 10 years to evaluate. So such products (interlayer) are not suitable for current patent system. I would like to know how to solve this problem. Manufacturers and concerned people needed too much energy to develop the market, gaining small money. Now let's go back to Japanese experience of my job. My 1st. application was done by Kawasaki municipal officer and the fabric was adhered by asphalt emulsions on 10th. Oct. 1977. The engineer recognized our fabrics to be superior to other ones which he had ever used. Thanks to my friend, Mr. S. Hayashi, I entered this field. He was a member of the study group in 1971.
Later, some operations were done by some organizations. Some operations were successful, but some failed. This fact and experience was good for me. I have assimilated much knowledge and I could become understood such fabrics supplier's comments as emulsions were not recommendable.
I myself have watched the Osaka prefectural operation for many years. This operation was done as the prior operation to the following 2 larger ones. It was called Yao-Domyoji line's front road of Kashihara High School. I think this operation was important, because it was soft pavement wide interlayer and that heavy traffic was suspected, although this was not milling and overlay. My personal opinion was that this test operation was successful, but, at the same time, some problems were included. Those problems were rather good for my knowledge. "10 years later, the road was rebuilt. Then fabrics were still tough." said the constructor. This rehabilitating operation was done in Jan. 1980 and rebuilding (milling and overlay) was done in Oct. 1990.
This test operation had a problem because it was done by asphalt emulsions. An Osaka prefectural engineer pointed out that operators sprayed very few asphalt, and then the quantity was increased, though

overlay was already started. So irregularities appeared there. Naturally some troubles happened, but I considered the operation was successful, however some nagative opinions existed. Anyhow 2 larger operations were done. Those were not wide interlayer operation in strict meaning.

In 1981, Osaka municipal engineers started to use wide interlayer overlay rehabilitation technique.

In 1978, our fabrics were initially used in Kumamoto Pref. and the quantities have reached to 200,000m^2, thanks to one of our largest distributors, Messrs. Nisshin Chemical Company.

Recently I asked some constructors for their reputations. Their answer was "Very good". Fabrics can be useful for more than 12 years. Of course, some constructors failed or were not so satisfied.

In Japan, more important techniques are milling and overlay, or preferably immediate overlay after milling. Many roads were started by thin ACC, and afterwards few overlays on these roads. Road surfaces often exceed door levels. That is why we need milling and overlay.

These were started by Osaka municipal engineers and later in Hyogo Pref. At 1st. stage, I recommended such time schedule of fabric laydown and overlay as follows; after milling, traffic has to be released for some days, and next, overlay.* This method is good for Japanese road maintenance. Because highways and most urban streets require rapid rehabilitation. Quite many operations had been done so, and now Hyogo Pref. adopts this technique as a common rehabilitation one. Observations must be carefully done and evaluated. I hope this technique will be adopted by Japan Highway Public Corp.

In 1986, Prof. Yamaoka started evaluation of interlayer. His 1st. announcement had been done at 1st. RELEM here. In 1990, Mr. Yasuda et al. reported 27 months result. Unfortunately the sheet used was not mine, though the results were satisfactory. (*5)

In 1987, our largest distributor, Nichireki Chemical Ind., Co.,Ltd. willingly started this field and now is developing widely in Japan.

In 1989, the Public Works Research Institute, Ministry of Construction, Japan started interlayer test again, and preliminary test shows our mat is good. Since 1989, applications for air fields have been also done. Parking area is also next good candidate.

3 Difficulties

Evaluation of wide interlayer overlay has many difficulties, so it is not easy to standardize rehabilitation techniques. Interlayer is not panacea. They have a limitation. It is quite natural.

At first, to make sure, 5 categories should be considered as I talked before. To study interlayer, I would list up many difficulties, but I don't think so seriously.
(1) Interlayer materials: Too many products, structures and polymers.
(2) Limitations:For the thermal crack, our interlayer is not recommendable until now. Common applications require that base roads must be prepared as specified and overlay thickness must be sufficient according to traffic volume and character. Wrong condition needs dig out.

(3) Adhesives: Asphalt emulsion is not recommendable, especially in high humid area like Japan. Because it often flows to outside area, it takes a long time to evapolate. If not evapolated, it causes slippage. In Japan, hot asphalt distributor is not easy to specify. Because asphalt emulsion distributors are the main trend. Hot asphalt distributor needs a certain quantity, say 2 tons minimum, which means small operations are not suitable.
(4) Road width: There are many widths of roads. Of course, we can slit the fabrics very easily. Now we have 4 widths in stock: 1.6, 1.9, 3.2, 3.8 meters respectively.
(5) Curvature: Slow curves are useful by one piece with no cut, though steep curvature needs cuts of fabrics.
(6) Incline: I don't recommend inclined roads in the beginning. To test the fabric, in Japan, the mountain area is easier to be appointed as a test road, but I don't recommend as a first stage. It had better be flat and straight.
(7) Desultory operations: Rehabilitation requests minimum operation and often forces to cut and cut overlay.
(8) Rehabilitation length: In Japan, the police often request overlays less than 100 meters length for safety and smooth traffic due to heavy traffic.
(9) In operation: Crew must be expert. They should know about such a interlayer system that especially no wrinkle is necessary. In Japan, workers become good operators after 1 roll used, and 2nd. roll may be the last one.
(10) Roll weight: 60 kilograms are too heavy for common Japanese workers. It can be solved by the shorter roll or laydown equipment.
(11) Laydown equipment: There is not enough equipment in Japan. If we could get more orders, it would be resolved.
(12) Officials: In Japan, most overlay operations are not private, but public. For prevention of injustice and educational purposes, officials are periodically transferred to other offices every 2 - 4 years.
 This system makes us difficult to follow up the operations.
(13) Improper evaluation: It will be caused by improper preparation, wrong operation and unbalanced overlay thickness against traffic volume. We should take care of these conditions. And also, impermeable membrane prevents water and air from intruding into sub bases, which contribute the extension of a road life. This effect is also difficult to be proved. Please refer to Mr. Raymond A. Forsyth's comment again.

4 Proposal to the audience

My experience shows that wide interlayer overlay rehabilitation system is a good one. So you may believe in it. Please try more operations.

To prevent default of interlayer, one idea is that, even in Europe, you may have an agent of "Petromat" like me. Or you may have a joint venture here. Please start first like in U.S. way according to your conditions. Secondly, please investigate what material is good for pavement. Then next, you will produce more sophisticated product. Maybe such product has been already produced here. If so, I know about it. Othe ideas to always succeed in overlay are prediction of results. One of them is to use "Roadrator". The other is to use Benkelmann's Beam or Nichireki's Romen Catcher System. You have to know limitations of interlayer.

Anyhow careful watching and right operation will result that interlayer system will become a standard overlay technique. Please accumulate data and announce the results to many people. You have a good chance.

5 References

(*1) "Report on Performance of Fabrics in Asphalt Overlays"
   US Department of Tranportation,Federal Highway Administration
   NEEP 10 (National Experimental and Evaluation Program Project No. 10) Sept. 1982.
(*2) M. Fukuoka p265 "Reflective cracking in pavements" RILEM 1989.
(*3) "Manual for road maintenance" 1978 Japan.
(*4) Raymond A. Forsyth "Paving Fabric applications by CALTRNS"
   Nov. 1989,Osaka Japan.
(*5) K. Yasuda et al. "Applications of geotextiles in paving(overlay)"
   Geotextile Seminar Mar. 1990, Tokyo Japan.

* This time schedule method went well. Later Hyogo's engineers applied immediate fabric laydown after milling and then overlay.

36 THE APPLICATION OF A GEOTEXTILE MANUFACTURED ON SITE ON THE BELGIAN MOTORWAY MONS–TOURNAI

R. DUMONT
Walloon Ministry of Infrastructure and Transport, Mons, Belgium
Y. DECOENE
Screg Belgium, Brussels, Belgium

Abstract

This paper describes the renovation of the continuously reinforced cement concrete pavement of the E42 motorway between Mons and Tournai (Belgium). After seventeen years of traffic this pavement was severely cracked; the cause of this phenomenon has not been clearly defined to date.

Instead of tearing up the pavement and laying a new one, the Authority decided in 1990 to test a system based on a geotextile impregnated with a polymer-bitumen on the carriageway in the direction of Mons, where the damage was less serious. The Engineers opted for a technique in which the geotextile manufactured on site with a suitable machine and the "membrane" (thick surface dressing) is subsequently overlaid with a 5cm thick porous asphalt course containing a bitumen with polymers and cellulose fibres.

Because of the success of this experiment made on more than 155,000 m2, the Authority decided in 1991 to apply the same system on the carriageway to Tournai, where more than 150,000 m2 of cracked cement concrete pavement had to be overlaid.

This paper gives details on the road works : degree of deterioration of the concrete pavement, principle of the adopted solution, characteristics of the different materials (geotextile, membrane, porous asphalt) and the machines used. The aspect of the overlay after two or three years will also be discussed either in the paper or at the Conference in Liège.

Reflective Cracking in Pavements. Edited by J.M. Rigo, R. Degeimbre and L. Francken.
© 1993 RILEM. Published by E & FN Spon, 2–6 Boundary Row, London SE1 8HN. ISBN 0 419 18220 9.

1 Introduction

The E42 motorway between Mons and Tournai in Belgium was constructed around 1973 with a continuously reinforced concrete pavement in its section between Péruwelz and Tournai. After seventeen years of traffic this pavement had severely deteriorated in both directions of travel.

With the traditional repair techniques it would have been necessary to demolish and replace the continuously reinforced concrete. Because of the high costs involved in such works the awarding Authority proposed tooverlay the concrete with an anticracking system composed first of a geotextile manufactured on site and impregnated with elastomer-bitumen and then of a wearing course in porous asphalt. The Authority thus hoped to extend the service life of the pavement and postpone the time when it may have to be demolished.

2 Description of the site

Since 1983, i.e. less than ten years after they were constructed, the continuously reinforced concrete pavements of the carriageways in the Péruwelz-Tournai section of the E42 motorway had been subject to severe longitudinal and transverse cracking.

Fig.1. "Paving" blocks

This abnormal cracking probably resulted from a conjuction of factors, namely :
- the 18mm diameter of the longitudinal reinforcing bars;
- the laying of the longitudinal reinforcement at one-third-depth in the concrete pavement, i.e. ± 6cm under the running surface;
- alkali-aggregate reaction in a Portland cement concrete with 50% of its stone skeleton composed of crushed limestone.

The increase in traffic volume and seasonal cycles promoting water entry into cracks gradually caused the concrete to crake and disintegrate into "paving" blocks which loosened from the reinforcement (figure 1).

Up to 1985 the damage was limited to a few stretches along the 16km of the section; after that date it expanded to the whale section, thus requiring permanent intervention by maintenance gangs to remove the concrete "paving" blocks and fill the resulting holes with bituminous material.

3 Choice of solutions

3.1 Preliminary studies

As reported before (Tessonneau and Samanos (1991)), many studies have been devoted to anticracking systems with polyester threads laid on site, especially at the Belgian Road Research Centre (BRRC) and at the Regional Laboratory of Autun in France. Their experiments have shown among other things that the excellent behaviour of such systems can be ascribed to a "debonding" effect which greatly reduces - in comparison with other systems - the maximum stress in the overlying asphalt course as the crack opens.

In addition, during simulation tests conducted with the equipment of BRRC (Clauwaert and Francken (1989)) it was observed that cracks, if any, may appear in the wearing course without affecting the imperviousness of the nonwoven geotextile impregnated with elastomer-bitumen. In porous asphalt such cracks do not interfere with the properties of the course since water is normally allowed to enter it, but in dense asphalt they can have adverse effects in times of frost.

3.2 Application to the Mons-Tournai motorway

On the one hand, since water entering cracks was one of the main causes of the damage, the solution to the problem was sought in a technique that would make it possible to waterproof the surface of the continuously reinforced concrete pavement and to maintain this waterproofness in the course of time, in spite of the large number of cracks.

On the other, the increase in traffic volume on a continuously reinforced concrete pavement with a transversely grooved surface having led to numerous complaints by frontagers, provision had to be made for the laying of a wearing course minimizing the noise generated by tyre-road contact.

A waterproof anticracking system with a wearing course in porous asphalt seemed to meet these two requirements without entailing excessive costs.

To adapt this waterproof anticracking system to the state of deterioration of the existing pavement surface, several tests were made during the two phases of execution.

1. First phase - carried out in May 1990.
The waterproof "membrane" (thick surface dressing) was laid all over the continuously reinforced concrete pavement of the Tournai-Mons carriageway, i.e. over a surface area of 158,000 m2.
This membrane was reinforced with interlaced polyester threads spread at the rate of 100g/m2 over the full width (11.25 m) of the carriageway in the three most severely deteriorated stretches of the motorway, i.e. over 75,000 m2.

2. Second phase - carried out in May 1991.
The waterproof membrane was laid all over the continuously reinforced concrete pavement of the Mons-Tournai carriageway, i.e. over a surface area of 158,000 m2.
This membrane was reinforced with interlaced polyester threads spread at the rate of 150g/m2 over a width of only 9m, i.e. over 127,000m2. The reduction in width resulted from the mild deterioration of the emergency stopping lane, which did not make it necessary to reinforce the waterproof membrane.
On the other hand, because of the more severe deterioration of the Mons-Tournai carriageway the rate of application of the polyester threads was increased to 150g/m2, in order to maintain the loose "paving" blocks which had not been removed in position.

4 Description of the technique

4.1 The selected anticracking system consists of two components : a reinforced membrane and a porous asphalt course, and a technological component : a layer of chippings (figure 2).

Fig.2. From right to left : the modified bitumen, the spreading of the geotextile, the precoated chippings.

The waterproofing membrane is constructed in two phases:
1. a thick "surface dressing" obtained by spreading a recycled elastomer-modified bitumen at a rate not lower than 3kg/m2;
2. a nonwoven geotextile composed of interlaced polyester threads, which were spread at the rate of 100g/m2 in the direction of Mons and 150g/m2 in that of Tournai. This geotextile is laid on the cooled thick surface dressing, using a thread-laying machine.
The machine used to lay polyester threads on the E42 motorway has a working width of 3m. It is fitted with two roller beams (big spools) each carrying 700 kg of continuous threads on a support of 1.50m.
It is equipped with a spray bar for polymer-bitumen emulsions which was not used, however, as the binder for the membrane had to be a highly viscous mixture of bitumen and recycled elastomers.

It should be noted that more recent versions of this machine have a 6,000 litre emulsion storage tank (Tessonneau and Samanos (1991)) in order to spray 800 to 2,200 g/m2 of binder; their working width can be adjusted from 1.25 to 3.70m.

4.2   A layer of 7/10mm chippings precoated with bitumen is spread at a rate of about 5kg/m2 immediately after the laying of the geotextile.
This layer is to maintain the geotextile in position on the thick surface dressing and to enable the subsequent porous asphalt paving equipment to travel on the newly laid reinforced membrane without damaging it.

4.3   The membrane is overlaid with a 5cm thick wearing course in porous asphalt 0/14 containing 5.4% of bitumen admiscend with new elastomers and cellulose fibres.
This wearing course is laid with two asphalt finishers working in parallel and three nonvibrating smooth-wheeled rollers, to avoid forming a cold construction joint in the porous asphalt.

5   Conclusions

The experiment carried out in 1990 and 1991 on the E42 motorway between Mons and Tournai (Belgium) has shown that it is technically feasible to overlay severely deteriorated continuously reinforced concrete with an anticracking system composed of an elastomer-bitumen impregnated geotextile manufactured on site and a wearing course in porous asphalt. The success achieved is due among other things, to the design of a special machine for laying continuous polyester threads in place on the road.

After two years of motorway traffic no reflection of cracks or other defects from the underlying pavement is observed in the section where this system was applied. The oral presentation of this paper at the RILEM Conference Exhibition will be an opportunity to report observations on this site after three years of traffic.

It may also be added that the application of the anticracking system on the continuously reinforced concrete pavement has made it possible to preserve the infrastructure in place and extend its service life, as well as to increase the comfort of road users by improving surface evenness, lowering the noise level, eliminating the risk of aquaplaning and reducing splash and spray in rainy weather.

Références

Clauwaert C., Francken L. (1989). Etude et observation de la fissure "réflective" au Centre de Recherches routières belge, in : Reflective Cracking in Pavements (eds. J.M.Rigo and R.Degeimbre), C.E.P. - L.M.C. - University of Liège, pp.170-181.

Tessonneau H., Samanos J. (1991). Un nouveau système antifissures : Filaflex, la membrane armée fabriquée en place, in. Revue Générale des Routes et Aérodromes, 681, pp.57-52.

37 BELGIAN APPLICATIONS OF GEOTEXTILES TO AVOID REFLECTIVE CRACKING IN PAVEMENTS

Y. DECOENE
Screg Belgium, Brussels, Belgium

Abstract

During the first Conference on "Reflective Cracking in Pavements" (Liège, 1989), P. Silence and J. Estival presented a paper about three Belgian experiments with geotextiles to avoid reflection cracking of pavements, more particularly in overlays on cement concrete pavements. In these experiments geotextiles impregnated with polymer-bitumen were covered with a porous asphalt wearing course.

This paper reports on pavement behaviour on these three test sites after five to seven years of traffic, especially the rate of any cracking which may have occurred.

Moreover, there is a brief description of further experiments with other geotextiles for the renovation of cement concrete roads: woven geotextile, nonwoven geotextile, geotextile manufactured on the job site, ... In most of those cases too the wearing course is in porous asphalt. The description includes the characteristics of the road, the adopted solution, and the materials used (geotextile, bitumen, type of asphalt, ...). The aspect of the overlays under traffic is also dealt with in the paper and will be discussed during the new Conference.

<u>Keywords:</u> Reflective cracking, Geotextile, Bitumen, Membrane, Polymers, Porous asphalt, Cement Concrete.

1 Introduction

During the first RILEM Conference on "Reflective Cracking in Pavements" (Liège, 1989), P. Silence and J. Estival presented a paper on Belgian applications of geotextiles to control reflective cracking; included in this paper was a description of three experimental jobs carried out in Belgium between 1985 and 1987. The present paper will describe current pavement behaviour on those three sites and then present the jobs which Screg Belgium, to which J. Estival belonged, has carried out with geotextiles in

Belgium since 1989.

2 Jobs from 1985 to 1987

The three jobs described in the 1989 paper (Silence and Estival (1989)) involved the application of an anticracking system with a woven geotextile; this system (see table 1) was laid on :

Cement concrete slabs at Erquelinnes (1985) and Tourpes (1987).
A layer in bituminous mix at Wépion (1987).

More details can be found in the aforesaid paper.

Table 1.   Applications from 1985 to 1987

Place	Year of renovation	Surface (m2)	Apparent cracks in 1992
Erquelinnes	1985	2,500	36 (°)
Tourpes	1987	19,200	0
Wépion	1987	20,600	0

(°) 30 in "rue des Bonniers" (1,180 m^2) and 6 in "rue Couroye" (1,320 m^2).

2.1 Works carried out

It may be recalled here that the cement concrete slabs at Erquelinnes were overlaid first with a woven geotextile impregnated with bitumen modified with new elastomers, and then with 4 cm of 0/10 mm Fixtone (doubly coated) porous asphalt.

At Tourpes the cement concrete slabs were overlaid successively with a bituminous slurry, a 3 cm thick type IIIC regulating course (Belgian Ministry of Public Works (1979)), a woven geotextile impregnated with an emulsion of bitumen and new elastomers, 5 cm of type IIIB bituminous concrete, and 4 cm 0/14 mm Fixtone porous asphalt.

The wearing courses in bituminous mix of the "rue Lecomte" and "rue du Suary" at Wépion were overlaid with a type IIIC regulating course about 4 cm thick, a geotextile impregnated with an emulsion of bitumen and new elastomers, and 3 cm of 0/10 mm Fixtone porous asphalt.

2.2 Behaviour

In 1992 no reflective cracking is observable in the 39,800 m^2 of overlay at Tourpes and Wépion.

The overlay at Erquelinnes (2,500 m^2) performs well when considering the facts that its total thickness is very limited (about 4 cm), that the renovated streets are densely inhabited, that the cement concrete slabs were unstable and in very poor condition (Estival (1988)), especially the "rue des Bonniers" which carries very heavy traffic leaving the nearby French-Belgian custom-house. Many double cracks have appeared at the surface of this street, reflecting the joints between the underlying slabs; they are spaced 2 to 3 cm apart, thus indicating vertical movement ("rocking") of the slabs at the wide joints. Four similar cases can be observed in the "rue Couroye", which was treated in the same way.

3 Jobs since 1989

3.1 Description

Since 1989, Screg Belgium companies have applied geotextile-based anticracking systems on several sites in Belgium (see table 2), mainly on cement concrete roads (excepted in Châtelet) and in built-up areas where the available thickness for overlay was very limited.

Table 2. Applications since 1989

Place	Year of renovation	Surface (m2)	Bitumen (°)	Geotextile (°°)	Wearing course	Apparent cracks in 1992
Hermée	1989	1,300	a	A	Porous asphalt (4 cm)	0
Genappe	1990	1,060	b	B	Porous asphalt (4 cm)	2
E42-South	1990	75,190	b	C	Porous asphalt (4 cm)	0
Woluwé-S-P	1990	1,820	b	D	Porous asphalt (4 cm)	0
N5-Bruxelles	1990	1,110	b	A	Porous asphalt (4 cm)	0
Quevaucamps	1991	2,580	a	A	Ultra-thin layer (1,5 cm)	0
E42-North	1991	141,200	b	C	Porous asphalt (5 cm)	0
R3-Châtelet	1991	22,000	b	C	Porous asphalt (2,5 cm)	0
N527-Irchonwelz	1991	3,600	a	D	Asphalt with a high mastic content (3 cm)	0
Lasne	1992	13,500	c	C	Porous asphalt (3 cm)	0

(°) a with new elastomers
    b with recycled elastomers
    c emulsion of bitumen with new elastomers

(°°) A nonwoven polyester
     B woven polyester
     C polyester manufactured on the jobsite
     D nonwoven polypropylene

Tests conducted with the simulation equipment of the Belgian Road Research Centre (Clauwaert and Francken (1989)) on various geotextiles with an overlying layer in porous asphalt had shown that, after the asphalt had cracked, the membrane consisting of a nonwoven geotextile impregnated with elastomer-bitumen remained imprevious, as opposed to that with a woven geotextile. That is why we have opted for nonwoven geotextiles ever since

The geotextile manufactured on site on motorway E42 and ring road R3 (figure 1), using a machine capable of laying continuous polyester yarns on site, was also of the nonwoven type (Samanos et al. (1991)).

Fig. 1 Application of a geotextile manufactured on site on ring road R3 at Châtelet

Provided lateral drainage of infiltrating water could be ensured, Screg Belgium preferred to use porous asphalts mainly because laboratory tests had shown that cracks, if any, first appear in the wearing course and not in the membrane. Now porous asphalt naturally allows the entry of water into those cracks, which may be dangerous with dense bituminous concretes - especially during frost.

All jobs between 1989 and 1992 were carried out after the awarding authorities had consulted with Screg Belgium; they wanted the renovation of the streets involved to be economical and durable, while raising the existing street levels as little as possible and minimizing the appearance of cracks especially above the joints in the underlying cement concrete pavement.

3.2 Considerations on execution

The 1990 job at Genappe was preceded in 1989 by a first application, in which an emulsion of bitumen with new elastomers was sprayed before applying a nonwoven polyester geotextile and a 4 cm thick layer of 0/14 mm porous asphalt. Heavy rain between the spraying of the emulsion and the application of geotextile washed part of the binder away, which resulted in peeling of the new surfacing a few weeks later and made it necessary to repeat the procedure with an anhydrous mixture of bitumen and new elastomers. It should, therefore, be avoided using the binder in emulsion form if the geotextile is not laid immediately on the binder coat.

The use of a polypropylene geotextile raised no problem with an overlying layer bound with pure bitumen or bitumen to which new elastomers had been added. On the other hand, for an overlying layer containing a bitumen with recycled elastomers which had to be prepared at more than 200°C the experiment at Woluwé-Saint-Pierre has shown that the polypropylene may melt under the heat of the fresh mix. A polypropylene geotextile should, therefore be avoided under a layer which contains a bitumen that has been modified with recycled elastomers.

Details on the execution of the Belgian jobs carried out in 1990 and 1991 with a geotextile manufactured on site have been published (Decoene (1992)); in addition, a paper on the E42 motorway job (Dumont and Decoene (1993)) has been prepared for the second RILEM Conference in Liège.

It should be added that the application of an anticracking system on cement concrete slabs was always preceded by cleaning and filling of the joints; on some sites a regulating course was applied before overlaying, either locally or over the entire surface (Genappe).

3.3 Behaviour

In 1992 no cracks are observable on the sites listed in table 2, except in the "rue de Villers" at Genappe. On this road, which carries a considerable amount of heavy traffic from or to the sugar factory of Genappe, the wearing course was locally repaired where the underlying cement concrete was of very poor quality; in addition, the double cracks which have appeared are indicative of severe slab rocking in the underlying cement concrete, just like at Erquelinnes (2.2).

4 Conclusions

The use of an anticracking system composed of an elastomer-bitumen-impregnated geotextile and a layer in bituminous mix is an economical solution, which makes it possible for example to prevent the destruction of an underlying structure in cement concrete. This has been demonstrated by the behaviour of the renovated pavements on the thirteen Belgian sites fitted with such a system between 1985 and 1992.

Observations on the roads thus treated have, however, shown the limits to the applicability of the technique : it should be avoided on cement concrete slabs which are subject to considerable vertical movements at the joints. Contract-awarding authorities are, therefore, recommended to measure slab rocking before making any decision.

It would be interesting to try and determine, as a function of traffic volume, the maximum permissible slab rocking value which is still compatible with the use of the technique.

Tests in the laboratory have shown that nonwoven geotextiles should be preferred and that in particular such geotextiles manufactured by laying continuous polyester yarns on site should be encouraged.

With a geotextile manufàctured on site it is possible to spread the emulsified elastomeric binder and the continuous yarns virtually at the same time. In other cases, and especially with a plant-manufactured geotextile, it is preferable to spray an anhydrous elastomeric binder.

By the time this paper is presented in Liège, additional information will be available from observations made in 1993 on overlays constructed in Belgium since 1985 by firms affiliated to Screg Belgium.

5 References

Belgian Ministry of Public Works (1979) **Cahier des charges-type 150** (eds Fonds des Routes), Administration des Routes, Bruxelles.

Clauwaert, C. and Francken, L. (1989) Etude et observation de la fissure "réflective" au Centre de Recherches routières belge, in **Reflective Cracking in Pavements** (eds J.M. Rigo and R. Degeimbre), C.E.P.-L.M.C.- University of Liege, pp. 170-181.

Decoene, Y. (1992) Gebruik in België van een speciale machine voor het in situ fabriceren van een geotextiel tegen het voorkomen van reflectiescheuren, in **Wegbouwkundige Werkdagen 1992** (eds Stichting Centrum voor Regelgeving en Onderzoek in de Grond-, Water- en Wegenbouw en de Verkeerstechniek), Ede, pp. 729-736.

Dumont, R. and Decoene, Y. (1993) Application of a geotextile manufactured on the jobsite of the Belgian motorway Mons-Tournai, in **Reflective Cracking in Pavements**, C.E.P.-L.M.C.-University of Liege.

Estival, J. (1988) Utilisation de géotextiles pour lutter contre la remontée des fissures dans les revêtements hydrocarbonés. **Bituminfo**, 55, 21-25.

Samanos, J. Roffé, J.C. Tessonneau, H. and Serfass, J.P. (1991) New Systeme for Preventing Reflective Cracking: Membrane Using Reinforcement Manufactured on Site (MURMOS), in **Highway Maintenance Operations and Research 1991** (eds Transportation Research Board), Transportation Research Record No. 1304, Washington, D.C., pp. 59-65.

Silence, P. and Estival J. (1989) Applications de tissus de polyester pour lutter contre la remontée des fissures, in **Reflective Cracking in Pavements** (eds J.M. Rigo and R. Degeimbre), C.E.P.-L.M.C.-University of Liège, pp. 281-287.

38 PREVENTION OF CRACKING PROGRESS OF ASPHALT OVERLAYER WITH GLASS FABRIC

ZHONGYIN GUO
Tongji University, Shanghai, P.R. China
QUANCAI ZHANG
Taiyuan Highway Administration, P.R. China

Abstract
Cracking is the principal type of failure of asphalt overlayers. In order to prevent the cracking progress of asphalt overlayers, glass fabric (grid) is used to enhance the resistance of asphalt overlayer to reflective cracking or fatigue cracking. Glass fabric (grid) is a very cheap textile material with high tensile strength and low deformation ratio in China. In this paper, both laboratory experimentation and field road test were carried out to study the effectiveness of the application of glass fabric (grid) to prevent the cracking progress of asphalt overlayer. The research results show that the application of glass fabric (grid) at the bottom of an asphalt overlayer can significantly prolong the life of the asphalt overlayer against cracking progress, but it hardly delays the initial cracking.
Keywords: Asphalt Overlayer, Cracking, Prevention, Glass Fabric

1 Introduction

Up to 95 percent of highway pavements in China are asphalt pavements. The typical asphalt pavement construction for the second and lower classes of highways comprises a thin asphalt surface of about 60mm with a underlaid strong semirigid subbase of about 300mm. This structure is so called weak surface pavement with strong subbase in China. Based on the design specification for highway flexible pavements, the design life of this type of pavement ranges from 8 to 10 years. Unfortunately, in four or five years at most since being opened to traffic, pavements must be overlaied. Normally, the overlayer is made with 30mm asphalt concrete. For this type of overlayer, the cracks in the existing pavement will very soon reflect through it. Two or three years later, the pavement needs a new overlayer. For economical limits, pavements can not be overlaid often and so can not be kept in a satisfactory surface condition and structure load bearing capacity. Glass fabric or grid is a cheap material (about 0.2 us¥ per square meter) and has high tensile strength and low deformation ratio compared with other types of geotextile. So that glass fabric or grid were chosen to be used as reinforcing materials at the bottom of asphalt overlayer.

The application of fabric to prevent reflective cracking progress has been investigated by the research group of Texas University at Austin and other researchers. But more research work has been done

Reflective Cracking in Pavements. Edited by J.M. Rigo, R. Degeimbre and L. Francken.
© 1993 RILEM. Published by E & FN Spon, 2–6 Boundary Row, London SE1 8HN. ISBN 0 419 18220 9.

on the application of geogrid.

2 Laboratory test

2.1 Experimentation of glass fabric (grid)
In this research project, four types of glass fabric (grid) were tested, which are 12×10, 10×10, 10×8 woven glass fabric and one type of glass fabric grid. It is meant that the number of the strands of the glass fabric in the weft is 10 and that in the warp is 12 by 12×10 (or 10×10, 10×8).

The unit weight of the glass fabric (grid) was weighed in a precision balance and shown in table 1.

Table 1. Unit weight of glass fabric(grid)

Type of glass fabric or grid	12×10	10×10	10×8	Grid
Unit weight (g/m^2)	115.1	101.7	93.9	320.5

For the narrow strip tensile test, the specimen size of glass fabric (grid) is 50mm in width and 150mm in length. Based on the ASTM Standard D1515, the specimens were loaded through controlling the axle deformation rate. Different deformation rates were used during the tests which are 50mm/min. for all materials, 210mm/min. and 5mm/min. for grid only.

In this experimentation, the maximum extension ratio (maximum deformation per unit length) and tensile strength of glass fabric were measured with the MTS load cell and LVDTs. The observed data are shown in Table 2.

Table 2. Maximum extension ratio and tensile strength of glass fabric (grid)

Type of glass fabric (grid)	Tensile strength (N/50mm)		Max. extension ratio (%)	
	Warp	weft	warp	weft
10×8	503	167.5	2.44	1.32
10×10	570	312.6	1.66	1.3
12×10	606	378	1.68	1.3
Grid	3125	2333	0.76	0.81

2.2 Fatigue test

Bending fatigue tests on asphalt concrete beams with or without 12×10 glass fabric at the bottom with rubber and polywood support and cracking inducer were carried out to study if the application of glass fabric could enhance the resistance of asphalt concrete to reflective cracking.

The specimen size for fatigue test is 50mm×50mm×240mm (height× weidth×length), which was compacted by static load. The type of asphalt mixture used for fatigue test is LH—15I. This mixture is a typical one used as asphalt surface (wearing course) mixture with the air content of 4% and the asphalt content of 5.5% by the weight of aggregate. The maximum size of the aggregate with continuous grade is 15mm.

Before doing fatigue test, the bending tensile strength and modulus of the same mixture were measured with four point bending test. The test results are shown in Table 3. From these results, it can be concluded that the application of glass fabric of 12×10 at the bottom of asphalt beams can significantly increase the bending tensile strength of asphalt concrete by 1.42 times, but has few effects on the modulus of asphalt concrete. That may be because the the deformation ability of glass fabric is much greater than that of asphalt concrete.

Table 3. Bending tensile strength and modulus

Type of specimen	Bending tensile strength(Mpa)	Bending tensile modulus(Mpa)
with glass Fabric (12×10)	5.384	2115.07
without glass Fabric	3.782	2295.02

For the fatigue test, the load of 5Hz. and sign wave is generated by the MTS 458-90 function generator and applied at the top-middle of the beam specimen. The fatigue test temperature is 15°C and under 7 stress levels.

The loading cycles to initial cracking and failure condtion of the tested specimen was recorded automatically. As for the failure condition, it means cracking progressing up to 10mm below the top surface for specimen with glass fabric at the bottom or cracking progressing up to the top surface for the specimen without glass fabric at the bottom. Through the test results, the relationship between the load and loading cycles were analyzed with data regression method using the following equation:

$$N = a(1/P)^b$$

where: N = loading cycles;
    P = means of load(KN);
    a, b = regression content (being shown in Table 4).

Table 4. Regression Analysis Results of Fatigue Tests

Cracking progressing condition	Type of Specimen	a	b	R
To Failure	with Glass Fabric	9.8177×10^9	3.2785	0.88
	without Glass Fabric	8.7371×10^8	2.8174	0.86
Initial Cracking	with Glass Fabric	1.1759×10^9	2.9090	0.89
	without Glass Fabric	2.3470×10^9	3.2485	0.85

Two way classification statistical method was used to analyze the effects of the application of glass fabric on the fatigue resistance. Through this result analysis, it can be concluded that the application of glass fabric at the bottom of asphalt concrete beams has effects on the resistance to reflecting cracking progress at a significance level of 0.1. But the application has hardly any effects on the resistance to the initial reflective cracking. On average, the fatigue life to failure condition can be prolonged 1.8 times, where glass fabric is used at the bottom of asphalt concrete beams.

3 Field road test

Test roads were constructed in Shandon and Shanxi Provinces respectively in 1987 and 1990. The field road test in Shandon province was carried out before the laboratory test work above, that in Shanxi province was done after.

3.1 Field road tests in Shandon Province
Asphalt overlayer was paved over two sections of existing highway asphalt pavement at No. 104 national highway and Tai-Xu Highway respectively. The thickness of asphalt overlayer are 30mm and 15mm with or without glass fabric at the bottom of the overlayer. Prior to paving the overlayer, the cracking condition, deflection and subgrade moisture of the existing pavement were measured.

The type of glass fabric used here is 10×10. The tack coat of $0.8kg/m^2$ was spread over the existing pavement surface, then the glass fabric was paved. After the asphalt mixture of LS-15 I was spread, an 8 ton steel roller was used to compact the overlayer 4 times.

The cracking area and deflection at the surface of the overlayer were surveyed in the first winter and spring since the test road was opened to traffic. The lowest air temperature in the winter is $-8°C$. The observation results of cracking area percent of Taixu highway test section delivered an unsatisfactory conclusion. From this test data, it can be concluded that the application of glass fabric at the bottom of asphalt overlayer could hardly enhance the fatigue resitance of the overlayer to reflective cracking and thermal cracking. However, the test results of No. 104 highway section show that the application of glass fabric can significantly reduce the cracking area of the overlayer. Comparing the cracking area of the overlayer with glass fabric at its' bottom with that without glass fabric, it can be seen that the cracking area can be reduced 50%. This observation data of No.104 national highway test section were shown in Table 5.

At that time, glass fabric had been used by many province highway administrations in pavement overlaying projects as a new material to prevent the progress of overlayer cracking. Their experiences also yielded contradictory conclusions. Some successfully utilized glass fabric to reduce the cracking area of overlayers, but some did not.

Considering the above condition, both Shandon and Shanxi province highway adiministrations are willing to offer finantional support for Tongji University to do laboratory investigation and also field road test again to study if the application of glass fabric could prevent the cracking progress of asphalt overlayer. As stated before, the laboratory research results yielded a satisfactory conclusion, that is, the application of glass fabric can increase the fatigue life of asphalt overlayer by 1.8 times.

Table 5. Cracking area of test pavements in No. 104 national highway test sections

Location	East side(with glass fabric)			West side(without glass fabric)		
	Cracking area of existing pavement (m^2)	Cracking area of overlayer		Cracking Area of existing pavement (m^2)	Cracking area of overlayer	
		Winter (m^2)	Spring (m^2)		Winter (m^2)	spring (m^2)
517+750	126.8	31.9	51.6	62.1	30.8	39.5
800	95.3	27.1	32.4	56.8	40.3	71.7
850	49.1	32.2	37.8	57.3	42.8	45.9
900	115.9	44.5	85.9	60.9	55.4	70.9
950 518+000	91.9	14.7	39.9	99.6	60.7	68.5

3.2 Field road tests in Shanxi Provience

In 1990, a test road of 500m in length was developed in No.208 national highway near Taiyuan City. The existing pavement was developed in 1982 and overlayed in 1987 with asphalt concrete of 25mm. The lowest air temperature here in winter is $-20°C$. The test section plane is shown in Figure 1. The 5000m^2 area test pavement was divided into 10 parts. Different overlayer thicknesses, asphalt mixtures and types of glass fabric were considered to be used for each part as shown in Figure 1. Before paving glass fabric and the overlayer, the cracking area, deflection and subgrade moisture were measured.

(1)	(2)	(3)	(7)	(8)
One Layer Glass Fabric(12×10) 30mm overlayer	No Glass Fabric 30mm overlayer	One Layer Geotexile 30mm overlayer	One Layer Glass Fabric(12×10) 15mm overlayer	No Glass Fabric 15mm overlayer
One Layer Glass Fabric (10×10) 30mm overlayer	One Layer Glass Fabric(10×10) 30mm overlayer	Two Layer Glass Fabric(10×10) 30mm overlayer	No Glass Fabric 30mm overlayer	One Layer Glass Fabric(10×10) 30mm overlayer
(4)	(5)	(6)	(9)	(10)

Figure 1. The plane of test pavements

As in Shandon, the overlayer's cracking condition and pavement deflection were surveyed for each month and each three month respectively. The results are shown in Table 6 and Table 7. Some specimens were sawn and cut out of the overlayer in order to see if the paved glass fabric had been broken during construction by the compacting roller, then the glass fabric and asphalt mixture were seperated at the temperature of 150°C. The glass fabric taken out of the overlayer was found to have been destroyed to some extent during the construction. The compacting roller weight, asphalt overlayer thickness and maximum aggregate size are the main factors breaking glass fabric.

There is the greatest cracking area in the spring. Considering the cracking area and deflection value of the existing pavement before being overlayed, it can be seen that the cracking area of the overlayer can be reduced by an average of 100% where the glass fabric was used at the overlayer's bottom, compared with the cracking area of the overlayer with glass fabric with that without glass fabric. So it can be concluded that the cracking progress of asphalt overlayer can be prevented to a certain extent with glass fabric (grid). The measured cracking area of the 15mm overlayer is much more than that of the 30mm overlayer. So it can be proposed that 15mm asphalt overlayer would be too thin and weak to resist cracking and prevent cracking progress.

4 Remarks

1. Glass fabric (grid) is a cheap and effective material in China to be applied at the bottom of asphalt overlayer to prevent its' cracking progress.

2. Laboratory test results show that the application of glass fabric at the bottom of asphalt overlayer can sinificantly enhance the resistance of asphalt overlayer to cracking progress. Fatigue life can be increased by 1.8 times and the bending strength by 1.42 times.
3. Through field road tests, it can be proposed that the cracking progress of asphalt overlayer can be effectively prevented where glass fabric (grid) is used at the bottom of asphalt overlayer.
4. Construction method is the principal factor affecting the effectiveness of the application of glass fabric (grid) to prevent cracking progress. It is proposed that the compaction roller weight should be lighter than 8 tons for initial compaction.
5. The maximum size of aggregate should be less than 15mm and the thickness of overlayer should be thicker than 30mm for keeping glass fabric(grid) from being seriously broken during construction.

Table 6. Cracking area of test pavements

Date	Cracking Area Ratio (%)									
	NO. of Parts									
	1	2	3	4	5	6	7	8	9	10
May, 91 (before overlaying)	51.0	24.4	29.7	65.5	34.0	17.4	14.8	31.3	12.9	8.1
Jan., 92	0	0.45	0.96	0.54	0.57	0.96	2.22	1.62	1.5	1.2
Mar., 92	0.4	1.15	0.78	0.51	1.38	1.02	3.2	4.09	2.14	1.44
Apr., 92	0.36	0.66	0.90	0.27	0.24	1.2	2.25	3.87	1.59	1.32

Table 7. Deflection of Test Pavements

Date	Means of Deflection(0.01mm)									
	No. of Parts									
	1	2	3	4	5	6	7	8	9	10
May, 91 (Before overlaying)	52.3	44.3	18.4	46.7	42.6	30.3	22.4	23.45	27.1	31.6
July., 91	50.9	56.7	21.7	48.4	41.5	29.2	20.4	23.51	25.5	26.8
Mar. 92	53.9	58.0	25.2	45.4	48.5	34.9	28.4	32.7	37.8	37

6. References

The design specification for highway flexible pavemats. (1987), The Transportation Department of P.R.China, Peking.

Button, J.W. and Lytton, R.L. (1987) Evaluation of fabric, fibers and grids in overlays. Proceedings for sixth International Conference Structural Design of Asphalt Pavements, vol.1, 925–934.

Caltabiano, M.A. and Brunton, J.M. (1991) Reflection cracking in asphalt overlays. AAPT, 60, 310–312.

Brown, S. F. and Brunton, J.M. (1985) Polymer grid reinforcement of asphalt. AAPT, 54, 18–44.

39 LONG TERM PERFORMANCE OF GEOTEXTILE REINFORCED SEALS TO CONTROL SHRINKAGE ON STABILIZED AND UNSTABILIZED CLAY BASES

P. PHILLIPS
Geofabrics Australasia, Sydney, Australia

Abstract
In a large country like Australia with a relatively small population, it is often uneconomical to use conventional methods to seal large lengths of roads. Many of these roads are located in areas of expansive clay subgrades and, with some exceptions, are located where quality road building materials are scarce. As a result more emphasis is made in utilising the insitu clay soils.

The expansive nature of these clay soils develops intense cracking patterns which are readily reflected into any pavement or surface seal. Cracking provides a ready path for moisture ingress, leading to pavement deterioration and failure.

The use of non-woven, needle punched, continuous filament, polyester geotextiles to reinforce the bituminous seal and allow it to bridge cracking and thereby stabilise moisture in the clay pavement has become common practice in Australia.

Keywords: Geotextile reinforced seal, Paving fabric, Conventional seal, Expansive clays, Trial, Core permitivity test, Homogenous waterproof layer, Bridge cracks.

1 Introduction
With the increasing emphasis being put on cost benefit based systems to compete for road funding, and increased pressure to stretch each road dollar as far as possible, Road Construction Authorities need to look at every alternative available to them. One of these alternatives is utilizing readily available, natural, materials such as low quality gravel and clay. These do not meet the specification for base materials and are susceptible to shrinkage cracking. Conventional surfacing would tend to crack and allow water to ingress causing rapid pavement deterioration.

The use of paving fabrics provides a degree of reinforcement to the surface of the conventional seal and offers the following benefits:
    Prevents water ingress by providing a more flexible homogeneous waterproof layer.
    Provides a flexible surface that can tolerate deflections up to three times greater than conventional seals.
Bridges shrinkage cracks-retarding their propagation to the

Reflective Cracking in Pavements. Edited by J.M. Rigo, R. Degeimbre and L. Francken.
© 1993 RILEM. Published by E & FN Spon, 2–6 Boundary Row, London SE1 8HN. ISBN 0 419 18220 9.

surface.

Prevents cracking of the seal as a result of rutting caused by weak base layers.

Prolongs the fatigue life of the seal caused by cyclic loading and weather.

Reduces moisture loss through capillary rise during dry periods, resulting in a more stable moisture content.

2 Problem description

Australia is characterized by extensive areas of expansive clay sub- grades. Construction techniques using conventional gravel, over these expansive clay subgrades, can be prohibited due to:

Substantial depth required to satisfy the low soaked C.B.R. values.

Lack of availability of quality gravel.

Substandard quality of available gravel.

Permeable nature of the gravel and the loss of strength with moisture ingress.

Reflective cracking through gravel pavements originating from the expansive nature of the clay subgrade.

The use of non woven, needle punched, continuous filament, polyester geotextiles to reinforce the bituminous seal, has become a common practice in recent years.

However, some technical aspects need to be evaluated before this relatively new system can be adopted as a standard practice. These include:

Structural adequacy of the pavement.

Effects of moisture infiltration from shoulder and road surface.

Position of traffic loading and its effect.

3 Test equipment used

This has led to the initiation of a full-scale test program sponsored by the Australian Road Research Board and the Road and Traffic Authority of New South Wales. Pavement loading was achieved using an Accelerated Loading Facility (ALF).

ALF is a mobile testing machine that can simulate years of traffic in a matter of days. This can be achieved by applying loads on a pavement through a full scale rolling wheel (refer figure 1).

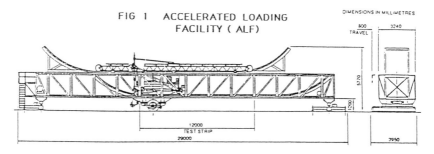

FIG 1   ACCELERATED LOADING FACILITY (ALF)

The performance of a pavement can be continually monitored, with data such as: surface deformation and deflection, moisture movements, crack development, load and number of passes recorded.

In house testing by Geofabrics Australasia was also performed on core samples to determine the infiltration of water into sprayed seal pave- ments. This was carried out using a modified permeability apparatus.

4 Design procedure

Clays are characterized as having soaked Californian Bearing Ratio (C.B.R.) in the order of 2%. However, if these clays are kept dry, the C.B.R. is in the order of 20-35%. By treating these clays with hydrated lime (3%) the C.B.R. increases to 51%. This is reasonable for lightly to medium trafficked roads.

The trial was conducted on a remote country road with an average annual daily traffic of 60 vehicles, or an Equivalent Standard Axles (E.S.A.) of 140,000 over a 20 year design life. This represents about 19 days of testing with the ALF at a standard load of 40 KN. The conventional type of pavement constructed in this area, is 300mm gravel built on clay subgrade.

The trial consisted of 3 sections:

Geotextile reinforced seal over clay.
Geotextile reinforced seal over 150mm and 300mm of lime tabilized clay.
Conventional seal over 150mm and 300mm of gravel and clay (control section).

As these areas sometimes flood, the formation was built so that the edges were 600mm above the natural surface (refer to figure 2).

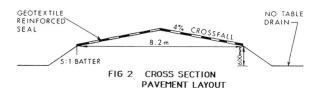

FIG 2 CROSS SECTION PAVEMENT LAYOUT

Dams were constructed along the edge of the formation to simulate the effects of flood at the edge of the seal. This allows testing with ALF under two conditions, flooded and dry.

5 Construction of trial

5.1 Geotextile reinforced seal over clay

The sequence of construction steps are as follows:

Surface preparation including shaping and compaction to relative densities (RD) specified.

```
        Geotextile Reinforced Seal
        -------------------------
            150mm clay                    RD > 100%
        -------------------------
            150mm clay                    RD > 100%
        -------------------------
            300mm clay                    RD >  95%
        -------------------------
        / / / / / / / / / / / / /     Natural Surface
```

The prepared surface is swept and given a light spray of water to promote penetration of the tack coat.

Application of tack coat or prime coat consisting of cut back bitumen C170 incorporating 15-20% cutter oil at a rate of 1.6L/m².

Placement of Bidim PF2 geotextile over tack coat, using a mechanical applicator.

Spreading of fine (7mm) stone over saturated geotextile.

Rolling by rubber multi tyred roller until sufficient bitumen has been squeezed out of the geotextile around the stones.

Application of C170 bitumen incorporating 8% cutter oil at a rate of 2.4L/m².

Spreading of 10mm stone.

Rolling of stone.

5.2 Geotextile reinforced seal over lime stabilized clay

Two types were constructed having depth of 150mm and 300mm.
The sequences of construction steps are as follows:

Surface preparation and treatment of clay with 3% hydrated lime. This is constructed in 150mm layers.

```
                    Geotextile Reinforced Seal
----------------------                        ------------------------
150mm clay treated                            150mm clay treated
with lime      RD = 100%                      with lime      RD = 100%
                                              ------------------------
                                              150mm clay treated
                                              with lime      RD = 100%
                                              ------------------------
450mm clay     RD = 100%                      300mm clay     RD >  95%
----------------------                        ------------------------
/ / / / / / / / / / /  Natural Surface / / / / / / / / / / /
```

Geotextile reinforced seal constructed in a similar manner to section 5.1.

5.3 Conventional seal over gravel

Two pavements were constructed having gravel depths of 150mm and 300mm respectively.
Construction sequences were as follows:
  Subgrade preparation including importation of gravel and compaction in 150mm layers.

```
                    Conventional Seal
------------------------          ------------------------
  150mm Gravel    RD > 100%         150mm Gravel    RD > 100%
------------------------          ------------------------
                                    150mm Gravel    RD > 100%
                                  ------------------------
  450mm clay      RD > 100%         300mm Clay      RD >  95%
------------------------          ------------------------
```
/ / / / / / / / / / / / Natural Surface / / / / / / / / / / /
Application of prime coat as in section 5.1 at a rate of 1.0L/m².
Placement of seal coat having an application rate of 1.35L/m².
Spreading of 10mm stone.
Rolling stone.

6 Trial evaluation

Summary of test results are listed in table 1 and plots of permanent deformation versus number of 40KN load cycles are listed in figure 3

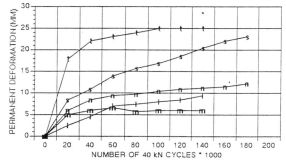
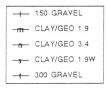

FIG 3
BREWARRINA ALF TRIAL RESULTS

Table 1   Summary of Trial

Section		Description of Pavement	Offset (M)	Water	No. of Cycles
A	1	Geotextile reinforced Seal over clay	3.4	nil	180,088
	2	Geotextile reinforced Seal over clay	1.9	nil	146,000
	3	Geotextile reinforced Seal over clay	1.9	Edge	173,906
	4	Geotextile reinforced Seal over clay	On angle	Edge	173,165
B	1	Geotextile reinforced Seal over 150mm lime stabilization clay	On angle	Edge	153
	2	Geotextile reinforced Seal over 300mm lime stabilization clay	On angle	Edge	20,466
C	1	Conventional seal over 150mm gravel	1.9	nil	145,965
	2	Conventional seal over 300mm gravel	3.4	nil	137,954

6.1 Geotextile reinforced seal over clay

The sections with no water (A1 & A2) performed extremely well, far exceeding the design traffic of 140,000 E.S.A. Performance of the section furthest away from the edge was by far the best with an an average deformation of 12mm at 180,088 E.S.A. Trial A3 was generally the weakest. This was due to the presence of water at the pavement edge for approximately three months. Moisture penetrated 1.2 metres from the edge of the pavement and resulted in severe swelling of the clay under the geotextile reinforced seal. Despite this a relative low deformation of 23mm was recorded at 173,906 E.S.A.

Trial A4 was conducted at an angle travelling close to the pavement edge with water dammed for 3 months. It resulted in immediate failure.

6.2 Geotextile reinforced seal over lime stabilization clay

Two trials were conducted, (B1 & B2) the first having 150mm of stabilized clay and the other 300mm stabilized clay. They were both tested at an angle to the edge of the seal with water to the edge. Both sections performed in a similar manner to the geotextile seal (A4) with severe rutting.

6.3 Conventional seal (control section)

Two trials were conducted, the first (C1) on 150mm gravel and the other (C2) on 300mm gravel. Obviously C2 was expected to be the strongest. However, this did not eventuate with substantial deformation taking place on the 300mm gravel section even after a few thousand E.S.A. Eventually it stabilized, indicating insufficient compaction of the gravel.

The performance of the 150mm gravel was far superior to the 300mm control section.

No experiments on gravel pavements were conducted with the presence of water as it was already known that this would allow water to penetrate into the pavement causing rapid deterioration.

7 Determination of the infiltration of water into the pavement

Summaries of test results are listed in figure 4 Core Permitivity Test.

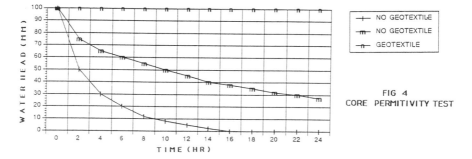

FIG 4
CORE PERMITIVITY TEST

geotextile reinforced seal pavement constructed from clay up to 40 years. This has to be looked at with great care. Even though the geotextile is a polyester that is chemically inert, its long term biodegradability needs to be verified in the field.

10.3 Weaknesses of this system

The main weakness of the geotextile reinforced seal pavement constructed from clay is the risk of damage in the outer wheelpath due to moisture ingress through the formation batter. The ALF trials concluded that moisture had penetrated up to 1.2 metres from the edge.

To reduce the risk of moisture penetration table drains should be eliminated to stop water ponding against the batter toe.

10.4 What is the best way of extending the life of this system

The width of the geotextile reinforced seal should be widened by 1.2 metres and guide posted to discourage trucks from travelling near the edge where all the damage occurs.

Avoid the use of painted centrelines. This is to encourage vehicle travel towards the centre of the road in which the pavement is the strongest.

11 Conclusion

The ALF trials on geotextile reinforced seals over clay bases has yielded two major benefits:

The development of guidelines for the design and management of all weather low cost pavements.
The ability of Road Building Authorities to use geotextile reinforced seals with confidence.

12 Acknowledgement

The assistance provided by the Road Transit Authority of New South Wales, the Australian Road Research Board Ltd. and Brewarrina Shire Council is acknowledged and appreciated. The views expressed are those of the author.

References

Sutherland M. and Phillips P. 1990 Geotextile reinforced sprayed seal roads in rural Australia.
Roads and Traffic Authority of New South Wales Brewarrina ALF Trial Workshop.
Roads and Traffic Authority of New South Wales Use of geotextile reinforced seals in the upgrading of the Cobb Highway to an all weather standard.
Australian Road Research Board Accelerated Loading Facility.

40 IN SITU BEHAVIOUR OF CRACKING CONTROL DEVICES

M. LEFORT and D. SICARD
Laboratoire Régional de l'Ouest Parisien, Trappes, France

Abstract
Many experimental sites have been set up in France to supplement the research work conducted on means of combatting the harmful consequences of shrinkage cracking in hydraulic binder treated bases.
This paper covers the state of the art in mix designs, pre-cracking and various crack control techniques.
Keywords: France, Cracking, Hydraulic shrinkage, Insertion, Geotextile, Geogrill, Special bituminous mix, Crack sealing, Precracking.

1 Introduction

Shrinkage cracking in materials treated with a hydraulic binder is an unavoidable phenomenon in the French climate. Nevertheless, between one third and two thirds of pavement structures, depending on the economic context of the moment with respect to petroleum products, include a base treated with a hydraulic binder. The development of the technique may be associated with that of the use of blast furnace slag in highway engineering and the implementation of the coordinated overlay program on the French national network during recent decades.
 Today, the usage quality level of pavements has risen particularly from the cracking standpoint. The highway agency or operator, as a result of user requirements, is increasingly unwilling to accept defects in appearance (aesthetics) or evenness (comfort) represented by a crack, especially if it is filled.
 The mechanisms by which cracks are transmitted to the surface of pavements are numerous and complex, but their study has brought out three main principles forming the basis of methods used to retard the transmission of shrinkage cracks to the pavement surface:
  Improve the ability of the wearing course to resist reflection cracking.
  Reduce the stresses on this wearing course by modifying the characteristics of the treated base.
  Set up an obstacle to the transmission of cracks between the treated base and the wearing course.

Reflective Cracking in Pavements. Edited by J.M. Rigo, R. Degeimbre and L. Francken.
© 1993 RILEM. Published by E & FN Spon, 2–6 Boundary Row, London SE1 8HN. ISBN 0 419 18220 9.

As concerns the first two points, the present paper simply reviews the state of the art. It focuses basically on the third area, in particular against the background of 10 years of observations on an experimental site.

It looks into the main insertion techniques (principle, application, effectiveness) developed in France to date.

2 Improve the ability of the wearing course to resist reflection cracking

Bituminous concrete, with regard to the cracking phenomenon, is stressed, firstly, in the process of crack propagation in the layer that it forms and, secondly, when the crack has gone through this layer, there is an effect on the stability or invariability of the material at the lips of and around the crack.

2.1 Crack propagation

Very little is known regarding the effect of the composition of the bituminous mix on the parameters of the crack propagation law. It is likely that favorable factors are the same as for fatigue behavior. Compact bituminous mixes with a high binder content are thus favorable, as are polymer modified bituminous mixes not exhibiting high moduli at the most critical service temperatures. The use of polymer bitumen may also be valuable, in particular for heavily travelled pavements, because it allows an increase in the binder content and hence in crack propagation resistance without any risk of instability.

2.2 Behavior of crack layer

It is difficult to use a theoretical approach to this problem, which occurs essentially during cold periods when the bituminous mixes are fragile and cracks are more open. It is however conceivable that these stresses justify properties in the bituminous mix comparable to those giving it its crack propagation resistance: good fatigue behavior; high binder or mortar content; low-susceptibility bitumen; particle-size recomposition offering optimum compactness; special mixes for heavy traffic.

3 Reduce the stresses applied to the wearing course by the hydraulic binder treated base

The objective is to obtain cracks as fine as possible with a small opening on each thermal cycle.

3.1 Act on the quality of the hydraulic material

The rules of the art and the precautions to be taken to limit the cracking of hydraulic binder treated bases are set forth in different documents [1][2][4].

Among the practical rules to be observed, the following may be considered:

Whenever possible, choose an aggregate with a low expansion coefficient.
Provide a good bond between the base and its subgrade.
Avoid high mechanical properties when cracking occurs (choice of binder, hardening processes, etc.).

3.2 Control the cracking rate

The opening and the advance of cracks result from hydraulic and thermal shrinkage: to obtain cracks which open little means that the displacement must be distributed between a large number of cracks, closer to each other than "natural cracks". This is achieved by providing, in the material before it hardens, joints or incipient cracks at close intervals.

The principal studies and experiments conducted were based on the cutting of notches at the top of the base after its compacting. These notches are made either by means of a heavy vibrating plate equipped with a knife or by a double-roller vibrating compactor, one of whose rollers has a cutting disk.

The results were conclusive with notches about 10 cm deep at intervals of 3 m. The layer must be slightly recompacted on the surface after the notches have been made.

3.2.1 "CRAFT" technique (Création Automatique de Fissures Transversales)

The process consists in making a groove, before compacting, over the entire height of the base using a plow which allows the projection of a fast breaking bitumen emulsion on the walls of the groove. The groove is immediately closed off and the placing of the material is pursued normally. The bituminous phase creates a discontinuity which allows the localization of the crack.

3.2.2 "ACTIVE JOINT" technique

This process consists in placing a groove every 2 m over the entire thickness of the spread layer, slightly compacted. A vertical joint element with an undulating profile in plastic is introduced. The groove is closed off and the placement of material continues normally. The joint element, 2 m wide, is placed transversely in the axis of each lane. Its height is about two thirds of the thickness of the layer and it is placed at the bottom of the layer. The joint constitutes an incipient crack and, owing to its undulating form, allows load transfer by the interlocking of the lips of the crack whatever the base material.

4 Set up an obstacle to crack transmission between the treated base and the wearing course

The concept of a "crack control membrane" or "stress distribution layer" has a threefold objective:

From the viewpoint of the transmission of stresses at the head of the crack, coming from slow thermal cycles, it plays a role similar to separation.
By providing a good bond between the bituminous mix and its substrate, it enables the structure to withstand fast traffic loads.
It preserves the imperviousness of the structure in spite of the thermal opening cycles.

The material used to form a crack control membrane must be sufficiently deformable under the slow stresses of thermal cycles to allow a dissipation of stresses appearing at the head of the cracks of the support, but it must conserve sufficient rigidity for fast loads due to traffic. It must not exhibit a high vertical compressibility so as not to increase excessively the bending stresses in the wearing course nor constitute the scene of permanent deformations leading to rutting on the surface. It must adhere to both sides in order to ensure proper bonding of the wearing course on the base and must also conserve its impermeability even over or below an open crack.

These techniques, which are being developed increasingly today, may be placed in three families of processes which are distinguished by the nature of the membrane or the crack control layer:

Fine bituminous mix with high binder and fines content,
Thick surface dressing or membrane with rubber bitumen or with elastomer bitumen,
Bitumen-impregnated geotextile.

The development and the definition of the different techniques are the result of observations made on many experimental sites.
One of them has been followed since 1982. The first results were presented during the first Liège conference in March 1989. They are reviewed in this paper and supplemented by the latest observations and measurements so as to constitute a complete overview of this work.

4.1 Ten years of observation on an experimental site

4.1.1 Characterization of site

The site was set up in June 1982. It is located on a major national highway RN20 about 50 km from Paris. In the section considered, there are two lanes in each direction over a total width of 13.5 m. The experiment involved the direction towards Paris and essentially the slow lane, where all the observations were carried out.

The traffic there is class To (17,372 vehicles/day in 1979 with about 1,000 commercial vehicles/day on the heavy traffic lane, and 17,700 vehicles/day in 1987).

The structure consists of a former flexible pavement overlaid in 1971 with:

25 cm of slag-treated 0/20 porphyry granular material,
6 cm of 0/10 semi-granular bituminous concrete.

The maintenance study called for a layer of 8 cm in 0/14 bituminous concrete. Prior to the placement of the test sections, the pavement exhibited regular transverse cracking at average intervals of 7.5 m and very good bearing capacity (Table 1).

Table 1 - Values of deflection and radii of curvature on the roadway intended for the test sections, before their setup.

	On cracks	Between cracks
Average D (1/100 mm)	8	7
Average Rd (m x 1/100 mm)	36070	50389
Average R (m)	5351	9411

In the roadway section reserved for the tests, the theoretical maintenance thickness (8 cm) was reduced to a total thickness of 6 cm and, depending on the direction of traffic, there are the following sections in the order given:

Bituminous mixes in two layers, with two sections, the first comprising a modified bitumen in the first layer and the second only pure bitumen.
Prior sealing of the cracks with a bituminous mastic.
Laying of a bitumen-impregnated geotextile.
Laying of a thick surface dressing.
Reference section.

4.1.2 Description of sections
Bituminous mixes in two layers
The system includes a first layer of from 1 to 2 cm with a fine bituminous mix and a second layer of from 4 to 5 cm in 0/10 gap-graded 2/6 bituminous concrete.
The fine bituminous material consists of a mix of 81% 0/2 crushed sand and 15% 0/4 river sand supplemented with filler in order to obtain a total fines content of 15%. The binder content is 10%.

1) The first section overlaid with this system has a fine bituminous mix with a binder consisting of an 80/100 bitumen modified by an SBS-type polymer. In the second layer, in 0/10 gap-graded bituminous concrete, the binder is a pure 60/70 bitumen.
The length of this section is 230 m and covers 34 cracks.

2) The second section overlaid with this system has a fine bituminous mix with a pure 80/100 bitumen. The second layer is the same as in the first section.
Its length is also 230 m and covers 29 cracks.

Prior sealing of cracks
Sealing was carried out with a bituminous mastic obtained with a bitumen modified by an SBS-type polymer. The width of application of the mastic over the cracks was about 10 cm, with a proportioning of the order of 0.35 kg/m. The placement of the mastic was followed by the spreading of lacquered porphyry chips.

This was all covered with 6 cm of 0/10 semi-granular bituminous concrete containing pure 60/70 bitumen.

The length of this section is about 200 m and covers 32 cracks.

Laying of impregnated geotextile
The geotextile used is a nonwoven polyester sheet calandered during manufacture and weighing 0.210 kg/m^2.

The impregnation was carried out by spreading, on the support, a cationic emulsion tack coat with 68% pure bitumen of the fast-break type. It was proportioned at about 1 kg/m^2.

The geotextile was unwound and placed after the breaking of the emulsion. As of this phase, the impregnation began and was completed during the placement of the 6 cm of 0/10 semi-granular bituminous concrete containing 60/70 pure bitumen.

The length completed with this technique is about 150 m and covers 16 cracks.

Placement of a thick surface dressing
The binder used for this surface dressing has the following composition:

 80.5% of a mixture containing 50% 80/100 bitumen and 50% 180/220 bitumen
 19.5% recovery rubber powder.

It was prepared on the spot and then spread in the proportion of 3.6 kg/m^2. This was followed by spreading with 6/10 precoated chips in the proportion of about 8 to 9 l/m^2, or practically to refusal.

After sweeping and vacuuming, this was covered with about 5 cm of 0/10 semi-granular bituminous concrete containing 60/70 pure bitumen.

This system was used over a length of about 190 m and covers 21 cracks.

Reference section
The control section consisted of a conventional cationic emulsion tack coat containing about 0.300 kg/m^2 of residual bitumen. It was covered with about 6 cm of 0/10 semi-granular bituminous concrete containing 60/70 pure bitumen.

Its length of about 220 m covers 30 cracks.
4.1.3 Reflection cracking trends
Until 1986, the surface condition of the test zone was checked twice a year. These observations were then linked

with the trends in reflection cracking. A detailed inventory of transverse cracking was made following each of these inspections so as to classify the cracks according to geometrical criteria using the method described in Table 2.

Table 2. Classification of transverse shrinkage cracks according to geometrical criteria

IDENTIFICATION PARAMETER	CLASSIFICATION OF CRACKS			
Spacing between cracks	Classification according to mean (m) and reduced standard deviation (σ/m)			
Width of cracked pavement (or lane)	A L < 1/3	B 1/3<L<2/3	C L > 2/3	
Opening – Appearance of lips	A fine crack	B Spalled crack (dislodging of agregates)	C Spalled crack (dislodging of pieces of mix)	
Branching	A no ramification	B doubled part.	C tripled part.	D crazed
Tortuousness L : width of theoretical strip encompassing the crack	A L < 20 cm	B 20<L<40	C 40<L<60	D L > 60
Discontinuity Number of pieces constituting the crack	A crack in 1 piece	B crack in 2 pieces	C crack in 3 or more pieces	

The essential part of the results obtained is shown in the graphs of figures 1a, 1b, 1c and 2. It shows the trends in some geometrical characteristics of the cracks, among those best characterizing their severity.

The example of these results shows that all the systems tested enabled the retarding of reflection cracking in relation to the reference control section, for at least 5 years, and even much more in certain cases. In addition, all the systems brought about a significant modification of the geometrical characteristics of the reflection cracks in relation to those of the control. In other words, it may be considered that all the systems tested on this site demonstrated effectiveness, on the one hand, in the crack reappearance time and, on the other, in the severity of the reflection cracking.

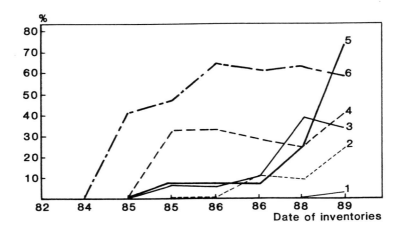

Fig.1a. Cracked width of class C on truck lane

Section No.
1. Fine bituminous mix with elastomer bitumen
2. Fine bituminous mix with pure bitumen
3. Crack sealing
4. Geotextile
5. Thick surface dressing
6. Control

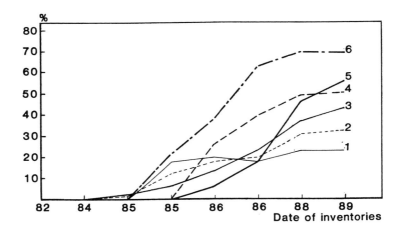

Fig.1b. Opening class B + C

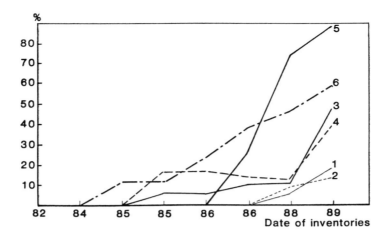

Fig.1c. Branching class C + D

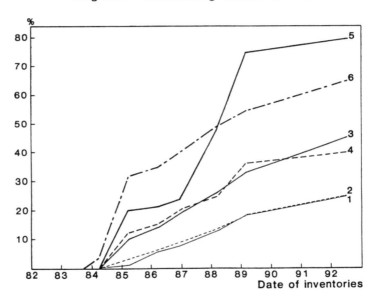

Fig.2. Percentage of reflection cracking

The reflection cracking was calculated in relation to the initial number of cracks on the section and by assigning the coefficient:
 0.25 to an incipient crack,
 0.50 to a discontinuous crack affecting the lane,
 1 to a crack which is continuous over the entire width of the lane.

Finally, permeability measurements were carried out during the last inspection of 1992, both in situ and on core samples located over the reflection cracks. Although there was a small number of measurement points, the two procedures yielded similar results. The conclusions are that when the crack reappears on the bituminous mix in two layers, it is permeable, whereas other systems which concentrate a relatively large amount of binder at the interface make it possible to maintain a relatively satisfactory imperviousness.

5 General assessment

The completion of pavement structures making wide use of hydraulic binder treated materials has led French engineering to orient its research and experimentation so as to reduce as much as possible the harmful consequences of reflection cracking.

New pavement structures are conceived and new mixes are designed with hydraulic binders specially developed for the road. It is however not always economically possible to use minimum-cracking hydraulic binder treated materials nor to overlay them with bitumen-treated materials offering maximum cracking resistance.

Consequently, investigations are also oriented towards treated base precracking techniques. There have been innovations in techniques and in equipment for producing but also for designing a precracked treated base structure.

The development of devices for controlling cracks in surface layers is of a more general nature. If the field of observations conducted on French projects is extended, the following conclusions may be drawn with regard to their in-situ effectiveness.

Two-layer bituminous mixes such as those tested on highway RN20.

This is the most effective technique as regards the retarding of reflection cracking and not the severity of the cracking when it has risen to the surface. This effectiveness is moreover reflected in the fact that practically all French highway contracting firms have developed such a system and offer it in their product catalogue.

A modified binder must be used for the first part of the system, and even for the wearing course in the case of heavy traffic. The thicknesses of each layer must also be complied with (calling for profile quality in the support).

The thicker the wearing course, the more effective is the system. Finally, the system does not provide waterproofing when the crack has risen.

Prior filling of cracks. It is good to know that prior filling of cracks is a positive approach, but cannot be a technique developed as a crack control system because its application is not compatible with that of a wearing course and, moreover, it is not applicable to new pavements. The filling of cracks remains a crack maintenance technique only on very heavily travelled roads in France.

Impregnated geotextiles. The development of simple systems for unwinding geotextile materials has allowed faster and better placement, leading to the extension of this technique. The nature of the geotextile is not important as long as it has been treated to be incompressible. On the other hand, the binder and its proportioning (related to the characteristics of the geotextile) are of primary importance. The use of a modified binder is advantageous, but it has not been unquestionably demonstrated that the spreading of the binder in the form of an emulsion is problematical. The system constitutes an impervious interface.

Mention should be made of the excellent performance of surface dressings over an impregnated geotextile.

In the same family of processes, also may be mentioned a contractor's process consisting in "blocking" the binder not with a prefabricated geotextile but with a projection of continuous wires, all of which is protected with a light spreading of chips.

Also noteworthy are geogrills. With this type of geotextile, the experiments conducted in France with regard to the problem raised by shrinkage cracking have always been unsuccessful. Note may however be made of the appearance of fiberglass fabrics which may represent an interesting new approach.

The thick surface dressing (membrane) using rubber bitumen or elastomer bitumen

The process, not employed extensively in France, is widely used in particular in the United States under the name SAMI (Stress Absorbing Membrane Interlayer). Experiments on highway RN20 may not be regarded as representative of the technique as it may be conceived and implemented today. In addition, tests on RN20, distorted by excessive use of chippings, were penalized by the placement of a wearing course whose quality and placement were far from being optimum.

Considering the information drawn from the various French projects, there is every indication that this is a technique to be classified among the most effective with regard to reflection cracking and which can guarantee waterproofing.

6 References

Bulletin de liaison des Laboratoires des Ponts et Chaussées n° 156 et 157. "Fissuration de retrait des chaussées à assises traitées aux liants hydrauliques.
Note d'information SETRA-LCPC n 55 "Règles de l'art pour limiter la fissuration de retrait des chaussées à assises traitées aux liants hydrauliques. March '90.
Note d'information SETRA-LCPC n 56 "Limites et intérêt du colmatage des fissures de retrait des chaussées semi-rigides". March '90.
Note d'information SETRA-LCPC n 57 "Technique pour limiter la remontée des fissures à la surface des chaussées semi-rigides". March '90.

41 ASPHALT OVERLAY ON CRACK-SEALED CONCRETE PAVEMENTS USING STRESS DISTRIBUTING MEDIA

G. HERBST
Central Road Testing Laboratory, Road Administration of Lower Austria, St Pölten-Spratzen, Austria
H. KIRCHKNOPF
Road Construction – Department 2, Road Administration of Lower Austria, Tulln, Austria
J. LITZKA
Institute for Road Construction and Maintenance, Technical University of Vienna, Austria

Abstract
The technical and economic optimization of asphalt overlay thickness on old concrete pavements was the target of a test track constructed in Lower Austria in the years 1988 to 1990.

A conventional type of overlay was compared with thinner overlays using various stress absorbing interlayers. After a period of four years tentative conclusions about the behaviour and performance of the different solutions can be drawn. The best results so far show a conventional 120 mm overlay and the use of geogrids. The first will need a longer time of observation.

Regarding costs overlays with the use of stress absorbing interlayers vary between +/- 10% from those for a conventional overlay.

Keywords: Test track, asphalt overlay, crack-seated concrete pavement, geotextile, geogrid, steel grid, reflective cracks, costs.

1 Introduction

A number of secondary roads in the province of Lower Austria have old concrete pavements with poor serviceability needing resurfacing to reinstall adequate riding quality. Formerly thickness design of asphalt overlays on cracked and seated concrete pavements had to consider a minimum thickness of 150 mm to avoid reflection cracking. With regard to structural aspects, this thickness means an overdesign for these low trafficked country roads (Herbst and Kirchknopf, 1988).

So a test track was built in the years 1988, 1989, 1990 to investigate a possible reduction of the asphalt thickness using stress distributing media, (Litzka et al., 1991).

2 Description of the test track

The test track is part of a secondary road near Tulln in Lower Austria holding a traffic volume of 3,000 AADT with about 10% commercial vehicles.

Because of the narrow width of the carriageway the design traffic load will amount to 0.6×10^6 10-tons-standard axles for a design period of 20 years.

The existing pavement consists of an old undowelled concrete pavement with no steel reinforcement overlying a 300 to 400 mm unbound gravel sub-base. This concrete pavement was overlaid in 1973 with 110 to 130 mm asphalt concrete.

Serious cracking and material loss in the AC have been the reason for rehabilitation measures from 1988 to 1990. On this occasion various test sections have been constructed to investigate into stress distributing media. The main objective was to learn about the possible reduction in asphalt overlay thickness.

After milling off the old asphalt overlay and crack-seating of the concrete pavement the following types of overlay were used:

- 120 mm AC 16-bituminous road base without stress distributing interlayers (1988, 1989)·
- 60 mm AC 16-bituminous road base on 100 mm granulated old asphalt without additional binder reusing the milled-off old asphalt as an unbound interlayer (1988)
- 60 mm AC 16-bituminous road base on a polypropylene geotextile (Polyfelt PGM 14) over a 30 mm AC-levelling course (1988,1989)
- 60 mm AC 16-bituminous road base on a polymer prestretched-geogrid (Tensar) without an underlying levelling course (1989)
- 60 mm AC-16 bituminous road base on a polymer prestretched-geogrid with underlying levelling course (1990)
- 60 mm AC-16 bituminous road base on a steel grid (mesh-track) without levelling course (1989). A part of this test section was executed without crack-seating of the concrete pavement.

Thus 12 test sections with an average length of 300 to 400 m have been constructed.

The geotextile was placed after spraying 1.7 kg/m^2 emulsion tack coat. The geogrid was fixed directly on the supporting layer and protected against possible overheating at placing of the AC-overlay using a 8/11 surface dressing. The steel grid was bolted directly to the concrete pavement applying 2 to 3 bolts and straps/m^2.

The laying of a wearing course was postponed for the end of the test tracks survey period.

3 Test program

The testing included the following measurements and investigations

- static (optical method) and dynamic deflection (Falling Weight Deflectometer) to assess the structural capacity of the pavement before and after reconstruction
- settlement of the crack-seated concrete pavement on selected areas to evaluate the effectiveness of the crack-seating procedure
- survey of cracks and joints before and of cracks after reconstruction to evaluate the effectiveness of stress distributing media.
- cross profiles (permanent deformation)
- E-moduli of the pavement layers by FWD back-calculation
- technological and mechanical material analysis of the AC-overlay to evaluate mix design influences on reflective cracking

According to the main target of this test track the survey of cracking was regarded as the most important criterion for performance. The other investigations have been made to support a deeper analysis.

4 Results of the various test sections

After two to four years of observation one can give only tentative results and conclusions.

Table 1 shows the development of reflective cracking versus the different types of construction. The amount of transversal reflective cracks is expressed as a percentage of the total length of transverse joints and cracks in the previous old concrete pavement. To enable a comparison only those figures are listed which are available for comparable periods of observation. A graphic display of the crack development may be taken from fig. 1.

The greatest percentage of reflective cracking has been found at the steel-grid section. Cracks occurred relatively early in both sections and of course to a far greater extent on the not crack-seated concrete pavement. This early appearance of reflective cracks is in contrast to positive experiences in other countries (Molenaar et al., 1989). Yet, it is to be stated that there were problems with the proper fixing of the grid to the supporting concrete. Meanwhile the authors have been informed that the method of fixing the grid to the ground has been changed. The grid is embedded into a slurry layer of about 17 kg/m^2 instead of using bolts and straps.

The far best results so far showed the 1990-geogrid section applying a levelling course. Up to now no reflective cracks appeared. But in this case it is to be stressed that the AC 16-mix design was of a highly plastic character. This makes comparability with the 1988 and 1989 test sections slightly difficult.

All test sections applied asphalts with B 100 standard bitumen where high binder contents as at the 1990 test track will promote permanent deformation in the asphalt overlay. A counterbalance by using polymer modified binders may indicate improved performance regarding reflection

Table 1. Development of reflective cracking

Type of overlay construction	Percentage of transversal reflection cracks (%)		
	after 12 months	after 19 months	after 30 (32) months
120 mm AC 16 road base without stress absorbing media	0.5	0.9	1.8
60 mm AC 16 road base on 100 mm granulated asphalt	0.9	1.7	4.9
60 mm AC 16 road base on geotextile (PGM 14) with 30 mm AC-levelling course	2.5	3.9	7.2
60 mm AC 16 road base on geogrid (Tensar)			
a) without AC-levelling course	2.3	2.6	6.1
b) with 30 mm AC-levelling course	0.0	0.0	—
c) like b) but without crack-seating	0.0	0.0	—
60 mm AC 16 roade base on steel grid (Mesh Track)			
a) with crack-seating	5.5	14.3	27.3
b) without crack-seating	22.2	31.9	43.9

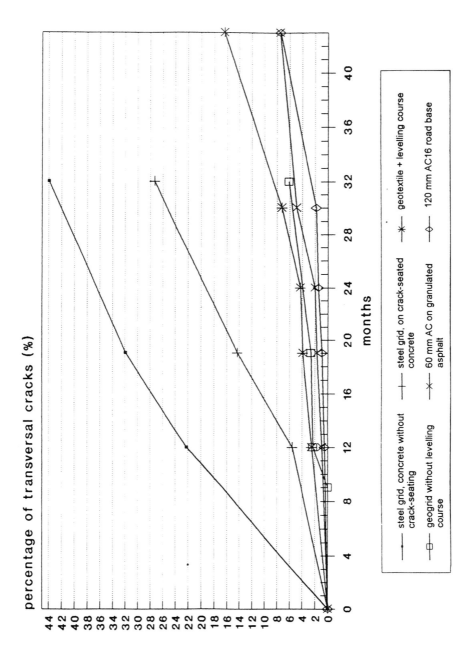

Fig. 1. Development of reflective cracking

cracking and resistance to permanent deformation but will render a far less applicable solution from the economic point of view.

From the other constructions so far the conventional 120 mm asphalt overlay showed the best results. The sections with the geotextile on a levelling course, with geogrid without levelling course and with the interlayer of granulated asphalt have more cracks but after 30 or 32 months respectively they are lying in about the same range.

The relatively good result of the 120 mm overlay without stress absorbing interlayer is to be evaluated with great care as it is known from practice that thicker overlays will need a longer time for crack appearance (see also fig. 1). Therefore final conclusions can only be made after a longer period of observation.

The observation of the test sections will be continued during the next two or three years to obtain final results.

It may be of further interest that the structural design proved to be sufficient in all cases, shown by design calculations using the moduli derived from the FWD-measurements (Litzka et al. 1991).

5 Economic aspects

To assess the economic efficiency of the various types of construction a comparison was made between their costs and the costs of a totally new construction following Austrian standards (RVS, 1986) (see table 2). The costs of overlays were about or less than 54% of those for a new construction. Within the overlays one may quote the following ranking of costs beginning with the cheapest solution:

1) 60 mm bituminous road base on polypropylene geotextile (PGM 14) over a levelling course
2) 120 mm bituminous road base without interlayer equal to 60 mm bituminous road base on polymer geogrid (Tensar) without a levelling course
3) 60 mm bituminous road base on granulated old asphalt interlayers or on polymer geogrid (Tensar) with levelling course or on a steel grid (mesh track) without levelling course.

Anyway, it must be stated that any comparison of construction costs is bound to the local conditions.

Table 2. Comparison of costs

Type of overlay construction	Costs (1990) AS/m2	compared to reconstruction (%)	comparison of overlay types (%)
new construction according to Austrian standard	665.00	100.0	—
120 mm AC 16 road base without stress absorbing media	330.10	49.6	100.0
60 mm AC 16 road base on 100 mm granulated asphalt	352.35	53.0	106.7
60 mm AC 16 road base on geotextile (PGM 14) with 30 mm AC-levelling course	292.70	44.0	88.7
60 mm AC 16 road base on geogrid (Tensar) a) without AC-levelling course b) with 30 mm AC-levelling course	316.15 357.00	47.5 53.8	95.8 108.3
60 mm AC 16 road base on steel grid (Mesh Track)	355.80	53.5	107.8

6 References

Herbst, G. and Kirchknopf H. (1988) Erfahrungen mit der Sanierung alter Betondecken durch Entspannen und Einbau dicker Asphaltschichten im Hocheinbau. **Bitumen**, Heft 2,

Litzka, J., Herbst, G. and Kirchknopf, H. (1991) Versuchsstrecke LH 112, Asparn-Pischelsdorf, Betondeckensanierung im Hocheinbau mit Asphalt, Schlußbericht. **Amt der NÖ. Landesregierung**, Straßenbauabteilung 2, Tulln.

RVS 3.63 (1986) Bautechnische Details, Oberbau. **Richtlinien und Vorschriften für den Straßenbau**, Forschungsgesellschaft für das Verkehrs- und Straßenwesen, Wien.

Molenaar, A., Moens, J., van der Walle, M. and van Drongelen, P. (1989) Steel reinforcement for the prevention of reflection cracking in asphalt overlays. In **Reflective Cracking in Pavements, Assessment and Control**, Rilem Conference Liege.

42 EXPERIMENTAL PROJECT ON REFLECTION CRACKING IN MADRID

R. ALBEROLA
National Highway Department, Madrid, Spain
A. RUIZ
Centro de Estudios y Experimentación de Obras Públicas, Madrid, Spain

Abstract
The paper reports on a project carried out in Madrid and describes the experimental stretch built for this purpose and the laboratory and field testing done on the performance of the joints and protection systems used.
 It also describes the tests run by CEDEX and reports on the results of the subsequent two-year monitoring period.
 Between Autumn of 1989 and Spring of 1990 an experimental stretch was built which was designed to stop cracking spreading from the bottom layer of roller compacted concrete (RCC) to the pavement made of asphalt concrete.
 The reconstruction of the right lane of the N-II national highway (Madrid to French border via Barcelona) at a point 35 km from Madrid was used to build this experimental stretch.
 In 1990 the traffic in the area was 32,000 veh/day, including 22% heavy vehicles, so the daily rate in the test section involved a minimum of 3500 industrial vehicles per day.
 The test section was 4500 m long in total and was divided into five 900-m sections in which the distances between joints varied and different processes were applied. All the joints were made cold in fresh concrete. They were spaced at 2.5, 3.5 and 6.0 metre distances.
 The pavement consisted of a 12-cm layer of asphalt concrete on a 25-cm base layer of RCC over a 20-cm layer of cement-treated soil sub-base.
Keywords: Reflected Cracks, Roller Compacted Concrete, Cold Joints, Sealing, Sand Asphalt.

1 Introduction

Roller compacted concrete began to be used in Spain in 1984 on roads bearing heavy and medium traffic. Since that date it has been used regularly both for strengthening and widening works and in new road building.
 Since the evenness achieved in the RCC is not very good, its use on roads with medium and heavy traffic has made

it necessary to lay an asphalt mix wearing course between 8 and 12 cm thick. Right from the initial works constructed with this technique its main problem proved to be the reflection at the surface of the cracks caused by shrinkage of the RCC. The problem is more acute in the central plateau zone of Spain where the daily temperature gradient can rise to over 30°C. Furthermore, while the cracks appearing in pavements under medium or light-class traffic (less than 1000 heavy veh/day) do not deteriorate to any appreciable degree, in pavements supporting heavy traffic (bear in mind that the legal single-axle weight allowed in Spain is 13 tons) the cracks evolve very quickly, in the form of ramifications and particles breaking off which leads to expensive maintenance work causing inconvenience to traffic.

As a result, right from the first applications, techniques have been studied which would be capable of minimizing the reflection of cracks. The stages of development of these techniques have been as follows:

- In an initial stage, RCC was laid without any contraction joints. The shrinkage cracks were irregularly reflected on the surface and pieces broke off at the corners of the places where longitudinal cracks crossed transverse ones. All this led to a complicated sealing of the joints and to generalized deterioration of the asphalt mix.
- Immediately afterwards contraction joints were sawn. These were initially spaced 15 or 10 m apart. In two works on the national highway grid in Spain known as the State Network of General Interest, in an area on the plateau with very heavy traffic, 0.85 to 1.70-m wide geogrids were placed over these joints, in between the binder layers and the asphalt aggregate wearing course. The cracks reflected onto the surface very quickly indeed (within two years). Ramifications also occurred which led to a rapid deterioration. Test stretches were built also on one work using asphalt mixes made with modified binders (approximately 5% SBS) and gap graded asphalt mixes containing fibres to be able to increase the bitumen content. The reflection of cracks occurred just as quickly as in the other stretches although here the cracks were less ramified and finer than the earlier ones. Other test stretches were prepared without geogrids but the same deterioration appeared as in the stretches with geogrid installed.
- The next stage of the technique, still the current state of the art, was to reduce the distance between the contraction joints. During this stage joints began to be made cold in fresh concrete in addition to the ones sawn in. In a parallel fashion the following systems were studied: sealing of the joints in the surface of RCC using elastomer-modified asphalt foil, interlayering

with sand-asphalt layers or membranes, use of coated materials specially designed to keep reflection cracking to a minimum.

This paper reports on the results obtained to date on a test stretch built according to these latest criteria.

2 Description of the work

During 1981 and part of 1982 a stretch of the Madrid-Barcelona-French border highway, the N-II, between km points 33 and 38, was improved by building dual carriageway and widening the right-hand edge. The slow lane was built of 12 cm of coated macadam on 20 cm of gravel-cement on a 40-cm sub-base of granular material. The traffic in this area was 18,590 veh/day of which 4100 were heavy vehicles.
This stretch was put into service in 1982 and five years later considerable deterioration had appeared.

A study of the problem revealed that the reasons for the unsuccessful performance of this stretch with the very aggressive traffic indicated were as follows:

- lack of thickness and quality in the gravel-cement layer
- lack of drainage capability in the sub-base
- lack of compaction in the subgrade with the result that the actual CBR was much lower than the assumed value.

The conclusions drawn from all the foregoing were that a strengthening of this kind was not to be recommended, that the slow lane needed to be torn up and rebuilt with a pavement suitable for the traffic to be withheld.

In view of these conclusions it was decided to take up the right-hand lane and hard shoulder on the outgoing carriageway, which was in bad shape, and to replace this by a suitable pavement using the following stages:

- Excavation of the pavement of the near-side lane and hard shoulder down to a depth of 50 cm.
- Supercompaction of the subgrade in order to provide appropriate bearing capacity in the formation level.
- Construction of a 20-cm thick soil-cement sub-base prepared in a mixing plant.
- Construction of a 25-cm thick RCC layer.
- Laying of a 12 cm asphalt mix wearing course

Different actions were tested in an effort to minimize the reflection of cracks.

The work was divided up into five sections which were 900 m long and the joints there were set at 2.5, 3.5 and 6.0 metres distances apart.

Three different types of protection systems were used for the joints: sealing, sand-modified asphalt and Supertecolan foil.

No protection was used in the sections where joints wereset 2.5 and 3.5 metres apart.
Figure 1 shows the division into sections and the different treatments applied.

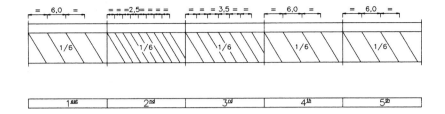

Length (m)	910	900		900		900		900
PROTECTION OF THE TRANSVERSAL JOINT	A/R	NO PROTECTION		NO PROTECTION		600m SF 5,2 kg/m²	300m SF 2,2 kg/m²	SA
PROTECTION OF THE LONGITUDINAL JOINT	A/R	450m A/R	450m SF 2,2 kg/m²	450m A/R	450m SF 5,2 kg/m²	600m SF 5,2 kg/m²	300m SF 2,2 kg/m²	SA

A/R = ASPHALT/RUBBER
SF = SUPERTECOLAN FOIL 10cm WIDE (kg/m²)
SA = SAND-ASPHALT

FIG. 1. TEST STRETCHES

3 Characteristics of the materials

The binder chosen was PUZ-II-350 cement both for the soil cement and for the RCC.

The soil-cement had to have a minimum 4% of cement and a simple compression strength of 25 kg/m² at seven days. The density had to be 97% of the Modified Proctor and the curing done with asphalt emulsion.

The grading of the RCC had to fall within the envelope below:

UNE Screens	Percentage Passing
20	100
16	88-100
10	70-87
5	50-70
2	35-50
0.4	18-30
0.08	10-20

The aggregate added came in two sizes, under or over 5 mm, and the CBR of the mix had to be over 65.

The binder proportion added had to be 10% in weight of the total dry materials and to have an indirect tensile strength over 33 kg/cm² at 90 days. An asphalt emulsion was used as curing compound, in the proportion of approximately 300 g/m².

4 Joint construction

The RCC was laid on site using a double vibrating rod spreader. Joints were cut cold in the fresh concrete after the compaction stage. A wedge-shaped cut with a vibrating tool was made, 1 cm thick and 12.5 cm deep.

5 Joint protection

5.1 Sealing
The joint was sealed (bridged) with a bituminous composition with a high rubber content.
The composition of this bitumen-rubber was as follows:
- bitumen content no higher than 55%
- natural rubber content not below 15% passing a 20 micron screen
- mineral filler content no higher than 35%, 100% passing a two tenths of a millimetre screen.

The joints were cleaned and heated using compressed air and a propane-fired nozzle and were subsequently sealed by the material chosen at a temperature 10° below the safety heat.

5.2 Supertecolam foil
This foil is made of asphalt film consisting of a polyester felt webbing of 130 and 180 g/m with both sides coated with a copolymer styrene-butadiene-styrene (SBS) sandwich, distilled asphalt bitumen and selected mineral filler.
The bottom side of the foil was coated by a polyethylene film as the non-stick material while the top side had a sand finish.

5.3 Sand-asphalt (SAMI)
The SAMI was made up hot and consisted of a modified hydrocarbon binder and an 0/5 grading aggregate in such a way that all its particles were coated in an even film of binder. The bitumen proportion was 10% in weight of the dry aggregate and the grading envelope was as follows:

Screen (mm)	Percentage in Weight
5	100
2.5	65-90
0.63	25-45
0.08	11-15

An automatic paver spread it over the entire lane width to a thickness of 1.5 cm. It was then compacted by pneumatic roller.

5.4 Asphalt concrete

Two types of asphalt concrete were used. For the binder layer a G-12 mix (asphalt concrete containing 6-8% voids) was used with a binder content in the proportion of 4.3% of the aggregate. The mix was laid in a thickness of 6 cm except in the section with the sand-asphalt where it was only 4.5 cm thick, corresponding to the thickness of the interlayer.

For the wearing course, 6 cm of an S-12 mix was used (asphalt concrete with 4 to 6% voids) with a binder content of 5.2%.

6 Laboratory tests

Eight cores of RCC were taken from each section. Part of these cores were taken from the contraction joints.

The cores were tested for indirect tensile strength at 28 days and a ratio was established between the density and the strength values. For the reference density (2274 T/m^3), a tensile strength was obtained for the material in the slabs of 1.5 MPa and for the material in the joints of 1.2 MPa, with a 78% ratio between the two.

Although the asphalt concretes used in the experimental stretch were not specifically tested for crack reflection, a general lab test was carried out on the mixes used in Spain (Ref. 1) with aggregates and bitumens similar to the ones used in this test. The study was based on a flexural strength test with three supporting points, a 400 μm amplitude displacement and 1 Hz frequency on prism-shaped cores 50x90x300 mm in size which were cut down to induce the appearance of a crack whose progress was monitored throughout the test. The result produced a crack spreading speed (IVF) expressed in μm per second. S-type mixes gave an IVF from 5 to 8 and G-type mixes one between 8 and 27.

7 Visual inspection

In the visual inspection work carried out so far (1990, 91 and 92) cracks have only been detected in this current year, distributed as follows:

	STRETCH	JOINT No.	CRACKS REFLECTED	%
1	Joints every 6 m Sealed protection	152	3	1.2
2	Joints every 2.5 m No protection	360	6	1.7
3	Joints every 3.5 m No protection	257	6	2.3
4	Joints every 6 m Supertelecolam foil	150	4	2.7
5	Joints every 6 m Sand-asphalt	150	16	10.7
	TOTAL	1,069	35	3.2

All the reflected cracks were located over joints what means that the system of cutting in fresh worked as expected. Probably not all the joints opened, at least in the case of the short spaced ones, but when the layer shrinkaged, it cracked in the joints.

The joints every 6 m performed the same or better than the joints every 2.5 m or 3.5 m except for the last section with a sand-asphalt interlayer. Cutting joints every 3.5 m have the adventage that even if only open one every two joints, the movement originated by this length of slab can be absorbed by the 12 cm thickness of asphalt layer. Joints every 6 m performed better when sealed or protected with the foil than with the sand-asphalt interlayer. The system of joints every 6 m and sealing for avoiding the entry of water even if the crack is reflected seems to be very adequate.

Figure 2 shows the progress of the cracking in comparison with another two similar roadworks where joints were set at 10 and 15 m apart.

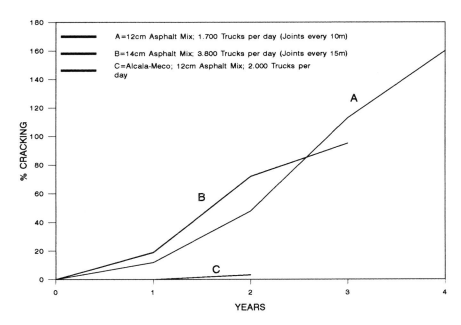

FIG. 2. CRAKING OF DIFFERENT SECTIONS

8 Conclusions

The technique studied to minimize the reflection of cracks in RCC whereby joints are cut cold in fresh concrete at 6-m intervals or under considerably reduces the cracking in comparison with the technique of cutting joints every 10 or 15 m.

Joints made in fresh concrete have had a good performance and are cheaper (about half the price) than sawn joints.

After two years no appreciable difference can be seen between the results of pavements where the joints are set at 2.5 or 3.5 m apart and without any protection devices or where the joints are set at 6-m intervals and protected by foils or sealings.

The foils or sealings have had a better performance up until now than the sand-asphalt layer.

9 References

García Carretero, J. and Martín González, N. (1991). Materiales y técnicas para minimizar el agrietamiento por reflexión de los pavimentos bituminosos, MOPT. Report TT-054/88.

43 TWO KINDS OF MECHANISM OF REFLECTIVE CRACKING

SHA QING-LIN
Research Institute of Highways, Beijing, P.R. China

Abstract
This paper discusses two kinds of mechanism of reflective cracking in pavement. The one is for thin asphalt surfacing over semi-regid roadbase or old cracked pavement. In this case, the reflective cracking in asphalt surfacing initiated at the bottom and propagated up to the surface of asphalt surfacing. The full scale experiments, including SAMI and polymer modified binder asphalt, constructed in Oct. 1988 has shown SAMI can reduce but can't eliminate reflective cracking. The another is for relatively thick asphalt surfacing, the thickness of which depends upon the local air temperature. In this case, the reflective cracking in asphalt surfacing initiated at the surface and propagated down to the bottom of asphalt surfacing. To reduce the reflective cracking, obviously, we should increase the crack resistance of the surface, and the various kinds of SAMI will not be successful. The two different mechanisms are proved by cracking survey, as well as the results of the thermal stress calculated with photo-electric analysis methed and finite element methed.
Keywords: Mechanism, Reflective Cracking, Semi-Rigid Roadbase, Thin, Thick, Thermal Stress.

1 Introduction

Cracking of various types is a major form of distress in bituminous roads in China and is an important problem concerned by most road engineers. It occures in asphalt surfacing constructed over both flexible and semi-rigid roadbase.

Cracking has been observed in all climatic zones of China, but is generally more severe in the north, where freeze-thaw conditions exist. Cracking usually commences during the first winter (poor quality bitumen) after the pavement is constructed and worsens each year.

It has been proved that the early cracks of the well designed and constructed new semi-rigid pavements are temperature induced or non-traffic related.

A feature of these cracks is that they are much wider at the surface than at the bottom of the surfacing.

It has been also proved that there are three different cases for the non-traffic related cracking as follows:

(a) When the thick asphalt surfacing was paved over an uncracked semi-rigid roadbase, the cracks of semi-rigid pavement are mainly thermal cracks of surfacing itself and the reflective cracking is very few.
(b) When the thin asphalt surfacing was paved over an uncracked semi-rigid roadbase, reflective cracking may occupy a marked part of surface cracks and in very unfavorable cases it may be more than 50%.
(c) When the asphalt surfacing was paved over a cracked semi-rigid roadbase and/or an old cracked pavement, reflective cracking is inevitable, and the higher the rate of reflective cracking the thinner the surfacing.

The asphalt surfacing thickness mentioned in first case (a) has a ralative meaning and depends on the local climatic condition, for example, in the north China where the lowest temperature in winter is about $-15°C$ the thickness is 7cm and more, and in the northeast China where the lowest temperature in winter is about $-27°C$ it is 12cm and more.

In this paper, the mechanisms of reflective cracking are discussed and the results of visual surveys of cracking, cores of bituminous pavements and photoelectric analysis method are reported.

Following this, the factors influencing the frequency of reflective cracking are listed and the measures for minimizing reflective cracking are suggested.

2 Reflective Cracking

When uncracked semi-rigid roadbase was covered with a bituminous surfacing it loses moisture as a result of evaporation and shrink, and tensile stress is induced. If the semi-rigid roadbase loses moisture slowly, it will shrink slowly and the induced stress will be low. The semi-rigid roadbase under a thin bituminous surfacing loses moisture quicker than under a thick asphalt surfacing. If roadbase was constructed with high quality semi-rigid material and located in favourable regions, it may not crack prior to the surfacing, especially when it was under thick asphalt surfacing. If roadbase was constructed with poor quality semi-rigid material or with high quality semi-rigid material but located in arid regions, it could crack prior to the surfacing, especially when it was under thin bituminous surfacing. If roadbase was constructed with poor quality semi-rigid material and located in frost, especially in heavy frost regions, it could crack prior to the surfacing owing to the low temperature contraction.

The cracks of semi-rigid roadbase cause the asphalt surfacing to crack, and reflection cracks result.

If semi-rigid roadbase was not covered in time with bituminous surfacing, it could crack owing to that semi-rigid material lost moisture as a result of evaporation and hydration.

On some sections of newly constructed freeways, the roadbases of which were constructed with high quality cement-treated and/or lime flyash treated granular materials, the 80-180mm thick asphalt surfacing took place about two months later than the completion of roadbases, consequently transverse drying shrinkage cracks of spacings between 6 and 10m were observed before paving asphalt surfacings. During the first winter after completion of the asphalt surfacing, prepared with poor quality bitumen, and/or during the second winter after completion of the asphalt surfacing, prepared with high quality bitumen, reflection cracks occured at the sruface even though the road was not yet opened to traffic. The reflection cracks observed therefore were due to low temperature-induced stresses in the surfacing.

The mechanism of reflective cracking in asphalt surfacing over semi-rigid roadbase is the same as that over existing cracked bituminous pavement.

Five 100m long test sections of overlay were constructed in 1986 on the 22m wide Beijing-Tanggu Class 1 bituminous road. The existing pavement consisted of 30mm of bituminous macadem, 80mm of bituminous penetration macadem and 300mm of soil lime. There were 227 cracks in the pavement, the great majority of which were transverse cracks of spacing 5-10m. The average crack rate was $231m/1000m^2$.

Five different thicknesses of overlay were constructed (40,60,80,100 and 120mm). The overlay was composed of an AC wearing course and a BM basecourse. A high quality bitumen was used in the left-hand carriageway and a relatively poor quality bitumen in the right-hand carriageway. During the first winter after the construction of the overlays, reflective cracking occured on each section. The results of cracking surveys, carried out in 1987 and 1990, are given in Table 1.

Table 1 shows that the reflection cracks increased with the decrease of the thichness of overlay and the rate of cracking in surfacing with high quality bitumen is much less than that with poor quality bitumen. Another section showed that even if the surfacing was 180mm thick, the reflection cracks still occured in it.

The cracks observed in asphalt surfacing, placed over uncracked semi-rigid roadbase were significantly less than that placed over cracked semi-rigid roadbase. In the former case, the cracks were mainly low-temperature contraction and temperature fatigue cracking of asphalt surfacing itself as proved by the results of cracking surveys and

Table 1. The rate of cracking in bituminous overlay (m/1000m²)

Thickness of overlay (mm)	40		60		80		100		120	
Quality of bitumen	Poor	High	Poor	High	Poor	High	Poor	High	Poor	High
1987.4	46.9	14.5	32.3	11.6	40.6	16.2	19.5	12.5	9.9	2.6
1990.6	142*	66.0	149*	49.5	102	42.9	116	23.1	29.7	9.9

* The alligator crazing cracks were not included.

laboratory experiments. In the latter case, there were not only the temperature cracks of asphalt surfacing itself but also the reflection cracks. One of the typical examples is the Zhengding test road, completed in October 1988. During the first winter (the lowest air temperature was -14°C) after construction no cracking was observed. Although the test road was not opened to traffic, over a two-winter period more or less cracks occured on most sections of it. The pavement structures, exposed time of roadbases and cracks observed in February 1990 are shown in Table 2. From Tab.2 it can be seen:

(a) Make good use of PMB seal coat to minimize cracking is successful.
(b) SAMI under thin asphalt surfacing may reduce but can't eliminate reflective cracking .
(c) The rate of cracks on section 16 is more than that on section 13 because the roadbase of section 16 had cracked before paving asphalt surfacing.

3 The mechanisms of reflective cracking

It was proved the reflective cracking occured mainly due to thermal stresses. The cores taken at locations of observed surface cracking during the cracking surveys in March 1987 had demonstrated that the mechanism of reflective cracking for thick asphalt surfacing, placed over cracked semi-rigid roadbase is different from that for thin asphalt surfacing over the same roadbase.

In thick asphalt surfacing the thermal stresses induced reflective cracking commenced at the surface and propagated downwards to connect with the cracking in roadbase.

Table 2. The pavement structures and cracks observed in surfacing on Zhengding Test Road

Section number	Pavement structure and layer thickness (mm)	Length of section	Exposed time of roadbase (m)	Rate of cracking (m/1000m^2)
1	PMB seal coat+60AC+120LFCS+430SL	100	22	0
2	60AC+120GCS+100LFCS+350SL	170	6	0
3	90AC+120GCS+100LFCS+320SL	200	44	0
4	120AC+120CTCS+390SL	162	31	0
5	150AC+120LFCS+360SL	170	7	0
6	120AC+120LFCS+390SL	205	12	3.0
7	90AC+120LFCS+420SL	225	45	3.5
8	150AC+120CTCS+360SL	200	55	5.0
9	60AC+120LFCS+450SL	190	94	6.4
10	40PMB AC+SAMI+155LFCS+430SL	100	54	11.3
11	50AC+SAMI+145LFCS+430SL	100	44	15.3
12	90AC+120CTCS+420SL	150	107	16.5
13	60AC+120SLCS+450SL	170	7	27.1
14	50AC+150LFCS+430SL	95	22	31.3
15	40PMB OGFC+SAMI+155LFCS+430SL	100	54	32.6
16	90AC+120LSCS+420SL	200	99	33.6

In Tab.2:
PMB : Polymer modified bitumen
SLCS: Soil lime crushed stone
LFCS: Lime flyash crushed stone
AC : Asphalt concrete
CTCS: Cement treated crushed stone
GCS : Graded crushed stone
SL : Soil lime

The features of cores, taken at the locations of cracks on some sections of foregoing Zhengding test road, are shown in following:

(a) Upper part of 38-82mm thick AC layers and roadbases cracked but lower part of AC layers uncracked, as shown in Fig.1 . The cores of this type have 16.7%.
(b) Upper part of 45-86mm thick AC layers cracked and its lower part uncracked, but cores of roadbase were damaged. The cores of this type have 48.9%.
(c) Both surfacing and roadbase cracked. The surfacing were 28-38mm thick. The cores of this type have 16.7%.
(d) 38-82mm thick surfacing cracked but cores of roadbase were damaged. These cores have 16.7%.

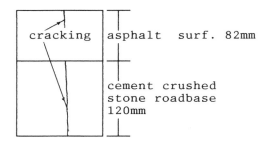

Fig.1. A typical core taken from Zhengding test road

The process of reflective cracking in thick asphalt surfacing demonstrated by cores is shown in Fig.2.

The mechanism of reflective cracking is proved also with the thermal stresses distribution within the surfacing calculated by the photoelectric analysis method and the finite element method.

The models used for investigating the thermal stresses within asphalt surfacing using photoelectric analysis method were shown in Fig.3 and Fig.4 (left hand).

The parameters of the two models for surfacing are $E_1 = 3,000$MPa, $\alpha_{T1} = 25*10^{-6}$ and for roadbase are $E_2 = 1,000$MPa, $\alpha_{T2} = 15*10^{-6}$.

When the surface temperature of models droped to a difference of 30°C, the thermal stresses occured within the surfacing were calculated and shown also in Fig.3 and Fig.4 (right hand). The Fig.3 and Fig.4 clearly demonstrated:

(a) When the asphalt surfacing was thick, the largest thermal stress occured at the surface directly above the cracking of roadbase. The surface stress of 1.231MPa is larger than that of 1.085MPa calculated for the same pavement structure but with uncracked roadbase. Therefore the reflective cracking occured above the cracking of roadbase and commenced at the surface, and then propagated down through the surfacing.
(b) When the asphalt surfacing was thin, the largest thermal stress occured at the bottom of surfacing. The bottom stress of 2.43MPa is much larger than that of 0.265MPa calculated for the same pavement structure but with uncracked roadbase. Therefore the reflective cracking inevitably occured above the cracking of roadbase and commenced at the bottom of surfacing, and then propagated up to surface. Once the thermal stress combined with the traffic induced stress the reflective cracking developed more rapidly.

The foregoing mechanisms of reflective cracking were proved also by the calculated thermal stresses in asphalt surfacing using the finite element method.

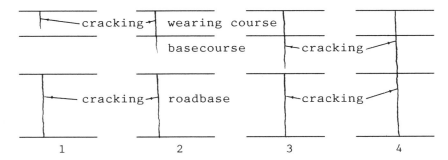

Fig.2. The process of reflective cracking in thick asphalt surfacing

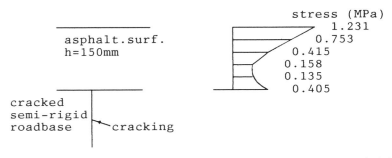

Fig.3. The distribution of thermal stresses (MPa) within thick asphalt surfacing over cracked semi-rigid roadbase

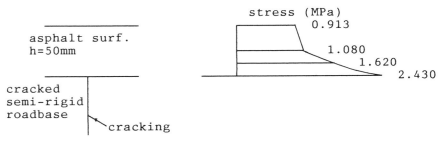

Fig.4. The distribution of thermal stresses (MPa) within thin asphalt surfacing over cracked semi-rigid roadbase

4 The main factors influencing the rate of cracking in surfacing over semi-rigid roadbase

There are four main factors that influence the rate of cracking in asphalt surfacing over semi-rigid roadbase.

4.1 The quality of bitumen
The quality of bitumen used for asphalt surfacing is the most important factor, which not only influences the temperature cracking of surfacing itself, but also the reflective cracking.

4.2 The types of semi-rigid roadbase material
The shrinkage and thermal contraction properties of various semi-rigid meterials are very different. The roadbases of different semi-rigid material therefore may crack to a different extent, and there are significant disparity in the rate of reflection cracks occured in asphalt surfacing over semi-rigid roadbase.

4.3 Moisture content of semi-rigid roadbase material
It is well known that the moisture content of semi-rigid material has significant effect on its shrinkage coefficient and maximum shrinkage strain. It was proved that if the moisture content of cement treated granular material is 1% more than optimum, its shrinkage strain will increase $40*10^{-6}-80*10^{-6}$ and more in dependence on the fine content of granular material and its plasticity.

4.4 Exposed time of semi-rigid roadbase
The exposed time of semi-rigid roadbase seriously affects the frequency of shrinkage cracking of roadbase. The longer the exposed time the more the shrinkage cracking.

5 Measures for minimizing reflective cracking

On the basis of foregoing cracking mechanisms, the following measures should be taken to minimize reflective cracking in semi-rigid pavement:

(a) High quality bitumen should be used in the surfacing. In addition, the use of modified bitumens to improve the low temperature cracking resistance of AC should also be considered.
(b) PMB seal coat may be used to increase cracking resistance of surfacing.
(c) High quality in terms of low coefficient of shrinkage cement and/or lime flyash treated granular materials should be used in the semi-rigid roadbase.
(d) The construction moisture content of the semi-rigid material should be reduced to a level consistent with achieving adequate compaction.
(e) The asphalt surfacing should be placed in time to avoide the shrinkage cracks of semi-rigid roadbase.
(f) For thin asphalt surfacings, make good use of a graded crushed stone interlayer or SAMI between the asphalt surfacing and semi-rigid roadbase should be considered.

44 MOVEMENTS OF A CRACKED SEMI-RIGID PAVEMENT STRUCTURE

A.H. de BONDT
Road and Railroad Research Laboratory, Delft University of Technology, Netherlands
L.E.B. SAATHOF
Road and Hydraulic Engineering Division, Ministry of Public Works, Transport and Water Management, Delft, Netherlands

Abstract
In this paper improvements for an existing overlay design procedure in case of reflective cracking in semi-rigid pavement structures are presented. These improvements are based on measured pavement movements at motorway A50 in the Netherlands. Movements due to temperature cycles and traffic (using FWD equipment and rolling loads) have been recorded. It became clear that at semi-rigid pavements the magnitude of the relative movements across the cracks caused by traffic is highly dependent on base temperature.
Keywords: Aggregate Interlock, Design Models, Fracture Mechanics, Load Transfer Measurements, Transverse Cracks.

1 Introduction

Transverse cracking in semi-rigid pavements is a serious problem for pavement authorities, because each year small maintenance measures such as filling of cracks are needed to prevent the ingress of water into the structure. Brooker, et al. (1987) stated that roadbase crack spacing is determined by the tensile strength of the roadbase material, its coefficient of thermal expansion and the temperature fall in the 1st or 2nd night after laying. After the laying of the asphalt concrete layers temperature cycles (daily and seasonal) and traffic loads will cause these cracks to grow to the pavement surface.
  Semi-rigid pavements show a wide variability in transverse cracking at the surface, because of the typical construction methods of cement stabilized bases and the fact that a wearing course mix always shows variability in its crack propagation resistance. Cement stabilized bases can be mixed in plant or in place. Especially the last option means that thicknesses and cement content (and thus tensile strength) can vary considerably. Bonnot (1983) has shown that in time not only more and more transverse cracks will reach the pavement surface, but that cracks already present will deteriorate. This is an important aspect if only local maintenance measures are applied. If overlay alternatives including soft interlayers and reinforcement are used it is essential to know the pavement movements which occur, because e.g. some alternatives are more successfull in resisting crack propagation caused by temperature cycles than propagation caused by traffic. In the next sections this topic will be discussed.

Reflective Cracking in Pavements. Edited by J.M. Rigo, R. Degeimbre and L. Francken.
© 1993 RILEM. Published by E & FN Spon, 2–6 Boundary Row, London SE1 8HN. ISBN 0 419 18220 9.

2 Superposition of Crack Propagation due to Temperature and Traffic

Goagolou et al. (1983) have theoretically analyzed the individual effects of daily temperature cycles and traffic on the reflective cracking process in semi-rigid pavements. They presented two figures showing crack length c within the overlay over overlay thickness h_0 versus actual number of load cycles N over number of cycles to failure N_{tot} for these individual effects (see Fig. 1). The curves are based on computations using elementary fracture mechanics principles.

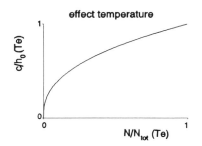
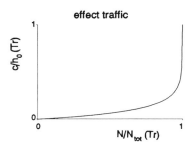

Fig. 1 Individual Effects of Daily Temperature Cycles and Traffic on Reflective Cracking in Asphalt Concrete Overlays

In the field both effects will occur, which means that in the overlay design superposition of both effects is necessary. The authors suggest superposition to be performed by assuming that crack propagation due to traffic occurs in day-time and crack propagation due to daily temperature cycles occurs at night. This means that in day-time a specific crack propagation will occur and pavement contraction during the night will insure further crack propagation. The next day this process goes on.

The consequence of this is that superimposing both parts of Fig. 1 means that one has to jump from one curve to the other using the actual crack length in the overlay. Fig. 2 shows an example of the superpositioning of the curves of Fig. 1 (In general the life of an overlay loaded by the effect of temperature $N_{tot}(Te)$ and traffic $N_{tot}(Tr)$ are not the same). It can be seen that superposition of the individual effects of temperature and traffic has a dramatic effect. The lifetime of the overlay (N_{tot}) is much shorter than expected on the basis of the curves of Fig. 1 on the first hand. This is caused by the fact that in case of the curves of Fig. 1 daily temperature cycles have a large effect at a small crack length and traffic has a large effect at larger crack length.

Analyzing the example of Fig. 2 it must be reminded that:

a) In the computations only smooth cracks have been applied (no aggregate interlock). In this way potential shear force transfer across a crack during a wheel passage is neglected.

b) No distinction has been made with respect to the time of the year. Only a constant daily temperature amplitude has been applied.
c) The lateral effect of crack propagation due to traffic has not been taken into account neither lateral wander. Only standard axle loads have been applied. Also no distinction has been made between the traffic lanes and the emergency lane.

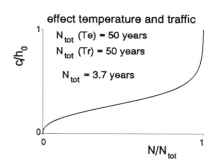

Fig. 2   Combined Effect

In section 4 future improvements of the method described above will be discussed. These improvements are based on the insight obtained by measuring pavement movements at the A50 in the Netherlands, see de Bondt, et al. (1992). These will be discussed in the next section.

3   Pavement Movements at the A50 in the Netherlands

3.1   General Information

The State Highway Administration (Rijkswaterstaat), the pavement manager of the A50, decided to construct test sections using several types of reinforcement in the overlay in the summer of 1992. This 3 km long section of this 2x2 lane motorway shows transverse cracking. Fig. 3 shows the number of cracks at the road surface in sections of a lane. The pavement consists of 0.20 m asphalt concrete on a 0.40 m cement stabilized base and is located on a sand subgrade. The road has been

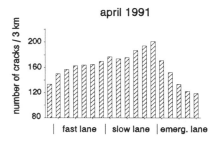

Fig. 3   Cracks at the Surface

constructed in 1971 and overlayed in 1981. Before the construction of the overlay information on the condition of the test sections, especially on the movements nearby the cracks, has been obtained.

3.2   Movements Caused by Temperature Cycles

Movements caused by temperature cycles have been analyzed by means of measuring the distance d between nails at two opposite points of 18 cracks twice a week from January until May. These points were located in the emergency lane at 1 m from the pavement edge. Because temperatures were recorded too, it was possible to get graphs such as Fig. 4. This picture shows d versus the temperature in the cement stabilized base for one of the cracks. Curling and warping of the pavement is probably the reason of the fact that at the same base temperature different values of d have been measured. The non-proportional relationship found can be caused by a lot of effects, e.g. the temperature dependent characteristics of the interface between the asphalt concrete layers and the base.

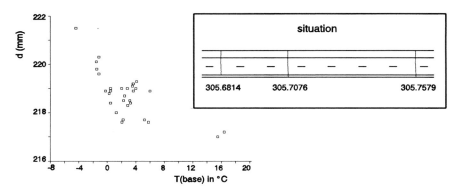

Fig. 4   Example of Crack Movement Caused by Temperature Cycles

3.3 Movements Caused by Traffic Loads

Fig. 5 and 6 show an example of surface deflections nearby a crack in summer and in winter conditions. Deflections have been measured by means of a falling weight deflectometer in the right wheel track of the slow lane (85 kN load). The effect of base temperature is very clear. In summer the surface deflection under the load is always the largest one, whereas in winter this is the case only if the load is far away from the crack (position a and b). From the curves at both sides of the crack it can be concluded that in winter two separate slabs exist, whereas this is not the case in summer. This means that after application of a thin overlay an extra tensile strain at the bottom of the overlay will occur in winter. Brown (1989) prefers position d for FWD testing at rigid pavements, because in this

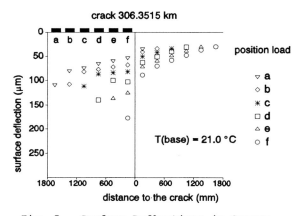

Fig. 5   Surface Deflections in Summer

way information is obtained on the deflected shape of the loaded and the unloaded slab.

Analysis of surface deflections measured at 55 cracks in position d (crack between d_{600} en d_{900}) showed that in general the deflected shape of the loaded and unloaded slab were parallel at a base temperature of 15.5 °C, whereas this was not the case at 3.5 °C. It also became clear that the effect of the magnitude of the falling weight load also depends on the crack width.

Fig. 7 shows the relative movement (d_L - d_U) at the same crack as shown before for several distances between load and crack (an FWD simulation of a moving wheel load). The geophones on both sides of the crack are denoted by L and U. Again the effect of temperature is clear.

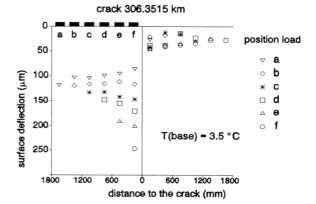

Fig. 6   Surface Deflections in Winter

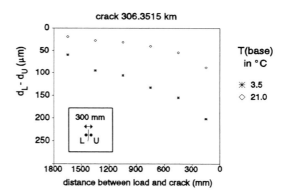

Fig. 7   Relative Movements

Load transfer measurements by means of LVDT's have been performed in order to get insight in the effects of a slow rolling wheel load. Fig. 8 shows the principle of this measurement (front axle 45 kN and rear axle 100 kN) and an example of the relative movements at a crack for a specific distance y between the measurement location and the centre of the wheel load. The maximum shear slip (P) occurs in this case at a wheel position of about 0.65 m from the crack. It can also be seen that the maximum shear slip before the crack passage (P) is slightly larger than the one after the crack passage (Q). It can be noticed that if the magnitude of the axle load increases by 2.2 the vertical relative movements do not increase to this extent in contrary to the horizontal movements. The pavement was very flat, which means that initial level differences could not have disturbed the results.

Measurements at this crack at a higher base temperature (13.5 °C) showed a decrease of the relative movements by a factor of about 100.

Interesting was also that the horizontal basin was deeper than the vertical basin. The small movements can be explained by the fact that construction of the cement stabilized base took place in fall 1971 (at a temperature between about 5 and 10 °C). This means that at a temperature of 13.5 °C the crack width will be very small.

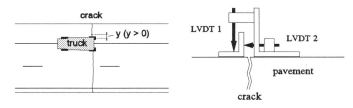

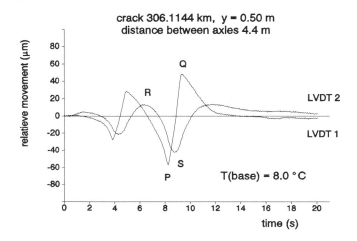

Fig. 8   Load Transfer Measurements at a Crack by means of LVDT's

Measurements at another crack and during a higher temperature showed that an increasing speed of the truck decreases the relative movements. At this stage it is not clear what the effect of the stiffness of the asphalt concrete is. Measurements elsewhere at a notched pavement (no friction possible) having an unbound granular base showed that a temperature increase of 6°C of the asphalt concrete layer reduced the relative movement PQ and RS by about 6. In both cases the horizontal basin was deeper than the vertical basin. Interesting was also that the curves between the basins caused by the front and the rear axle were more flat than the ones of Fig. 8.

Fig. 9 and 10 show the effect of the distance y between the measurement location and the wheel load centre on the relative movements PQ and RS. Because it is impossible to perform measurements under the centre of the load these values have to be extrapolated to obtain values under the loading centre. It can be seen that this extrapolation is difficult to perform.

In the field degradation of the crack faces (grinding) will occur due to shear slip caused by traffic. At the A50 this has been simulated by the application of 100 falling weight load applications of 163 kN. Fig. 10 shows that after the degradation procedure the vertical relative movements were higher, whereas the horizontal relative movements remained the same. During the degradation procedure also surface deflections have been measured by means of the geophones. The loading plate and the geophones were located in position f (see Fig. 5). Results showed that these values as well as the relative movements measured by the LVDT's remained constant.

Surface deflections have also been measured (position f) at the slow lane and the emergency lane in the driving as well as in the reverse direction at two base temperatures. For a specific crack and temperature the ratio d_0 over d_{300} did not show systematic differences.

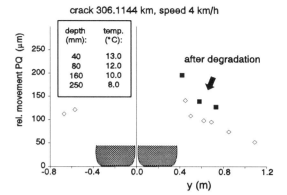

Fig. 9   Relative Movements

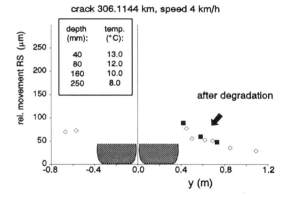

Fig. 10   Relative Movements

3.4 Comparison of Measurements of a Specific Crack

Fig. 11 shows a comparison of results for a specific crack. It can be noticed that the base temperature and in this way the crack width has a remarkable influence on the deflection ratio (The stiffness of the asphalt concrete also depends on temperature).

The differences shown in Fig. 11 can be explained if the mechanisms are analyzed which occur in a crack during a wheel passage. Fig. 12 shows that first a free shear slip without friction will occur, but soon this slip will develop friction because of the normal restraint caused by the crack dilantancy. This means that further shear slip will be restrained. All in all, it means that during a wheel passage a shear slip s will occur, while at the same time the crack width will increase from w_0 to w, due to dilatancy.

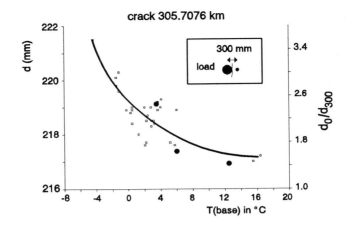

Fig. 11 Crack Movement and Load Transfer versus Base Temperature

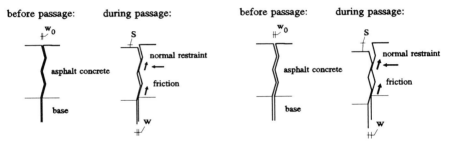

Fig. 12 Load Transfer under Different Climatic Conditions

Cores showed that the cracks in the asphalt concrete have grown around the mineral aggregates. It can thus be concluded that the amount of aggregate interlock in the asphalt concrete layers can be very good or very bad, depending on the actual crack width (w_0) and the applied load. The cores also showed that the crack faces in the cement stabilized base were very smooth, which means that no aggregate interlock can occur. This could be expected, because of its composition of cement and sand.

4 Conclusions

- At semi-rigid pavements relative movements across cracks due to traffic are highly dependent on base temperature.
- To get a better understanding of the strains thin overlays have to resist under traffic loadings, not only vertical movements should be measured at cracks, but also horizontal movements.

5 Recommendations for Further Development

Based on the measurements discussed in section 3 it is clear that the overlay design method described in section 2 can be improved by:

a) Taking into account the mechanisms which occur at a crack during a wheel passage. To be able to do this constitutive laws have to be determined by means of laboratory tests. After this the finite element code CAPA presented by Scarpas, et al. (1993) which has the proper finite elements, can be applied for pavement analyses.
b) Taking into account the fact that different daily temperature amplitudes occur. Temperature distributions during a year are necessary as input for CAPA to analyze the effect of these varying amplitudes.
c) The lateral effect of crack propagation in an overlay due to traffic and lateral wander can be examined by using load transfer results and CAPA analyses. The effect of the magnitude of a wheel load can be examined in the same way. Distinction between traffic lanes and the emergency lane can be performed by using different traffic volumes.

Finally, it should be possible that for a specific temperature profile during the design period the effectiveness of overlay alternatives using soft interlayers and reinforcements can be determined.

6 References

de Bondt, A.H. Saathof, L.E.B. and Steenvoorden, M.P. (1992) **Test Sections Asphalt Reinforcement A50 Friesland - Vol.1 General Information and Measurements before Overlaying (in Dutch) - Main Report.** Road and Railroad Research Laboratory, Delft University of Technology - Road and Hydraulic Engineering Division, Ministry of Public Works, Transport and Water Management, the Netherlands.

Bonnot, J. (1988) Fissuration de retrait des chaussées à assises traitées aux liants hydrauliques. **Bulletin de liasion des laboratoires des ponts et chaussées**, 156, 37-66.

Brooker, T. Foulkes, M.D. and Kennedy, C.K. (1987) Influence of Mix Design on Reflective Cracking Growth Rates through Asphalt Surfacings, in **Proceedings 6th International Conference on the Structural Design of Asphalt Pavements**, Ann Arbor, MI, 107-120.

Brown, S.F. Brunton, J.M. and Armitage, R.J. (1989) Grid Reinforced Overlays, in **RILEM Conference on Reflective Cracking in Pavements**, Liège, 63-70.

Goagolou, H. Marchand, J.P. and Mouraditis, A. (1983) Application à la fissuration des chaussées et au calcul du temps de remontée des fissures. **Bulletin de liasion des laboratoires des ponts et chaussées**, 125, 76-91.

Scarpas, A. Blaauwendraad, J. de Bondt, A.H. and Molenaar, A.A.A. (1993) CAPA: A Modern Tool for the Analysis and Design of Pavements, in **RILEM Conference on Reflective Cracking in Pavements**, Liège.

45 A CRACK-RESISTANT SURFACE DRESSING: THE RESULTS OF 8 YEARS OF APPLICATION, AND FUTURE PROSPECTS

J.P. MARCHAND
Cochery Bourdin Chausse, Nanterre, France

Abstract

A surface dressing of polymer bitumen, reinforced with a geotextile impregnated with the same polymer bitumen, has been developed in France since 1985. The results of this technique after eight years of monitoring are presented in this communication.

The points dealt with relate to the specifications laid down by the contractor so as to guarantee the performance characteristics of the product employed; the laying procedures developed in order to prevent creases in the geotextile; the internal procedures designed to ensure that the teams engaged in the application of the technique conform to standard practice; how the technique has been monitored so as to assess the efficacity of the surface dressing without neglecting its sealing of the underlay; the performance of the surface dressing in terms of surface characteristics; and the technical statement issued in 1991 on this technique, which is known as ARMACCO.

In conclusion, mention is made of the transposition of the technique to a system incorporating the same geotextile, but impregnated with an unfluxed emulsion of the same polymer bitumen, together with a a thin bituminous concrete wearing course incorporating polymer bitumen.

Keywords : Pavement, Crack, Surface Dressing, Wearing Course.

1 Introduction

The range of crack-prevention techniques developed up to the present includes:
* Those which limit the cracking of the course treated (pre-cracking <1>).
* Those which treat surface cracks intermittently (sealing).
* Those which combine crack treatment and surface maintenance. This category includes wearing courses having at their base a coated sand membrane with a high bitumen content, or a geotextile impregnated with binder.

Among these techniques, the ARMACCO crack-resistant surface dressing is an original solution. It has been developed since 1984 and at the present time is the only crack-prevention system which has been the subject of a technical statement.

Reflective Cracking in Pavements. Edited by J.M. Rigo, R. Degeimbre and L. Francken.
© 1993 RILEM. Published by E & FN Spon, 2–6 Boundary Row, London SE1 8HN. ISBN 0 419 18220 9.

2 Brief reminder of the process

It is a surfacing process, the surface consisting of:
* On the underside, a geotextile saturated with polymer bitumen, adhering to the underlay and acting as a membrane absorbing the movement of cracks.
* On the upper side, a two-layer polymer bitumen surface dressing constituting the wearing course.

The process was described at the first Conference on the Reflective cracking in Pavement <2>, and we confine ourselves here to an account of the results achieved after eight years.

3 The distinguishing features of the process

In order to ensure that the rise of cracks to the surface is effectively prevented, the contractor drew up rules and controls pertaining to the conformity of the geotextile and the binder, together with low-temperature specifications (-10°C) pertaining to the laboratory performance of the geotextile/binder complex.

The rules relating to conformity of the geotextile are given in table 1, and the specifications of the binder-impregnated geotextile are contained in table 2.

Table 1: Controls of conformity of the geotextile.

Characteristics	Extreme values	Geotextiles tested Polypropylene	Polyester
Mass per surface area (g/m2) [NF G 38-013]	> 120 < 150	132	142
Binder absorption capacity (g/m2) [IST 180-8-34]	> 700	930	680
Surface shrinkage at 150°C (%)*	< 10	8	1
Compressibility at 0.2MPa (%) [NF G 38-012]	< 60	45	17
Breaking force at - 10°C 2 mm/min. (kN) [DIN 53 857]	> 0.5	0.7	0.7
Breaking stress at - 10°C 2 mm/min. (%) [DIN 53857]	> 4.0	5.0	9.5
Breaking stretch at - 10°C 2 mm/min. (%) [DIN 53 857]	> 30	39	57

* Method developed by the contractor.

As can be seen from the two tables, the breaking stretches of the binder-impregnated polypropylene and polyester geotextiles are identical, whereas under the same operating conditions the polyester has a greater breaking stretch than the polypropylene. The performance characteristics of the geotextile alone are not sufficient to predict those of the complex.

Table 2: Specifications for binder-impregnated geotextiles (degree of impregnation: 0.8 kg/m2).

Characteristics	Extreme values	Geotextiles tested	
		Polypropylene	Polyester
Breaking force at - 10°C 2 mm min. (kN)[DIN 53857]	> 10	1.5	1.3
Breaking stress at - 10°C 2 mm min. (MPa)[DIN 53857]	> 8.0	8.8	10.5
Breaking stretch at - 10°C 2 mm min. (%)[DIN 53 857]	> 45	54	54

4 Détails of production and laying

The geotextile is laid immediately after the binder has been spread. The latter being a fluxed polymer bitumen, an interval of 30 minutes or even an hour is permissible because of the presence of volatile fractions. This is important, for the binder is spread at a temperature of 150 to 165°C, so that it is possible to wait a few minutes while the temperature of the spread binder drops, and so that the surface of the geotextile shrinks as little as possible (especially in the case of polypropylene).

The choice of the length of the rolls depends on the radius of curvature of the pavement (table 3). Rolls of width 0.50 m and 1.00 m are laid by means of manual unwinders (fig. 1). Wider rolls are laid mechanically with a wide unwinder (fig. 2)

Any creases that may occur are immediately corrected, either by raising the geotextile and laying it back down flat, or by brushing with a broom. If the crease subsists, it can be cut away.

Passing a heavy pneumatic-tyred compactor over the geotextile helps it to become impregnated with the binder.

Table 3: Choice of unwinder.

Radius of curvature of pavement	Width of geotextile	Type of unwinder
20 < R < 30 m	0.50 m	manual
30 < R < 50 m	1.00 m	manual
50 < R < 70 m	1.75 m	wide
70 < R < 100 m	2.50 m	wide
R > 100 m	3.80 m	wide

Fig 1 : Manual unwinder

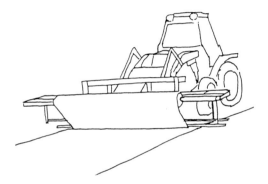

Fig 2 : Wide unwinder

5 Internal information procedures

All the procedures enabling the crack-resistant surface dressing to be laid in accordance with standard practice are contained in an internal document for the use of the supervisory staff on site. This document comprises sections on the resources to be made available, supplies, site preparation, the laying operation itself, and the controls to be carried out by the person in charge.

6 Monitoring and technical statement

The process was thought up in the context of a contest of innovative techniques organized in 1983 by the French Highways Department. The sites were monitored by both the contractor and the technical departments of the French Administration, and led to the the issue in February 1991 of a technical statement <3>, which stated:

6.1 Behaviour in terms of prevention of the rise of cracks

"The eariest applications date from 1985, and development on an industrial scale began in 1986. This process has been used mainly in rural areas, for traffic up to TO (750 to 2,000 heavy vehicles per day in each direction). Most applications were on continuously-graded aggregate underlays containing transverse cracks due to thermal shrinkage, and/or longitudinal cracks due to differential settlement or drying up of the underlying soil.

The monitoring of the sites, which in some cases lasted for five years, clearly showed that the technique fully serves its purpose, namely to prevent the rise of cracks and ensure imperviousness. Sometimes cracks can be detected without breakage of the membrane, which ensures that imperviousness is maintained".

6.2 Behaviour in terms of surface characteristics

"Bonding is satisfactory. The values of macrotexture and longitudinal coefficient of friction are comparable to those of a surface dressing.

Nevertheless, the surface aspect is not always satisfactory, and in some cases glazing is observable along the junctions between traffic lanes.

Note may be taken, in the development stage of this technique, of the problems of junctions of the surface dressing between traffic lanes, and of the behaviour of the surface due to creases in the geotextile, characterized by localized bleeding or plucking".

6.3 Behaviour in terms of formulation

"The single-layer double-chipping formula seems to be the trickiest to achieve successfully, having regard to the cumulation of two layers of binder (one for impregnating the geotextile and the other for the surface dressing itself). The two-layer technique apparently allows of greater flexibility of application because of the reduced batching of the first layer of surface dressing".

7 Development and transposition

With some thirty sites treated since 1985, this is now one of the oldest-established and most effective crack prevention techniques. But it suffers from the unfavourable prejudice held by local inhabitants and by contractors where surface dressings are concerned (noise, and displacement of chips), despite improvements made in the adhesive properties of binders and the technique of spreading.

Consequently, to meet clients' requirements, the contractor has transposed the geotextile-reinforced surface dressing to a bituminous surface reinforced with the same geotextile. This is known as ARMACCO E.(Fig.)

Fig 3 : Laying of ARMACCO E

This transposition meant relinquishing the impregnation-fluxed polymer bitumen and emulsifying the same polymer bitumen, but without solvent. This is a very important point, because the presence of solvent tends fo soften the bituminous concrete.

The proportion of emulsion (containing 70% bitumen) is adjusted so as to obtain ultimately an impregnation of 800 gr per square metre. The binder is the same polymer bitumen used to prepare the emulsion. It is applied to a thickness of 3 to 4 cm (equivalent to 75 to 100 kg per square metre) and has a ratio of $r > 1.17$ under the bending shrinkage test <3> of the Autun Regional Laboratory of the French Ponts et Chaussées (Road Research Laboratories), where

$$r = \frac{\text{cracking time of ARMACCO E}}{\text{cracking time of reference pavement}}$$

This rates it as a highly effective technique.

Performance in terms of the prevention of the rise of cracks alone is not enough. It must be accompanied by a rutting resistance test. This test gives a rutting value of 2.7 mm at 3,000 cycles at 60°C, which is markedly lower than the 20 mm required by the standard NF P 98-132 relating to thin bituminous concretes.

More than 300,000 square metres have been laid since 1990. Despite a shorter length of experience than with the reinforced surface dressing technique, the in situ behaviour of the reinforced mix is satisfactory.

8 References

<1> MARCHAND J-P (1989) L'enduit superficiel antifissure : de sa conception à sa réalisation, 1st int. conf. on Reflective Cracking in Pavements, Liège, PP 420-427.
<2> COLOMBIER G and al (1990) CRAFT, Travaux
<3> ARMACCO (1991) Technical Statement n° 36

46 THIN OVERLAY TO CONCRETE CARRIAGEWAY TO MINIMISE REFLECTIVE CRACKING

I.D. WALSH
Engineering Services Branch, Kent County Council, Maidstone, UK

Abstract

A cracked, jointed reinforced concrete pavement at Robin Hood Lane, Lydd, Kent was overlaid with 40 mm nominal thickness of bituminous material in 1989. Four materials, Dense Bitumen Macadam with a modified and unmodified binder and Hot Rolled Asphalt with a modified and unmodified binder were laid, half on the concrete directly, half incorporating a non-woven geotextile membrane, to produce 8 trial areas each approximately 160 m long. The concrete pavement was made good by vacuum grouting before overlay. After 3 years the Dense Bitumen Macadam has shown inferior performance to Hot Rolled Asphalt. The effect of using binder modified with Ethyl Vinyl Acetate polymer was negligible, the effect of geotextile was significant in reducing reflective cracking over cracks but insufficient to prevent reflective cracking over contraction joints.

Keywords: Polymer Modified Asphalt, Geotextile, Concrete Carriageway, Reflective Cracking.

1 Introduction and Description of the Site

A significant proportion of the new construction in Kent since the mid 1960's has included cement bound materials. This has been either lean concrete used as roadbase or PQ concrete. In addition there are numerous older concrete roads throughout the county.

Whilst concrete has been found to perform adequately in terms of its structural contribution, it has presented maintenance problems in that joints and cracks reflect through bituminous overlays and allow water into the construction. Thus causing damage to construction by chemical attack of the concrete and softening of unbound

---

Reflective Cracking in Pavements. Edited by J.M. Rigo, R. Degeimbre and L. Francken.
© 1993 RILEM. Published by E & FN Spon, 2–6 Boundary Row, London SE1 8HN. ISBN 0 419 18220 9.

layers and subgrade.

To date, this problem has been generally overcome by applying thick overlays (normally 180 mm) to suppress reflective cracking, even though this thickness of overlay is often not required for structural strength.

It will obviously be beneficial in terms of minimising maintenance, if materials or techniques can be found which will allow thinner overlays to be applied to pavements with cement bound layers. Following a trial in 1986 on M2[1] in Kent which investigated the effect of overbanding or geotextile Stress Absorbing Membrane Interlayers (SAMI), it was found that overbanding was counterproductive whilst geotextile may have a beneficial effect. There was little difference between SBS and EVA modified binder detected. A further comparative trial was installed in 1989 on a concrete pavement at Robin Hood Lane, Lydd, incorporating unmodified and EVA modified binder in two materials and with geotextile SAMI over half the width.

Robin Hood Lane is a two lane single carriageway road approximately 650 m long and 6 m wide with a 30 mph speed limit. The road was constructed during the war to carry tanks and other heavy equipment to a nearby army camp but is now carrying some 300 heavy commercial vehicles per day in each direction principally to and from gravel pits between Lydd and Dungeness.

A site location plan is given in Figure 1.

Figure 1. Site Location

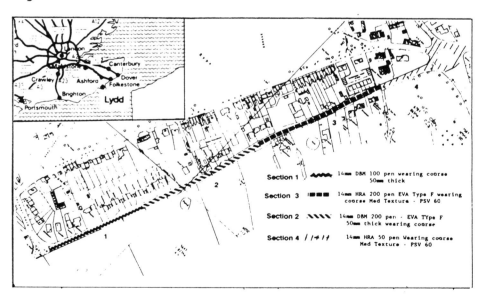

The pavement was identified as being in need of maintenance in 1987 to remedy extensive reflective cracking which had occurred at the joints and at mid slab cracks. In addition to the reflective cracking, local residents were complaining about `vibration' from the carriageway when heavy vehicles passed. On investigation this proved to be caused by rocking slabs. A road pavement appraisal identified the areas where slabs were rocking under traffic and identified surface defects in the surfacing.

The pavement construction is jointed reinforced concrete slabs approximately 250 mm thick laid straight on a sub grade of silty single sized flint gravel. The length of the slabs is somewhat variable but the average was just under 10 metres.

The concrete had been overlaid with 20 mm to 30 mm of Hot Rolled Asphalt wearing course, probably in the late 1960's, which had debonded locally particularly in the vicinity of joints and cracks.

2 Surface Treatments

Areas where slabs were shown to be rocking under traffic were vacuum grouted in advance of the Trial using a sand/cement grout. This work was carried out in mid April 1989 by a proprietary process. This also restored the correct longitudinal profile to the road surface, removing any `steps'.

According to current practice as outlined DTp Advice Note HA6/80, vacuum grouting should be carried out when the relative deflection between two adjacent slabs exceeds 0.2 mm or when the absolute deflection of a slab exceeds 0.4 mm. For this trial, however, it was decided to grout where the absolute deflection exceeded 0.3 mm. This was to avoid, as much as possible, the vertical movements under load becoming a major factor in causing reflective cracking through the thin overlays.

The old surfacing material was removed by cold planing.

The geotextile was a non woven 100% polypropylene fabric, complying with the requirements of USA Specification "Task Force 25"[2] laid by machine. It was stuck to the underlying concrete using 200 penetration grade straight-run bitumen, applied at approximately 1 litre/m^2 from a surface dressing tanker. The quantity is designed to saturate the geotextile after compression. It is deliberately insufficient to saturate the uncompressed geotextile to prevent binder picking up on the tyres of the public and contractor's vehicles.

The geotextile covered the whole of the eastbound lane and extended 300 mm into the westbound lane to include the longitudinal centre joint, four different materials were laid as described below to provide 4 sections as given in Figure 2.

Section 1 contained:
14 m nominal size dense bitumen macadam (DBM) wearing course to BS 4987[3] with 100 penetration grade unmodified bitumen laid 50 mm thick.

Section 2 contained:
A 14 mm dense bitumen macadam wearing course identical to Section 1 but the binder was 200 pen bitumen with 5% of EVA type F added as described below.

Section 3 contained:
Hot Rolled Asphalt (HRA) wearing course to BS 594[4] with nominal 40% stone content and 200 penetration bitumen binder with 5% of EVA type F as for Section 3, laid 50 mm thick.

Section 4 contained:
Hot Rolled Asphalt wearing course but with 50 penetration grade unmodified bitumen binder, laid 50 mm thick.

The combination of the four different materials, each with and without geotextile gave eight trial lengths. The sections of DBM with 100 pen binder and HRA with 50 pen binder, without geotextile were considered as control lengths since these materials are in common use in the country.

3 Materials Properties and Measurements on road pavements

The materials were supplied to conform with the gradation and binder content of BS 4987 1985, or in the case of hot rolled asphalt to BS 594 1983. In the latter case the target binder content is determined following the Marshall Stability and flow determination in accordance with BS 598 Part 3 1985.[5] In accordance with Council practice the test was carried out on a 30% stone content mix using 50 pen bitumen binder. The mix achieved 7.0 kN stability, 4.2 mm flow at a target binder content of 7.6%. Since a 40% stone content mix was required for this thickness of mat the binder content reduced to 6.6%. This binder content was also used for the EVA modified binder mix.

The tests carried out on the binder were as follows:

penetration, softening point, Fraas Breaking Point (IP80)[6], EVA content by Infra-red Spectrophotometry (if appropriate) storage stability test and Ductility at 4°C[7] both on the binder from the supplier's storage tank and from binder recovered from the mixed material using the Rotary Evaporator Method.[8]

EVA is variably soluble in Melthylene Chloride (CH_2Cl) therefore the percentage of EVA in the recovered sample may not be all of the EVA present in the mixed material.

After laying, 200 mm diameter cores were taken from the finished mat for measurement of rut resistance using the Dry Wheel Tracking Test.[9]

In Kent there are minimum requirements for texture depth in order to provide an adequate slip/skid resistance and these were measured by the laser Texture Meter[10] and Sand Patch Test. The aggregate in the Dense Bitumen Macadam had a Polished Stone Value (PSV) of 52, the precoated chippings applied to the Hot Rolled Asphalt, a PSV of 62.

Subsequently friction testing was carried out by the Pendulum Test.[11] The results of the Texture Depth and Friction Testing (Skid Resistance) are given in Table 1.

A Benkelman Beam[12] was used at 8 or 9 joints/cracks in each section to measure the benefits achieved in improved structural strength by both the grouting process and the effect of the geotextile overlay. The results of Deflection measurements are given in Table 2.

4 Observations during laying

The materials were laid between the 17 and 31 May 1989, during dry and sunny weather. The nominal thickness of material laid was 50 mm. However when the road was cored recently for investigation of crack propagation it was found that the macadam sections (1 and 2) were laid at a mean thickness of 41 mm whilst the asphalt sections were laid at a mean thickness of 47 mm.

No particular problems were encountered in laying the geotextile on a binder film nor laying the DBM or HRA materials, despite the very different penetration and softening point values on the binder with EVA compared to unmodified bitumen. No problems were experienced with applying and retaining precoated chippings to achieve the required texture depth. To prevent excessive heating of the geotextile, which melts at approximately 150°C, the maximum laying temperature of the blacktop overlay is specified at 145°C; this can cause problems with compaction in winter. This temperature was not exceeded by the DBM. However the HRA had occasional loads exceeding this temperature particularly the 50 pen bitumen material; some material was discarded, some was laid and removed as its delivery temperature exceeded 173°C and excessive binder hardening was feared; no damage to the geotextile was observed indicating that the existing road surface (temperature approximately 33°C) provided adequate temperature gradient to prevent damage.

After the pavement was completed both the horizontal and transverse profile was within UK standards for new roads.

Table 1. Material summary at time of works

Sect			EVA %	Pen dm	Binder Properties a Supplier b Tank c Recovered		IP Duct. cm	Binder Content		Wheel Track Rate mm/m	Texture Depth mm		Skid Resist. after 3 yrs.
					SPt °C	Fraas °C		Actual %	Target Mean %		Sand Patch	Laser	
1	100 pen 14mm DBM	a b c	-	- 95 105	- 48.3 44.0	- -9 -10	- 8.0 6.0	4.8	5.1	0.49	0.60	0.42	52
2	200 pen + EVA(F) 14mm DBM 200 pen	a b c b	4.8 -	171 141 123 182	42.0 43.1 42.2 39.6	-10 -	- 61.5 11.0* 24	5.4	5.1	0.33	0.55	0.40	54
3	200 pen + EVA(F) 40%/14 HRA	a b c	4.8	154 139	42.4 41.4 44.6	-14 -13 -16	65. 14.0*	7.0	6.6	4.78	1.02	0.82	57
4	50 pen 40%/14 HRA	a b c	-	35 42 38	57.3 66.7 62.8	-11 -11 -12	0.5 0.5	6.2	6.5	0.86	1.44	0.99	59

*Percentage of EVA in recovered binder unknown and may be variable.

Table 2. Deflections (mm X 10^{-2}) average for section

Sect	Before grouting Both	After grouting Geo	After grouting No Geo	After overlay Geo	After overlay No Geo	After 3 yrs Geo	After 3 yrs No Geo
1	34	18	-	12	-	15	-
2	26	20	18	7	11	15	25
3	34	15	17	11	10	8	25
4	22	18*	20	12	25	13	25

*Not all of the section required grouting

5 Observations of Crack Propagation

A survey was carried out prior to the overlay being placed to establish the location of transverse joints and cracks in the concrete. This information was then used as the datum against which all subsequent assessments of cracking were made.

Further surveys to establish the incidence of reflective cracking were carried out approximately 6 months, 1 year and 3 years after treatment. The results of the crack surveys are given in Table 4 below and summarised in Figure 2.

Figure 2

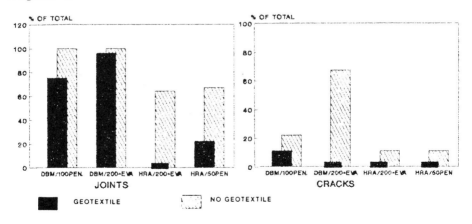

Table 4. Number of Transverse Cracks/Joints

DATE OF INSPECTION			DENSE BITUMEN MACADAM				HOT ROLLED ASPHALT			
			100 pen. SECTION 1		200 pen+ EVA SECTION 2		200pen+EVA SECTION 3		50 pen SECTION 4	
			GEO	NO GEO	GEO	NO GEO	GEO	NO GEO	GEO	NO GEO
BEFORE OVERLAY		J	16	16	22	22	22	22	9	9
		F	5	15	19	21	19	22	6	6
		P	4	2	1	1	1	1	0	0
19/12/89		J	7	14	6	21	0	5	0	6
		F	0	1	0	9	0	0	0	0
		P	0	3	0	1	0	0	0	0
20/ 3/90		J	11	16	14	22	0	5	0	6
		F	0	4	0	13	0	0	0	2
		P	0	0	0	2	0	0	0	0
23/ 4/92		J	12	16	21	22	0	14	2	6
		F	1	4	0	14	0	3	0	1
		P	0	0	0	1	0	0	0	0

NOTES: J = Joint   F = Full width crack   P = Part width crack

Joints are those arising at regular intervals from the bay construction process; it is very unlikely that a concrete train was used; alternate-bay construction being the norm at that time. Cracks were either full width or part width and were located generally close to the middle of the bays. This cracking was probably due to insufficient support from the underlying material; this support being provided subsequently by the cementitious grout which would, to a degree, also enter the wider cracks restricting movement. However at the small number of locations where multiple parallel cracks have appeared on the road surface this was found after coring, to be from a similar crack pattern in the underlying concrete. It is therefore important before thin overlaying to ensure all the concrete pavement is sound by appropriate thin bond or full depth repairs.

The main features of the trial in terms of reflective cracking were:-

   a) the high proportion of joints which have reflected through the dense macadam overlay, regardless of whether geotextile membrane was used. At least 75% of the joints reflected through the overlay in three years where geotextile was used and all of the joints reflected through where there was no

geotextile.
b) The large difference in the number of joints which reflected through the rolled asphalt overlay between the side with geotextile and the side without it. Only 2 joints (22% of joints in section 4 and 6% of all joints in the asphalt) reflected through where geotextile was used. This contrasts with 22 joints which reflected through on the side with no geotextile. These 22 joints represent approximately $^2/_3$ of all joints in the concrete.
c) the relatively low proportion of cracks compared to joints which reflected through the overlay. Overall, two thirds of the joints reflected through the overlay, whereas only 20% of the cracks reflected through. The vast majority of these cracks where on the side without geotextile and, as with the joints, more cracks reflected through the DBM lengths.

6 Discussion of Results

6.1 Effect of Macadam vs Hot Rolled Asphalt

Without geotextile 100% of the joints and a significant percentage of the cracks have reflected through the macadam overly, the performance of the Hot Rolled Asphalt is significantly better with only 60% joints reflected through and almost no cracks.

Macadam is essentially a close graded aggregate mixture with a low (4.8 -5.4%) binder content. The binder primarily acts as a lubricant during laying and compacting but the binder film thickness is quite small. The interlocking properties of the aggregate predominate and the material generally provides an excellent rut resistant structural layer. This is demonstrated by the rut resistance (low wheel tracking rate) (shown in Table 1).

Hot Rolled Asphalt is essentially a gap graded sand/filler mixture/bitumen matrix containing a percentage (40%) of coarse aggregate which has little interlocking capability. The matrix properties largely define the properties of the material and the bitumen is the major determinant of these.

Thin overlays to concrete pavements are subject to horizontal and vertical movements at joints and cracks. Because of the grouting technique, the vertical movements at joints had been largely eliminated, leaving the horizontal movement as a result of daily and annual thermal change. The extent of these movements were not measured, but elsewhere in Kent, concrete pavements made with similar materials have had joint movements in the region of 0.08 to 0.10 mm per °C temperature change i.e. 5 mm approximately movement per joint between summer and winter. Such movements may be exacerbated by the increased thermal gain from changing the white surface to black, but this is offset by the insulation value of the overlay; measurements of this effect are in progress in Kent. It cannot be predicted theoretically because of friction effects with the subgrade prevent all the joints moving equally.

The blacktop overlay in this case will be subject to tensile forces as the temperature increases above the 20°C at time of laying; maximum surface temperatures recorded exceed 58°C and to compression as the surface cools in winter, minimum temperature -10°C. The mode of failure in these circumstances is fatigue. On behalf of all County Councils in the UK, Birmingham University[13] carried out comparative fatigue testing on macadam mixtures complying with BS 4987 and Hot Rolled Asphalt mixtures complying with BS 594. Where deformations (deflections) are low a controlled stress criteria is applicable, however for high deformations (deflections) a controlled strain criteria was seen to be most applicable. Any fatigue loading must also take account of `rest periods' where delayed elastic recovery and `healing' occurs; the latter occurring as fresh cracks reseal themselves to give over 90% of their original strength in periods of the order of 3 days[4][5]. This was modelled in the laboratory using 4 mixtures made up into beams which were subject to bending stresses. Three mixtures were Hot Rolled Asphalts and 1 dense bitumen macadam wearing course. The results are summarised in Fig 3.

Figure 3

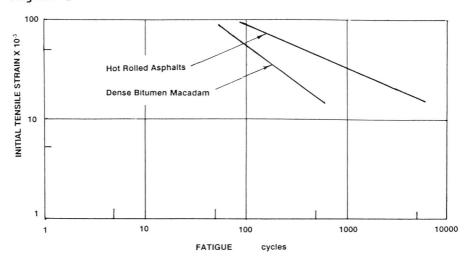

Two of the materials in this study were similar to those used on this trial, making the results valid for comparative purposes. In all cases the higher the binder content, the longer is the fatigue life. Pell[16] attributes this phenomenon to the dependence of fatigue life on the strain in the bitumen presence in the mix; the richer the mix the thicker is the film of bitumen round each particle and hence the lower is the actual bitumen strain. Furthermore the greater the bitumen film thickness the longer these properties will be maintained as the bitumen hardens in-service.

The binder content in the 100 pen DBM mixture was 0.3% less than target mean, that with 200 pen and EVA polymer was 0.3% above target mean; this may explain the slightly improved performance of the EVA modified mixture on the side with geotextile compared to the unmodified mixture. However, the CSS research found little difference in fatigue life for dense bitumen macadam wearing courses over a binder range of $^+0.5\%$. A reason for this different performance may be as a result of the binder properties themselves discussed later or the thickness of the laid material.

Hot Rolled Asphalt has a significantly higher binder content and higher binder film thickness and in the laboratory was found to have an order of magnitude larger fatigue life in cycles to failure, especially at low initial strains. The improved crack resisting performance of the EVA modified binder in the Hot Rolled Asphalt may also be explained, at least in part, by the 0.4% higher than target binder content in the mixture compared to the 50 pen unmodified mixture that was 0.3% below target.

The study also showed that for any single mix composition and binder content, the normal range of voids in a well compacted wearing course, has no significant influence on fatigue life.

The much better fatigue response of Hot Rolled Asphalt compared to dense bitumen macadam is clearly a most significant reason for its lower rate of reflective cracking.

6.2 Effect of thickness

It is recognised that increasing the thickness of the overlay improves resistance to reflective cracking by a combination of insulation effect and having any strains dissipated over a greater cross-sectional area of material. Whilst the nominal thickness was specified to be the same throughout, the DBM materials were laid thinner than specified. The 100 pen material (Section 1) had a mean thickness of 40 mm. Whilst the 200 pen and EVA material (Section 2) had a mean thickness of 42 mm, this difference is unlikely to have affected the comparative performance.

The hot rolled asphalt sections (3 and 4) had similar mean thickness of 47 mm - 48 mm respectively, however this disguises the fact that the geotextile side had a measured thickness of 44.5 mm whilst the non-geotextile side had a thickness of 50.5 mm. This makes the improved performance of the geotextile in ameliorating reflective cracking even more significant.

In comparing the two basic materials, DBM and HRA, it is not unreasonable to compare the geotextile side where the mean thicknesses were only 2.5 mm different. Therefore, the significance of the intrinsic material properties must be primarily responsible for the very significant better performance if the hot rolled asphalt.

6.3 Effect of EVA polymer vs unmodified binder

The EVA chosen was a 44 Melt Flow Index, 33% Vinyl Content Ethyl Vinyl Acetate known locally as EVA Type F. This was pre-blended with 200 pen bitumen in a 95%/5% blend in a low shear mixer. The binder is transported to the mixing plant and stored in a stirrer-tank, as the storage stability is poor, ie. the EVA will separate from the bitumen and rise if left. EVA does not react the same with all sources of bitumen and to be effective as a modifier should not dissolve but should form discrete globules or strands within the bitumen.[17]

These particles increase the stiffness of the binder and are commonly used to improve rut resistance and elastic stiffness of bituminous materials. The principal effect of EVA is to reduce the penetration and increase the softening point of the base bitumen and reduce its temperature susceptibility, the Penetration of 200 pen + 5% EVA (F), at 155-170 pen, is significantly higher than 100 pen bitumen and very much more than 50 pen bitumen. However, in terms of the rutting performance of the macadam mixtures made with the binders, EVA modified mixtures are very similar to those with the respective unmodified binders indicating that rutting of these materials is most significantly affected by aggregate interlock. For Hot Rolled Asphalt the rut resistance was much lower (i.e. higher wheel tracking rate) than for unmodified material but still well within the level required for the traffic on this road (10 mm max).

EVA (F) was used because it had performed as well as SBS modified bitumen on the M2 trial[1] and is considerably cheaper. Since reflective crack propagation is probably a low temperature performance problem it was thought to be measurable by the Ductility Test and Fraas Brittle

point both of which measure low temperature properties.
   Generally the Fraas Brittle point was slightly lower for the EVA mixture than unmodified bitumen; the EVA (F) had also significantly increased the Ductility at 4°C, particularly from the tank at the mixing plant. The ductility of 61 cm for 200 pen and EVA being similar to that quoted for 7% SBS modified bitumen (Pen 90, Softening Point 85°C.[18] Theoretically therefore improved low temperature performance would be expected without jeopardising the rut resistance texture or retention of in Hot Rolled Asphalt. Rut resistance and texture have been maintained and slightly improved resistance to reflective cracking is observable in the Hot Rolled Asphalt section, where binder properties are most significant; on the geotextile side, none of the 22 joints yet visible after 3 years, compared with 2 joints out of 9 being visible on the unmodified 50 pen side. However over the cracks, a similar proportion ($^2/_3$) reflected through. Set into the context of the difference in ductility between modified 200 pen and unmodified 50 pen bitumen, it would indicate that ductility is not a significant predictor of resistance to reflective cracking. As stated before, this improved performance may be simply as a result of the higher binder content in the EVA material rather than any modification of binder properties. EVA acts as a `thickening' agent permitting higher binder contents to be retained on the aggregate without drainage or introducing `instability'; this is beneficial in itself.
   Results on materials recovered from the samples taken from the road and reheating can be variable as the globules/strands may not dissolve in Methylene Chloride (CH_2Cl) (Dimethyl Chloride) and, depending upon their size, may be filtered out. With some sources of bitumen not used here the EVA completely dissolves [17], such bitumen should not be used with EVA as no improvement in properties is exhibited after blending. EVA content can be measured in bitumen using an infra-red spectrophotometer.
   Since the cost of adding EVA add about 10% to the laid cost of hot rolled asphalt compared to unmodified material, the benefits of using this polymer for reflective cracking control is as yet unproven.

6.4 Effect of Geotextile

The geotextile is a non-woven polypropylene membrane approximately 1mm thick. It is laid on a binder film also about 1 mm thick and it sticks down to it. A very small percentage of binder may be absorbed by the underlying road surface or run into cracks. However after the hot asphalt is laid the bitumen film is reheated, saturates the geotextile and bonds it both to the existing road and the overlay material. When cores were taken, the geotextile may be left bonded to the concrete (most common) or is lifted with the core. In the latter case, any poor concrete may be lifted with the core. Where the geotextile remained on the road surface the binder film was clean, bright and on the base of the core. Considerably larger force was required to remove the core on the geotextile side than where the blacktop was laid directly on the tack-coated concrete. The geotextile was not torn across the cracks.

It is suggested that the bitumen impregnated geotextile acts as a controlled slip plane at the joint/crack dissipating the movement over an unknown distance away. The M2 overbanding trial[1] indicated that movement must be dissipated over a width exceeding 150 mm as overbanding 300 mm wide led to cracks in the overlay at the edges of the overbanding and also in some cases within the overband. This caused serious disruption of the overlay, leading to rapid and significant deterioration.

The bitumen impregnated goetextile has little tensile strength and cannot play any reinforcing role. By providing this slip plane the geotextile therefore may be capable of reducing the strain in the overlay. However it may be hypothecated that if this degree of reduction is insufficient to reduce the strain in the overlay to an acceptable level for the material, cracking will occur.

It was observed that cracks reflected faster from joints where movement takes place and hence induced strain was greatest, than from cracks in the concrete. It was also observed that the geotextile reduced slightly the rate of reflective cracking with DBM, 43% of joints had come through in the first year with geotextile compared to 88% without, 70% compared to 100% after 2 years. DBM is said to have a lower resistance to tensile strain than with Hot Rolled Asphalt. Tensile strain testing is still to be carried out on these overlay materials to evaluate this property. However with DBM this modest improved performance with geotextile may not justify the expense unless movement in cracks is very small; this pertains, for example, following the `cracking and seating' process.

Hot Rolled Asphalt has a much larger resistance to cracking under tensile strain and the goetextile was very effective at maintaining the strain below that which will cause cracking of the material. Even after 3 years no cracks are visible in the modified section and only 22% on the unmodified section. This compares to $^2/_3$ of the joints on the side with no geotextile.

It appears therefore that the hypothesis of the

strain-reducing effect of geotextile may be valid.

It is now likely that further cracking of the overlay will occur through fatigue as the number of temperature cycles continues; and as the bitumen hardens with time.

There is possibly a slight beneficial effect on deflection at the pavement surface, measured by Benkelman Beam, of the presence/absence of geotextile, in that a 5×10^{-2} mm difference between eastbound and westbound reduced to 2×10^{-2} cm after geotextile and overlay as shown in Table 3. This would be too small to be of structural significance overall and the deflection levels are typical of what could be expected for a sound concrete pavement.

With this good performance, the cost of geotextile is justifiable in maintaining a crack-free road beneath Hot Rolled Asphalt. The cracks in the DBM now need attention by sealing to prevent water ingress into the surface of the pavement possibly leading to frost damage and accelerated fretting, particularly of the leaner DBM mixtures.

7 Alternative strategies for reducing reflective cracking

As an alternative to geotextile, or in combination with it, increasing the thickness of the pavement reduces the strain in the overlay and increases the thermal insulation reducing the movement. Trials elsewhere in Kent with 90 mm, 150 mm and 200 mm overlay to concrete roads are taking place as part of the USA strategic Highway Research Program (SHRP) and on other projects. 180 mm overlay to a cracked lean mix concrete road has remained uncracked indefinitely.

Other studies on the M20 motorway in Kent and elsewhere are in progress to investigate the benefits of using other polymer modified materials. On the initial evidence from this trial polymer modification of macadam type mixtures is ineffective, polymer modification of hot rolled asphalt seems to provide little benefit.

The cost of geotextile at about £1.30 sq.m is comparable to £4.00 sq.m for a 40 mm overlay; in addition to the latter cost has to be included increased costs of raising kerbs and footways and ironwork if applicable. Only when the Robin Hood Lane trial starts to crack over the geotextile treated joints on the HRA section, will the true cost-benefit of geotextile be known.

8 Skid Resistance and Texture Depth
Skid resistance is a combination of microtexture, provided by the Polished Stone Value (PSV) of the aggregate and macrotexture (rugocity) which provides

edges for tyres to grip, measured by Texture Depth. When surfaces are new the bitumen film provides the major microtexture, to give a pendulum value of 66 but the better performance of HRA long term at 58 compared to macadam 53 is a result of a higher PSV stone being used; both materials satisfy the County's requirements. The texture depth of the macadam sections had fallen slightly over the 3 years, the macadam was throughout below the County intervention level and was planned for surface dressing; this will probably be carried out using a geotextile/chipping dressing[19] which has proved excellent for sealing cracked pavements in Kent and restoring skid resistance.

9 Conclusion

1. Grouting beneath the concreted slab has reduced vertical deflection at joints to an acceptable level so that little vertical movement was translated to the overlay.
2. If an overlay is being designed for suppression of reflective cracking, rolled asphalt is significantly more effective than dense bitumen macadam.
3. Maintaining the binder content at or above the mean/target value increases binder film thickness and improves resistance to cracking.
4. Improving the ductility of the binder, as measured by the IP Ductility test does not seem to improve the resistance to reflective cracking.
5. The use of EVA (F) had little effect in suppressing the reflective cracking, however it does permit higher binder contents to be used without introducing a rutting problems. It is unlikely to be cost effective for reflective crack control alone.
6. The use of a non-woven geotextile paving fabric was very effective in conjunction with Hot Rolled Asphalt, in suppressing reflective cracking for at least 3 years with a 50 mm overlay at this site, and is probably cost/effective.
7. The optimal thickness of overlay to reduce the rate of cracking so that it matches the need to resurface for other reasons, e.g. loss of skid resistance or rutting, i.e. probably in the region of 10 - 14 years, has yet to be determined for Kent materials and temperature range.

10 Acknowledgements

This paper was produced with the data provided by the Materials Group, Technical Manager Ian Valentine, and Highway Maintenance Management Group, Project Engineer DT

Coupland, within the Engineering Services Branch. The paper is prepared with permission of the Director of Highways & Transportation, Mr A Mowatt.

11 References

1. Walsh, I.D (1989) Overbanding and polymer modified asphalt in overlay to concrete carriageways 1st Int Conf Reflective Cracking in Pavements RILEM Liege.
2. Task Force 25 Committee, `Specification for Paving Fabrics, Federal Highway Administration, Washington DC.
3. British Standards Institution (1988) Coated Macadam for roads and other paved areas, BS 4987 Parts 1-2 BSI.
4. British Standards Institution (1985) Hot Rolled Asphalt for roads and other paved areas, BS 594 Part 1, Specification for constituent materials and asphalt mixtures. BSI.
5. British Standards Institution (1985) Sampling and examination of bituminous mixtures for roads and other paved areas BS 598 Part 3 Methods for design and physical testing BSI.
6. Institute of Petroleum, Breaking Point of Bitumen-Fraas Method IP80/53. IP London.
7. American Society for Testing and Materials, Standard Test Method for Ductility of Bituminous Materials D113-79 ASTM, Philadelphia.
8. British Standards Institution (1990) Method of recovery of bitumen binders by dichloromethane extraction (Rotary Film Evaporator Method). Draft for Development DD2 BSI.
9. British Standards Institution (1990) Method for determination of Wheel Tracking rate of cores of bituminous wearing courses Draft for Development DD184.BSI.
10. Hosking, J.R. Roe, P.G and Tubey L.W. (1987) Measurement of Macrotexture of Roads Part 2; a study of the TRRL mini texture meter Transport and Road Research Laboratory Report RR 120.
11. Instructions for using the portable skid resistance tester (1969) Road Note 27 Road Research Laboratory, HMSO.
12. Kennedy, C.K. Ferie, P. and Clarke, C. (1978) Prediction and Pavement Performance and the design of overlays, Transport and Road Research Laboratory Report LR833.
13. Lees, G. (1987) Report on Testing and Properties of BS 594 and BS4987 Bituminous Paving Mixtures, County Surveyors' Society, UK.
14. Bazin, P. and Sawnier J. (1967) Deformability,

Fatigue and healing properties of asphalt mixtures. Proc 2nd Int Conf Struct Des of Asph Pavs, Univ Mich.
15. Raithby, K.D and Sterling A.B. (1970) The effect of rest period on fatigue performance of hot rolled asphalt under reversed axial loading. Proc Assoc Asph Pav Tech Vol.29.
16. Pell, P.S. and Hanson, J.M. (1973) Behaviour of Bituminous Roadbase Materials under repeated loading Proc Assoc Asph Pav.Tech.
17. Kent County Council (1989) EVA modified binders (Report on Blending EVA and different bitumens, KCC.
18. Whiteoak, C.D.(1989) Shell Cariphalte, DM. An SBS modified bitumen Shell Bitumen Review 64.
19. Walsh, I.D. (1986) Geotextile in highway surface dressing 3rd Int Conf Geotextiles, Vienna.

47 DESIGN AND FIRST APPLICATION OF GEOTEXTILES AGAINST REFLECTIVE CRACKING IN GREECE

A. COLLIOS
Edafomichaniki Ltd, Athens, Greece

Abstract

Rehabilitation of pavements constructed today in Greece has to be cost-effective and with a long-term satisfactory behaviour against reflective cracking. An appropriate method of design by means of geosynthetics, based upon semi-empirical formulas and cummulative knowledge of construction details, seems to provide an efficient tool for geotechnical engineers in their effort to pavement rehabilitation design. The first application of this method in Greece provided excellent results.

<u>Keywords</u> : Geosynthetic, Cost effectiveness, Cycles-to-failure, Design traffic number, Skilled construction.

1 Introduction

Pavements actually constructed in Greece consist of materials with origin and properties rather variable, depending on the accessibility and the climate conditions of the area of interest. All these structures are subjected to the combined influence of water and temperature variations, dynamic traffic loading, shrinkage and swell of the natural soil, resulting to a variable degree of pavement cracking.

Rehabilitation of those fractured structures presents a complex problem, accentuated by the repairing cost, since all cracks propagate almost immediately to the new layer for rehabilitation. This phenomenon, called reflective cracking, may be retarded or prevented and the degree of rehabilitation success is directly based upon the ability of the new layer to resist crack propagation.

The use of geosynthetics against reflective cracking in hot asphalt pavements has been proved rather effective and since its succesfull first application in Greece seems to offer a valuable reduction of the rehabilitation costs of old asphalt pavements. A reliable method of design, widely approved by the authorities and easily applied by the consultants will surely contribute to the official adoption of the use of geosynthetics in asphalt pavement rehabilitation in the near future.

2 Reflective cracking and fracture mechanics

The main reason for the development of reflective cracking in asphalt pavements is the horizontal and vertical relative displacements of the old pavement, mainly due to a combination of fatigue (cumulative effect of traffic), thermal influence, shrinkage (during pavement curing) and sub-soil movements (excessive saturation may cause swell, long-term consolidation, local landsliding, earthquakes etc). The effect of frost heave is also of concern, especially in Northern Greece, where pavement temperatures in winter may fall to -10° C, while the subbase conserves a higher temperature (at about the water freezing point). All those displacements result into tensile stresses within the body of flexible pavements, creating cracks and fissures, but also in rigid road constructions, where cracking is mainly generated by the existence of construction joints. The dynamic traffic loading over a simply repaved layer creates large shear stresses on the interface, largely constributing to the progagation of reflective cracking through the new pavement to the surface of the road. This propagation is presented under two differrent functions, either as vertical, creating a stable interface, or as horizontal movement creating a non-stable interface, before further vertical travel [1].

The rate of crack development in asphalt concrete due to traffic or thermal loading conditions (dynamic loading) may be estimated using the empirical relationship proposed by Paris - Erdogan [2] :

$$\frac{dc}{dN} = A\,(\Delta K)^n \qquad (1)$$

where dc/dN = crack development rate
       ΔK = stress intensity factor for the considered mechanism of dynamic loading
       A, n = fracture parameters, constant for each type of the asphalt overlay mix

The number of passages for the design vehicle (or load cyclic applications) N_f needed for the crack to propagate through the overlay may be computed by integrating (1) as follows :

$$N_f = \int_0^{h_o} \frac{dc}{A\,(\Delta K)^n} \qquad (2)$$

where h_o = overlay thickness
       c = crack length

The above formulas (1) and (2) allow an empirical graph to be established, concerning the type of overlay asphalt mix used in road rehabilitation in Greece, relating the maximal number of traffic loading repetitions to failure (thus the life expectancy of the project) to the thickness of the new overlay, provided that the stress intensity factor and the constant fracture parameters are determined.

The performance of creep tests on asphalt overlay cylindrical samples allow the determination of the diagram relating the stiffness E of the material to the loading time t. The slope m of the diagram log E - log t is related to fracture parameter n by :

$$n = \frac{2}{m} \qquad (3)$$

Using the results of experimental studies [3], [4], for a given loading frequency and mechanism, parameter A may be calculated as follows:

For traffic dynamic loading $\quad \log_{10} A = 2,19 - \dfrac{n}{0,42} \qquad (4)$

For thermal cracking $\quad \log_{10} A = 4,40 - \dfrac{n}{0,50} \qquad (5)$

The stress intensity factor is admitted to be influenced by the opening mode of the crack and is computed as follows [5] :

$$\Delta K = \frac{\mu}{\xi + 1} \cdot \sqrt{\frac{2\pi}{r}} \cdot [u] \qquad (6)$$

where [u] : relative displacement of crack limits
r : distance of crack to the considered load
$\xi = 3 - 4v$, v: Poisson's coefficient
μ = Lame's coefficient

The above procedure helps consulting engineers avoid costly and time consuming laboratory fatigue tests for the determination of the crack growth. The stress intensity factors and fracture parameters vary with the different mix temperatures for a given composition mix (according to the imposed Greek specifications) but may be relatively easily estimated for each specific project.

3 Geosynthetic contribution and design

To minimize the development of reflective cracking, consultants may try to :
   a) Determine and eliminate the original mechanism causing cracking which is a rather costly affair.
   b) Prevent or at least prolong the cracks once initiated, from propagating to the new overlay, that is either to increace the number N_f for a given overlay thickness or to reduce the thickness h of the overlay for a given number of cyclic loading (design traffic number or passages).

Geosynthetics placed at the interface of the old-new layer create a large anisotropy in the pavement structure and largely constribute to preventing or at least prolonging reflective cracking by :
  - Absorbing and relieving shear forces by their low modulus.
  - Relieving strains at vertically deforming construction joints or important cracks.

- Largely reducing the amount of water in-flow down to the fissured sub-grade, by creating an impermeable armed blanket together with the bitumen.
- Reducing the amount of bitumen necessary for repair of the "crocodile-type" fissuration, maintaining an optimum thickness.

One easily applied design method when using geosynthetics is based upon a variation of the proposed by the American Asphalt Institute method of the Design Traffic Number DTN by introducing the fabric efficiency factor FEF :

$$FEF = \frac{Nr}{Nm} \quad (7)$$

where Nr = number of load cycles to failure with geosynthetic.
       Nm = number of load cycles to failure without geosynthetic.

The FEF has been determined by laboratory tests and varies between 2,10 and 15,9, depending on the type and characteristics of the geosynthetic [6]. R.M Koerner proposes the use of a modified design traffic number DTN_r as follows :

$$DTN_r = \frac{DTN}{FEF} \quad (8)$$

where DTN = design traffic number without geosynthetic.
       DTN_r = modified design traffic number with geosynthetic.

Interesting financial comparisons for each type of project may be performed relating the value N_f (formula 2 - figure 1) to the DTN or to the DTN_r value and the thickness of the overlay for repaving. Such comparisons may lead to a reduction of approximately 20 - 30 % of the rehabilitation cost for the design life expectancy of the specific project.

4 First Greek application of design and construction

The first official application of the method in Greece (Ministry of Public Works) was performed at the rehabilitation project of the provincial road connecting Loutra Thermis to Vassilika, at the area of Thessaloniki prefecture in October 1988. Following an extensive soil investigation, the sub-soil was determined to be sandy clays with a few gravels (CL), with a I_{CBR} value equal to I = 2,5 - 4,0% and a plasticity index of Ip = 26 %. A general cracking of the last asphalt concrete layer was observed, due to multiple reasons involving ground saturation and swell (not adequately drained embankments) and large temperature variations ($\Delta T = 45°$ C), combined to heavy traffic.

Repaving was proposed by hot asphalt concrete mix, laboratory tested on creep tests upon cylindrical samples. Values of parameter m ranged between 0,75 and 0,85 with a mean value of m = 0,80, resulting to n = 2,50. The stress intensity factor ΔK was calculated for different values of the relative displacement of the cracks limits, ranging between 0,01 m and 0,05 m according to the site measurements. Equations (1) and (2) provided us with the curves of figure 1, relating the number of cyclic loading for failure N_f to the overlay thickness h, for each separate width of crack.

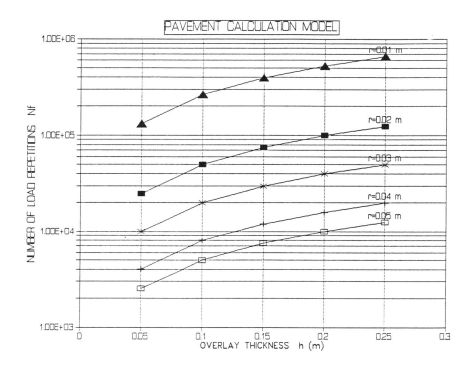

Fig 1 Correlation of the cyclic loading number N_f to the overlay thickness h.

The geosynthetic that was used consisted of a nonwoven needle-punched polypropylene geotextile with a mass/unit area of 140 gr/m² and a fabric efficiency factor FEF = 2,10. As deduced by figure 1 for a maximal crack width of 2,00 cm, the necessary overlay thickness was chosen to be h = 0,10m with an ultimate calculated N_f = 50.000 cycles. Comparing to the initial design traffic number of DTN = 10, the life expectancy of the rehabilitation was calculated to approximately 13,6 years. The reduction of the thickness due to the fabric, calculated by means of the modified traffic number DTN_r = 4,8 was calculated to be approximately 3 cm, offering a sensible cut-off of the initially projected rehabilitation cost (estimated at bi-annual base). During construction, all cracks having a width larger than 0,50 mm were carefully sealed with hot bitumen while the largest fissures (up to 2,0 cm) were filled with resins. Hot pure bitumen binder was uniformly sprayed at a temperature of 155° C with an average quantity of 1,5 lt/m² at an area slightly larger than the final to be covered by the geotextile. After having by hand laid down the geotextile with 0,20 m overlaps and smoothed all wrinkles, an additional binder 0,5 lt/m² was applied on the surface and then asphalt overlay was placed on top in two layers of 5 cm each.

Four years after the rehabilitation, the results are rather impressive. No trace of fissuration or reapparance of cracks is observed, although traffic loading has nearly doubled at the area. In spite of two successive "heavy" winters, the road behaviour is judged as rather satisfactory and the calculated life expectancy seems rather realistic, inducing that the above proposed procedure may be correctly reproduced for each project.

5 References

[1] COLOMBIER, G. (1989). Fissuration des chaussees-nature et origine des fissures moyens pour maitriser leur remontee. RILEM, Reflective Cracking in Pavements, Liege 8 - 10/3/89.
[2] PARIS, P.C. and ERDOGAN, F. (1963). A critical analysis of crack propagation laws, Journ. of Basic Engin, Series D, 85 - No 3.
[3] MOLENAAR A.A.A. (1983). Structural performance and design of flexible road constructions and asphalt concrete overlays, Ph. D. thesis, Delft University.
[4] JAYAWICKRAMA, P.W. SMITH, R.E. LYTTON R.L.. Development of asphalt concrete overlay design program for reflection cracking, RILEM, Reflective cracking in Pavements, Liege 8 - 10/3/89.
[5] IRWIN, G.R. (1957). Analysis of stress and strain near the end of crack traversing a plate, Journ of Appl. Mechanics, No 24.
[6] KOERNER, R.M. (1990). Designing with Geosynthetics, Prentice Hall, pages 242 - 244.

AUTHOR INDEX

Abd El Halim, A. O. 299
Ahmiedi, E. 193
Alberola, R. 282, 433
Alexander, W. 254
Antoine, J. P. 179
Blaauwendraad, J. 121
Campbell, D. M. 323
Caperaa, S. 193, 220
Cescotto, S. 146
Collios, A. 482
Colombier, G. 49, 273
Coppens, M. H. M. 200
Costa, C. 146
Courard, L. 146
Currie, B. 316
de Bondt, A. H. 121, 449
Decoene, Y. 384, 391
Di Benedetto, H. 179
Dumas, Ph. 246
Dumont, R. 384
Fonferko, L. 290
Francken, L. 75, 136, 206
Gordillo, J. 282
Graf, B. 159
Grzybowska, W. 290
Gschwendt, I. 129
Guo, Zhongyin 398
Hendrick, S. 146
Herbst, G. 425
Inoue, K. 378
Ishai, I. 343
Jacob, T. R. 169
Jaecklin, F. P. 100
Karam, G. 370
Kawamura, K. 237
Kief, O. 343
Kirchknopf, H. 425
Kirschner, R. 187
Kubo, H. 237
Kuck, P. J. 146, 323
Kunst, P. A. J. C. 187
Laurent, G. 353

Lefort, M. 413
Litzka, J. 263, 425
Livneh, M. 343
Majumdar, A. K. 334
Marchand, J-P. 273, 458
McKenna, R. 254
Michaut, J. P. 193, 220
Molenaar, A. A. A. 21, 121
Neji, J. 179
Nunn, M. E. 360
Pandey, S. N. 334
Pasquier, M. 179
Petit, C. 193, 220
Phillips, P. 406
Poliaček, I. 129
Potter, J. F. 360
Qing-Lin, Sha 441
Razaqpur, A. G. 299
Rigo, J. M. 3, 146
Rikovský, V. 129
Ruiz, A. 433
Saathof, L. E. B. 449
Samanos, J. 307
Sasaki, K. 237
Scarpas, A. 121
Serfass, J .P. 353
Sicard, D. 413
Silfwerbrand, J. 228
Sprague, C. J. 323
Stanzl-Tschegg, S. E. 263
Tessonneau, H. 307
Tschegg, E. K. 263
Vanelstraete, A. 136, 206
Vecoven, J. 246
Walsh, I. D. 464
Werner, G. 159
Wieringa, P. A. 200
Wojtowicz, J. 290
Woodside, A. R. 316
Woodward, W. D. H. 316
Zhang, Quancai 398

SUBJECT INDEX

This index has been compiled from the keywords assigned to the individual papers by the authors, edited and extended as appropriate. The numbers refer to the first paper of the relevant paper.

Active joint technique 413
Aggregate interlock 449
AMIR compactor 299
Anticrack systems 246
Asphalt 169
Asphalt concrete 290, 353
      overlay 254, 378, 398, 425
      reinforcement 200
Assessment 3
Australia 254

Beam model 228
Belgium 384, 391
Bending test, 3 point 187
Binder effects 246
Binder, modified 360
Binder quality 100
Binders 75, 273
Bitumen 391
Bituminous concretes 193
Bituminous felts 334
Bituminous sprayed reseal 254

CAPA 121
Case histories 3, 159, 370
Cement bound roadbases 28

Cement concrete 391
China 441
Chip seal 323
Chopped fibre reinforcement 169
Coated bitumen 49
Cold joints 433
Cold microsurfacing 282
Compaction 299
Composite overlay 316
Concrete carriageway 464
Conference, 1989 3
Core permittivity test 406
Core testing 441
Cost comparisons 425
Cost effectiveness 482
Cracks
      bridging 406
      causes 49, 100
      classification 413
      control method 129
      growth, analysis 146
      initiation 136, 206
      interface shear transfer 121
      length evolution 193
      measurements 237, 449
      movements 21

origin 3
propagation 3, 146, 159, 193, 220, 464
repair techniques 169
resistance 179
  overlay materials 21
retarding interlayer 353
sealing 413
CRAFT 273, 413
CRC pavements 384
Cyclic testing 179

Damage
  accumulation 159
  classification 100
  characterization 179
Deflections 449
  measurements 21, 100
Design 3, 75, 159, 307
  curves, asphalt overlays 187
  methods 129
  models 449
  overlays 21, 100, 121, 482
  traffic loading 482
Dynamic testing 200

Economic aspects 425
Elastic modulus 75, 169
Elastomer bitumen emulsion 49, 307
Evaluation procedures 100
Expansive clays 406

Fabric 246
  glass 398
  reinforced chip seal 323
  system 159
Fatigue 49, 129, 200, 334
  cracking 220
  crack propagation 220
  life 228
  testing 193, 398
Fibre modification 353
Fibres 282
  temperature limitations 3

Field trials 129, 237, 254, 353, 360, 370, 391
  glass fabric 398
Finite element methods 75, 121, 146, 220, 228, 299
Fissurometer 179
Flexibility test 282
Fracture behaviour 263
Fracture mechanics 21, 121, 159, 220, 449, 482
France 413
Full depth excavation 169
Full scale testing 129, 406
Future trends 75

Geofabrics 49
Geogrid 100, 299, 360, 425
Geosynthetics 482
Geotextiles 129, 179, 290, 353, 360, 384, 391, 413, 425, 464
  asphalt overlays 100
  binder impregnated 458
  felts 334
  reinforced seal 406
Glass fabric 398
Glassfibre grid 200
Greece 482

Index testing 3
Initial stiffness modulus 200
Installation,
  chip seal 323
  overlays 100
Interface system 136, 206
  classification 136
Interlayers 3, 75, 246

Japan 237, 378
Jute geotextiles 334

Laboratory testing 3, 75, 169, 179, 187, 246, 290, 307
  glass fabric 398
  precoated geotextile felts 334
  roller compacted concrete 433

Laying techniques, geotextiles 458
Load testing 316
Load transfer measurements 449
Long-term performance 458
    interlayer 353
Low temperature cracks 237
Low temperature fracture behaviour 263

Mechanism, reflective cracking 441
Microasphalt 316
Microsurfacing 3, 282
Mode I fracture behaviour 263
Modelling 75
    cracking 121
    crack propagation 146, 159
    numerical 136
    pavements 21
Modified binder 360
Modified bitumen emulsions 282
Modulus 169
    initial stiffness 200
    ratio 220
Monotonous testing 179
Mortar content 246
Multilayer linear elastic model 75
Multilayer materials 220
MURMOS 307

Netherlands 449
Non-woven geotextiles 334
Non-woven polypropylene fabrics 370
Numerical modelling 136

On site manufacture 384
On site reinforcement 307
Overlay 159, 187, 290, 299, 334, 378
    asphalt 254, 425
    design 21, 121
    modelling 75
    polypropylene bituminous composite 316
    reinforcement 121

Pavement
    design 100
    evaluation 21
    modelling 21
    retrofitting 100
Paving fabric 406
    specification 254
Performance testing 3
Ply adhesion strength 254
Poland 290
Polyester geotextiles 406
Polyester threads 307
Polymer modified asphalt 464
Polypropylene bituminous concrete overlay 316
Polypropylene fabric 370
Pore size determinations, jute geotextiles 334
Porous asphalt 391
Pre-cracking 413
    underlays 273
Reflection cracking 75
Reinforcement 75, 169, 187, 254
    asphalt 200
    overlay 121
Repair 169, 237
Resurfacing 323
Retardation 316
Roadbases, cement bound 228
Roller compacted concrete 433
Rupture 193

SAM membrane 282
SAMI 169
Sand asphalt 353, 433
SBS modified asphalt 353
Sealing 433
Semirigid pavement 282
    design 129
Semirigid roadbase 441
Shear coupling 159
Shear testing 21, 290
Shear transfer 121
Shrinkage-bending test 246
Shrinkage cracking 413

Simulation 146
    thermal cracking 206
Site trials 273, 316, 406, 413, 425, 433, 464
Skid resistance 316, 313, 464
Soft pavement 378
Spain 282, 433
Specifications 100
    geotextiles 458
    paving fabric 254
Specimen shape 263
Splitting test 263
Sprayed reseal 254
Standardization 3
Strain calculations, interfaces 136
Stress absorbing membrane 360
Stress intensity factors 159
Structural evaluation 136
Surface dressing 49, 323, 458
Surface initiated cracking 360
Surface treatment 290
Surveys of visual condition 21
Sweden 228
Synthetic reinforcement 187

Tack coat materials 75
Temperature effects 75
Tensile fatigue life 228
Tensile strength, asphalt 206
Tensile testing 316
Tensile yield 228

Testing 3, 206, 316
    dynamic 200
    fatigue 193
    jute geotextiles 334
    laboratories 75, 246
    splitting 263
Test track 425
Thermal effects 100, 136, 146, 159, 237, 441, 44
Thermal stresses 206
Thermal stress transfer 169
Thermorheological properties 206
Thin overlays 464
Thin surfacings 228
Traffic loadings 75, 136, 146, 159, 220, 449
Transverse cracks 449

Ultrasonic transmission 179
Underlays, pre-cracking 273
Underseal 169
United Kingdom 360, 464

Visual condition surveys 21
Visual inspection 433

Waterproof layer 406
Waterproof structure 323
Wearing course 458
Wheel tracking device 334
Wide interlayer 378

Yield strain 360

RILEM, The International Union of Testing and Research Laboratories for Materials and Structures, is an international, non-governmental technical association whose vocation is to contribute to progress in the construction sciences, techniques and industries, essentially by means of the communication it fosters between research and practice. RILEM activity therefore aims at developing the knowledge of properties of materials and performance of structures, at defining the means for their assessment in laboratory and service conditions and at unifying measurement and testing methods used with this objective.

RILEM was founded in 1947, and has a membership of over 900 in some 80 countries. It forms an institutional framework for cooperation by experts to:

* optimise and harmonise test methods for measuring properties and performance of building and civil engineering materials and structures under laboratory and service environments;

* prepare technical recommendations for testing methods;

* prepare state-of-the-art reports to identify further research needs.

RILEM members include the leading building research and testing laboratories from around the world, industrial research, manufacturing and contracting interests as well as a significant number of individual members, from industry and universities. RILEM's focus is on construction materials and their use in buildings and civil engineering structures, covering all phases of the building process from manufacture to use and recycling of materials.

RILEM meets these objectives though the work of its technical committees. Symposia, workshops and seminars are organised to facilitate the exchange of information and dissemination of knowledge. RILEM's primary output are the technical recommendations. RILEM also publishes the journal *Materials and Structures* which provides a further avenue for reporting the work of its committees. Many other publications, in the form of reports, monographs, symposia and workshop proceedings, are produced.

Details of RILEM membership may be obtained from RILEM, École Normale Supérieure, Pavillon du Crous, 61, avenue du Pdt Wilson, 94235 Cachan Cedex, France.

Details of the journal and the publications available from E & F N Spon/Chapman & Hall are given below. Full details of the Reports and Proceedings can be obtained from E & F N Spon, 2-6 Boundary Row, London SE1 8HN, Tel: (0)71-865 0066, Fax: (0)71-522 9623.

Materials and Structures

RILEM's journal, *Materials and Structures*, is published by E & F N Spon on behalf of RILEM. The journal was founded in 1968, and is a leading journal of record for current research in the properties and performance of building materials and structures, standardization of test methods, and the application of research results to the structural use of materials in building and civil engineering applications.

The papers are selected by an international Editorial Committee to conform with the highest research standards. As well as submitted papers from research and industry, the Journal publishes Reports and Recommendations prepared buy RILEM Technical Committees, together with news of other RILEM activities.

Materials and Structures is published ten times a year (ISSN 0025-5432) and sample copy requests and subscription enquiries should be sent to: E & F N Spon, 2-6 Boundary Row, London SE1 8HN, Tel: (0)71-865 0066, Fax: (0)71-522 9623; or Journals Promotion Department, Chapman & Hall, 29 West 35th Street, New York, NY 10001-2291, USA, Tel: (212) 244 3336, Fax: (212) 563 2269.

RILEM Recommended Practice

Autoclaved Aerated Concrete - Properties, Testing and Design
   Technical Committees 78-MCA and 51-ALC

RILEM Reports

1 **Soiling and Cleaning of Building Facades**
   Report of Technical Committee 62-SCF. *Edited by L. G. W. Verhoef*
2 **Corrosion of Steel in Concrete**
   Report of Technical Committee 60-CSC. *Edited by P. Schiessl*
3 **Fracture Mechanics of Concrete Structures - From Theory to Applications**
   Report of Technical Committee 90-FMA. *Edited by L. Elfgren*
4 **Geomembranes - Identification and Performance Testing**
   Report of Technical Committee 103-MGH. *Edited by A. Rollin and J. M. Rigo*
5 **Fracture Mechanics Test Methods for Concrete**
   Report of Technical Committee 89-FMT. *Edited by S. P. Shah and A. Carpinteri*
6 **Recycling of Demolished Concrete and Masonry**
   Report of Technical Committee 37-DRC. *Edited by T. C. Hansen*
7 **Fly Ash in Concrete - Properties and Performance**
   Report of Technical Committee 67-FAB. *Edited by K. Wesche*

RILEM Proceedings

1. **Adhesion between Polymers and Concrete. ISAP 86**
   Aix-en-Provence, France, 1986. *Edited by H. R. Sasse*
2. **From Materials Science to Construction Materials Engineering**
   Versailles, France, 1987. *Edited by J. C. Maso*
3. **Durability of Geotextiles**
   St Rémy-lès-Chevreuses, France, 1986
4. **Demolition and Reuse of Concrete and Masonry**
   Tokyo, Japan, 1988. *Edited by Y. Kasai*
5. **Admixtures for Concrete - Improvement of Properties**
   Barcelona, Spain, 1990. *Edited by E. Vazquez*
6. **Analysis of Concrete Structures by Fracture Mechanics**
   Abisko, Sweden, 1989. *Edited by L. Elfgren and S. P. Shah*
7. **Vegetable Plants and their Fibres as Building Materials**
   Salvador, Bahia, Brazil, 1990. *Edited by H. S. Sobral*
8. **Mechanical Tests for Bituminous Mixes**
   Budapest, Hungary, 1990. *Edited by H. W. Fritz and E. Eustacchio*
9. **Test Quality for Construction, Materials and Structures**
   St Rémy-lès-Chevreuses, France, 1990. *Edited by M. Fickelson*
10. **Properties of Fresh Concrete**
    Hanover, Germany, 1990. *Edited by H.-J. Wierig*
11. **Testing during Concrete Construction**
    Mainz, Germany, 1990. *Edited by H. W. Reinhardt*
12. **Testing of Metals for Structures**
    Naples, Italy, 1990. *Edited by F. M. Mazzolani*
13. **Fracture Processes in Concrete, Rock and Ceramics**
    Noordwijk, Netherlands, 1991. *Edited by J. G. M. van Mier, J. G. Rots and A. Bakker*
14. **Quality Control of Concrete Structures**
    Ghent, Belgium, 1991. *Edited by L. Taerwe and H. Lambotte*
15. **High Performance Fiber Reinforced Cement Composites**
    Mainz, Germany, 1991. *Edited by H. W. Reinhardt and A. E. Naaman*
16. **Hydration and Setting of Cements**
    Dijon, France, 1991. *Edited by A. Nonat and J. C. Mutin*
17. **Fibre Reinforced Cement and Concrete**
    Sheffield, UK, 1992. *Edited by R. N. Swamy*
18. **Interfaces in Cementitious Composites**
    Toulouse, France, 1992. *Edited by J. C. Maso*
19. **Concrete in Hot Climates**
    Torquay, UK, 1992. *Edited by M. J. Walker*
20. **Reflective Cracking in Pavements - State of the Art and Design Recommendations**
    Liege, Belgium, 1993. *Edited by J. M. Rigo, R. Degeimbre and L. Francken*
21. **Conservation of Stone and other Materials**
    Paris, 1993. *RILEM/UNESCO*
22. **Creep and Shrinkage of Concrete**
    Barcelona, Spain, 1993. *Edited by Z. P. Bazant and I. Carol*